# Integrated Circuit Test Engineering

Ian A. Grout

# Integrated Circuit Test Engineering

## Modern Techniques

With 149 Figures

 Springer

Ian A. Grout, PhD
Department of Electronic and Computer Engineering
University of Limerick
Limerick
Ireland

British Library Cataloguing in Publication Data
Grout, Ian
   Integrated circuit test engineering: modern techniques
   1. Integrated circuits - Verification
   I. Title
   621.3'81548
ISBN 978-1-84628-023-8

Library of Congress Control Number: 2005929631

ISBN 1-84628-023-0          ISBN 1-84628-173-3 (eBook)   Printed on acid-free paper
ISBN 978-1-84628-023-8

Typesetting: Camera ready by author
Production: LE-TeX Jelonek, Schmidt & Vöckler GbR, Leipzig, Germany

9 8 7 6 5 4 3 2 1

Springer Science+Business Media
springeronline.com

This book is dedicated to Jane.

**Test**

*A procedure intended to establish the quality, performance or reliability of something, especially before it is taken into widespread use.*

New Oxford English Dictionary

# About the Author

Ian Grout is a lecturer within the Department of Electronic and Computer Engineering at the University of Limerick, Ireland. He was born in London, UK in 1967, and earned his PhD from Lancaster University in 1994. He has worked within the microelectronics field for several years, in particular Integrated Circuit test research and education. He has been a lecturer at Limerick since 1998.

# Preface

The extensive use of electronic products in everyday life has only been made possible with the advent of, and substantial advances in, the field of microelectronic circuit engineering. The types of microelectronic circuits that can be developed today provide for a complex circuit behaviour within a physically small package. The advent of high-value, high-functionality portable electronic systems such as the laptop computer, mobile phone and Personal Digital Assistant (PDA), are testament to this. Since the successful demonstration of the transistor back in 1947 by Bardeen, Brattain, and Shockley at Bell Laboratories (USA), in just over half a century the ability to create microelectronic circuits containing tens of millions of transistors is a remarkable statement of achievement. The trend towards increases in design complexity and speed of operation, coupled with the need for improvements in the engineering processes utilised in order further to reduce product costs, is leading to improvements in all aspects of device design, fabrication and test. It is the field of test engineering that this book aims to identify and discuss.

With the increased demands placed on test engineering activities within the electronic, and in particular microelectronic, circuit and system development and production (manufacturing) environments, the need to consider the importance of developing the right circuit/system test procedure is critical to the commercial success of a developed product. Test and evaluation activities provide an essential input to the product development at a range of stages within a product development cycle in that they:

- Provide insight into a design development activity from concept through to implementation in order to ensure that the design can be adequately tested once it has been manufactured/fabricated.

- Ensure devices that contain circuit faults due to problems with either the design or fabrication are not perceived as "good devices".

- Provide the means to evaluate the operation of the design over a range of operating conditions that may be encountered in a final application.

- Determine and guarantee the operating specifications that will be required by the end-user.

- Allow for failures to be identified and the causes of the failures to be fed-back to design and fabrication for appropriate corrective action to be undertaken in order to eliminate the sources of the identified failures, and reduce the risk of faulty devices being passed onto the customer.

The important role of test engineering is now widely acknowledged, such that the test procedure (program) development process is now considered of equal importance to that of design. Test requirements, once seen in many cases as an afterthought and considered only once a design had been generated, are now in many cases a prerequisite to the design activity, and test is now an active process from design concept identification and specification through to high-volume fabrication of the final product. This is due to a number of factors including the:

- Rapid increase in design complexity, particularly for digital systems (more functionality per mm$^2$ of silicon area).

- Utilisation of new fabrication processes (reduced geometries allowing for the creation of smaller devices but which may suffer from fabrication process variation problems leading to new device failure modes and an increase in the fabricated device parameter variability).

- Increase in design performance (higher operating frequencies, greater demands on design specifications).

- Need for more competitive products (the customer receiving devices with superior performance at lower cost, given the availability of a greater number of suppliers of the necessary technology).

- Need to test the operation of the fabricated device in a cost-effective manner with the available test equipment, without the requirement to acquire and utilise prohibitively high-cost test equipment.

- Need to enable the device to run self-test algorithms once in the final application using additional circuitry built into the design (BIST: Built-In Self-Test), and for this type of self-test to be independent of external circuitry.

Over the last few years, the evolution of test has brought it closer to design, bridging the *"traditional"* gap between design and test. Here, where once the design and test activities were separate and distant in many aspects ranging from the ultimate *"end product"* to the terminology used, the gap has been bridged with a unified *"Design for Testability"* (*DfT*) – sometimes referred to as *"Design for Test"* - approach. Specialist engineers in design and test are supported with a more generalist DfT engineer providing the ability to bridge the gap, but not necessarily required to be a specialist in either field – the need for specialists is based on the need for in-depth knowledge of specific design and test issues, roles which a single person could not realistically be expected to undertake. The gap between design and test expertise, leading to potential problems in a product development process, is being replaced with this *"Design for Testability"* (DfT) link, providing the bridge between the experts. This is coupled with the traditional boundary between digital and analogue electronic circuits, a nice and neat separation of behaviour into logic and voltage/current for design and analysis purposes, being eroded by the requirements for true mixed-signal device operation. Digital, analogue and mixed-signal DfT for devices ranging from a single operational amplifier through to the latest generation microprocessor is a complex task requiring substantial knowledge of design, test and the implications of both aspects on the other.

The purpose of this book is to discuss the range of requirements placed on the test engineering activities and the test engineer within the rapidly changing microelectronic circuit engineering environment, and to identify the methods available to solving a range of test problem scenarios.

The book is presented as follows:

**Chapter 1** provides an introduction to the role of test within the electronic/microelectronic engineering environment and highlights a number of key trends. The trends in test are highlighted, along with an overview of key actions.

**Chapter 2** provides an overview of the key fabrication processes. Whilst not aimed at being an exhaustive review of the field, it is aimed at identifying key aspects relating to test program development, in particular those relating to process induced circuit failure modes and test program development following a defect oriented approach.

**Chapter 3** overviews test program development for digital logic, both combinational logic and sequential logic. The principles identified here are relevant for later discussions into System on a Chip (SoC) test program development.

**Chapter 4** introduces semiconductor memory structures and test procedures. Traditionally, memories have been discrete packaged devices which, when utilised with a suitable processor, develop the basic operation of a computer system. Increasingly, the move towards higher levels of integration is leading to memories being provided as macro cells for placement alongside other circuitry on the same silicon die, leading towards *System on a Chip* (SoC) and *System in Package* (SiP) solutions that require extensive use of memory.

**Chapter 5** introduces analogue test. The key differences between digital and analogue test are presented, along with the challenges that exist in the need to develop suitable analogue circuit stimulus, to capture signals and to analyse results.

**Chapter 6** provides a discussion into the combining of digital and analogue test methods to provide the ability to test mixed-signal designs. In particular, data converter test requirements and solutions are considered.

**Chapter 7** discusses test procedures for digital and analogue Input/Output cells which interface the core of the silicon die to the package. Such cells have particular test requirements.

**Chapter 8** further develops digital test concepts and introduces structured approaches to test program development and linkages to design with the "*Design for Testability*" (*DfT*) approach.

**Chapter 9** discusses the problem with, and need for solutions to, the testing of System on a Chip (SoC) devices. Essentially, the previous concepts are revisited, limitations identified and the need to adopt structured DfT approaches in product development identified.

**Chapter 10** introduces test pattern generation and fault simulation techniques. These activities are required as an input to the development of structural test programs. Fault simulation provides an important input to the development of structural test programs that aim to detect circuit faults rather than confirm the circuit functionality. Digital fault simulation is well established for specific fault models, whereas analogue and mixed-signal test still require more refinement.

**Chapter 11** discusses the need for, and issues relating to, the use of Automatic Test Equipment (ATE) for semiconductor devices. The need to understand the structure of the ATE, along with the limitations, are key to the success of test program application in a production environment, are identified.

**Chapter 12** highlights the role of Test Economics as an aid to the development of effective and cost efficient test programs.

The text is supported with examples of how the concepts may be demonstrated on modest sized designs suitable for laboratory exercises. The text is aimed at providing an insight into the area of modern test engineering for those who would like to learn more of the subject area as well as those who are entering the area for the first time. Electronic material for use with this text can be downloaded from *springer.com*

# Acknowledgements

Thanks must be expressed to the following people. First, to JJ O'Riordan (Analog Devices Inc., Limerick, Ireland) for his work within the test engineering community nationally and his continued support for the test engineering teaching and research within the University of Limerick. Second, thanks to the following postgraduates within the DfT Group: Joseph Walsh, Thomas O'Shea, Michael Canavan, Jeffrey Ryan and Jason Murphy, for their enthusiasm and interesting discussions.

# Contents

# Abbreviations

| | |
|---|---|
| ABM | Analog Boundary Module |
| AC | Alternating Current |
| ADC | Analog(ue) to Digital Converter |
| ALU | Arithmetic and Logic Unit |
| AMS | Analog(ue) and Mixed-Signal / Austria Mikro Systems |
| ANSI | American National Standards Institute |
| ARM | Advanced RISC Machines |
| ASIC | Application Specific Integrated Circuit |
| ASSP | Application Specific Standard Product |
| ATE | Automatic Test Equipment |
| ATPG | Automatic Test Program Generation |
| AWG | Arbitrary Waveform Generator |
| | |
| BiCMOS | Bipolar and CMOS |
| BICS | Built-In Current Sensor |
| BILBO | Built-In Logic Block Observer |
| BISR | Built-In Self-Repair |
| BIST | Built-In Self-Test |
| BSDL | Boundary Scan Description Language |
| BST | Boundary Scan Test |
| | |
| CAD | Computer Aided Design |
| CAT | Computer Aided Test |
| CDM | Charged Device Model |
| CF | Coupling Fault |
| CIF | Caltech Intermediate Format |
| CISC | Complex Instruction Set Computer |
| CMOS | Complementary MOS |
| COTS | Commercial Off The Shelf |
| CPLD | Complex PLD |
| CPU | Central Processing Unit |

| | |
|---|---|
| CTFT | Continuous Time Fourier Transform |
| CTL | Core Test Language |
| CUT | Circuit Under Test |
| | |
| DAC | Digital to Analog(ue) Converter |
| DBM | Digital Boundary Module |
| DC | Direct Current |
| DDR SDRAM | Double Data Rate SDRAM |
| DfD | Design for Debug |
| DFF | D-Type Flip-Flop |
| DfA | Design for Assembly |
| DfM | Design for Manufacture |
| DfR | Design for Reliability |
| DfT | Design for Testability / Design for Test |
| DFT | Discrete Fourier Transform |
| DfY | Design for Yield |
| DIB | Device Interface Board |
| DIL | Dual In-Line |
| DIP | DIL Package |
| DL | Defect Level |
| DNL | Differential Non-Linearity |
| DPM | Defects Per Million |
| DR | Data Register |
| DRAM | Dynamic RAM |
| DRC | Design Rules Checking |
| DRDRAM | Direct Rambus DRAM |
| DSM | Deep Sub-Micron |
| DSP | Digital Signal Processor / Digital Signal Processing |
| DUT | Device Under Test |
| | |
| ECL | Emitter Coupled Logic |
| EDA | Electronic Design Automation |
| EDIF | Electronic Design Interchange Format |
| EEPROM | Electrically Erasable PROM |
| EMC | Electromagnetic Compatibility |
| EMI | Electromagnetic Interference |
| EOS | Electrical Overstress |
| EPROM | Erasable PROM |
| ESD | Electrostatic Discharge |
| ESIA | European Semiconductor Industry Association |
| | |
| FA | Failure Analysis |
| FET | Field Effect Transistor |
| FFM | Functional Fault Model |
| FFT | Fast Fourier Transform |
| FMC | Fault Model Coverage |
| FPGA | Field Programmable Gate Array |

| | |
|---|---|
| FRAM | Ferromagnetic RAM |
| FSM | Finite State Machine |
| | |
| GaAs | Gallium Arsenide |
| GAL | Generic Array of Logic |
| GALPAT | Galloping Patterns |
| GDSII | GDSII Stream File Format |
| | |
| HBM | Human Body Model |
| HBT | Hetrojunction Bipolar Transistor |
| HDL | Hardware Description Language |
| | |
| IC | Integrated Circuit |
| $I_{CC}$ | Power supply current (into $V_{CC}$ pin for Bipolar circuits) |
| ICT | In-Circuit Test |
| $I_{DD}$ | Power supply current (into $V_{DD}$ pin for CMOS circuits) |
| $I_{DDQ}$ | Quiescent power supply current ($I_{DD}$) |
| $I_{EE}$ | Power supply current (out of $V_{EE}$ pin for Bipolar circuits) |
| IEE | The Institution of Electrical Engineers |
| IEEE | Institute of Electrical and Electronics Engineers |
| IFA | Inductive Fault Analysis |
| IIC ($I^2C$) | Inter-IC Bus |
| IMAPS | International Microelectronics and Packaging Society |
| INL | Integral Non-Linearity |
| I/O | Input/Output |
| IP | Intellectual Property |
| IR | Instruction Register |
| $I_{SS}$ | Power supply current (out of $V_{SS}$ pin for CMOS circuits) |
| $I_{SSQ}$ | Quiescent power supply current ($I_{SS}$) |
| ITRS | International Technology Roadmap for Semiconductors |
| | |
| JEDEC | Joint Electron Device Engineering Council |
| JEITA | Japan Electronics and Information Technology Industries Association |
| JETAG | Joint European Test Action Group |
| JETTA | Journal of Electronic Testing, Theory and Applications |
| JTAG | Joint Test Action Group |
| | |
| KGD | Known Good Die |
| KSIA | Korean Semiconductor Industry Association |
| | |
| LFSR | Linear Feedback Shift Register |
| LSB | Least Significant Bit |
| LSI | Large Scale Integration |
| LSSD | Level Sensitive Scan Design |
| LVS | Layout vs Schematic |

| | |
|---|---|
| Matlab | Matrix Laboratory (from The Mathworks Inc.) |
| MATS | Modified Algorithmic Test Sequence |
| MBIST | Memory BIST |
| MCM | Multi-Chip Module |
| MEMs | Micro Electro-Mechanical Systems |
| MISR | Multiple Input Signature Register |
| MM | Machine Model |
| MOS | Metal Oxide Semiconductor |
| MOSFET | Metal Oxide Semiconductor Field Effect Transistor |
| MPU | Microprocessor Unit |
| MPW | Multi-Project Wafer |
| MS-BIST | Mixed-Signal BIST |
| MSAF | Multiple Stuck-At-Fault |
| MSB | Most Significant Bit |
| MSCAN | Memory Scan |
| MSI | Medium Scale Integration |
| MUX | Multiplexer |
| | |
| NDI | Normal Data Input |
| NDO | Normal Data Output |
| nMOS | n-channel MOS |
| NRE | Non-Recurring Engineering |
| NVM | Non-Volatile Memory |
| | |
| OEM | Original Equipment Manufacturer |
| OO | Object Oriented |
| Op-Amp | Operational Amplifier |
| ORA | Output Response Analyser |
| OS | Operating System |
| OVI | Open Verilog International |
| | |
| PAL | Programmable Array of Logic |
| PC | Personal Computer |
| PCB | Printed Circuit Board |
| PDA | Personal Digital Assistant |
| PDF | Portable Document Format |
| PGA | Pin Grid Array |
| PI | Primary Input |
| PIPO | Parallel In Parallel Out |
| PLA | Programmable Logic Array |
| PLD | Programmable Logic Device |
| PLL | Phase-Locked Loop |
| pMOS | p-channel MOS |
| PMU | Precision Measurement Unit |
| PO | Primary Output |
| PODEM | Path Oriented Decision Making |
| PPM | Parts Per Million |

| | |
|---|---|
| PROM | Programmable ROM |
| PRPG | Pseudo-Random Pattern Generator |
| PSF | Pattern Sensitive Fault |
| | |
| QTAG | Quality Test Action Group |
| | |
| RAM | Random Access Memory |
| RF | Radio Frequency |
| RISC | Reduced Instruction Set Computer |
| RMS | Root Mean Squared |
| ROM | Read Only Memory |
| RTD | Round Trip Delay |
| RTL | Register Transfer Level |
| RWM | Read Write Memory (also referred to as RAM) |
| Rx | Receiver |
| | |
| SA0 | Stuck-At-0 |
| SA1 | Stuck-At-1 |
| SAF | Stuck-At-Fault |
| SDRAM | Synchronous DRAM |
| SFDR | Spurious Free Dynamic Range |
| SIA | Semiconductor Industries Association |
| SiGe | Silicon Germanium |
| SiP | System in Package |
| SIPO | Serial In Parallel Out |
| SISO | Serial In Serial Out |
| SISR | Serial Input Signature Register |
| SLDRAM | Synchronous-Link DRAM |
| S/(N + THD) | Signal to Noise Ratio plus Total Harmonic Distortion |
| SNR | Signal to Noise Ratio |
| SoB | System on Board |
| SoC | System on a Chip |
| SOF | Stuck Open Fault |
| SOI | Silicon on Insulator |
| SPI | Serial Peripheral Interface |
| SPICE | Simulation Program with Integrated Circuit Emphasis |
| SPLD | Simple PLD |
| SRAM | Static RAM |
| SSAF | Single Stuck-At-Fault |
| SSI | Small Scale Integration |
| STC | Semiconductor Test Consortium |
| STIL | Standard Test Interface Language |
| | |
| TAB | Tape Automated Bonding |
| TAP | Test Access Port |
| TCE | Thermal Coefficient of Expansion |
| TCK | Test Clock |

| | |
|---|---|
| TDI | Test Data Input |
| TDO | Test Data Output |
| TF | Transition Fault |
| TMS | Test Mode Select |
| TPG | Test Program Generation |
| TRST | Test Reset |
| TSIA | Taiwan Semiconductor Industry Association |
| TSMC | Taiwan Semiconductor Manufacturing Company |
| TTL | Transistor-Transistor Logic |
| Tx | Transmitter |
| | |
| UART | Universal Asynchronous Receiver Transmitter |
| ULSI | Ultra Large Scale Integration |
| UTP | Unit Test Period |
| UUT | Unit Under Test |
| UV | Ultraviolet |
| | |
| $V_{CC}$ | Power Supply Voltage (positive - for Bipolar circuits) |
| $V_{DD}$ | Power Supply Voltage (positive - for CMOS circuits) |
| $V_{EE}$ | Power Supply Voltage (negative - for Bipolar circuits) |
| VDSM | Very deep sub-micron |
| VHDL | VHSIC Hardware Description Language |
| VHSIC | Very High Speed Integrated Circuit |
| VLSI | Very Large Scale Integration |
| $V_{SS}$ | Power Supply Voltage (negative - for CMOS circuits) |
| | |
| WALKPAT | Walking Pattern |
| WSI | Wafer Scale Integration |

# Chapter 1

# Introduction to Integrated Circuit Test Engineering

*Acknowledgement of the important role that test engineering related activities undertake within an electronic/microelectronic circuit development process is leading to the evolution of the role of test engineering and the skills set required by the test engineer within a product development team. This chapter will introduce the role of the test engineer and the need to consider the test issues throughout a design development process from design concept through to high-volume production. A seamless integration of the design, fabrication and test activities is the only realistic way forward in order to ensure that a design concept is brought through to a successful production and delivery to the end-user (customer).*

## 1.1 Introduction

Whenever an Integrated Circuit (IC) product is developed and manufactured, the three key engineering disciplines required are:

- Design
- Fabrication
- Test

Each discipline is primarily concerned with solving a particular problem and traditionally, these have been considered in the main as separate activities with a limited interface boundary:

- The *design* [1] activity is primarily concerned with taking an IC specification (high-level, word based description of the IC operation) and translating this into a working circuit that meets a set of performance targets. The completion of this is a fully developed design that has been evaluated in simulation, and is in a design database format that the fabrication activity can utilise.

- The *fabrication* activity is primarily concerned with taking a completed design and translating this into the physical IC using the available fabrication process. The completion of this is a fabricated circuit, to be provided at the wafer level, bare die (for Multi-Chip Module (MCM) or System in Package (SiP) products), or a fully packaged device.

- The *test* [2-6] activity is primarily concerned in ensuring that the fabricated circuit actually meets the required set of performance targets, and that it has been fabricated without fault and within the defined fabrication process tolerances.

This *traditional* separation of these tasks has led to a number of barriers developing between the disciplines that have the effect of enforcing limitations in the capabilities of the interactions that must however be undertaken. These basic interactions between the disciplines are shown in Fig. 1.1.

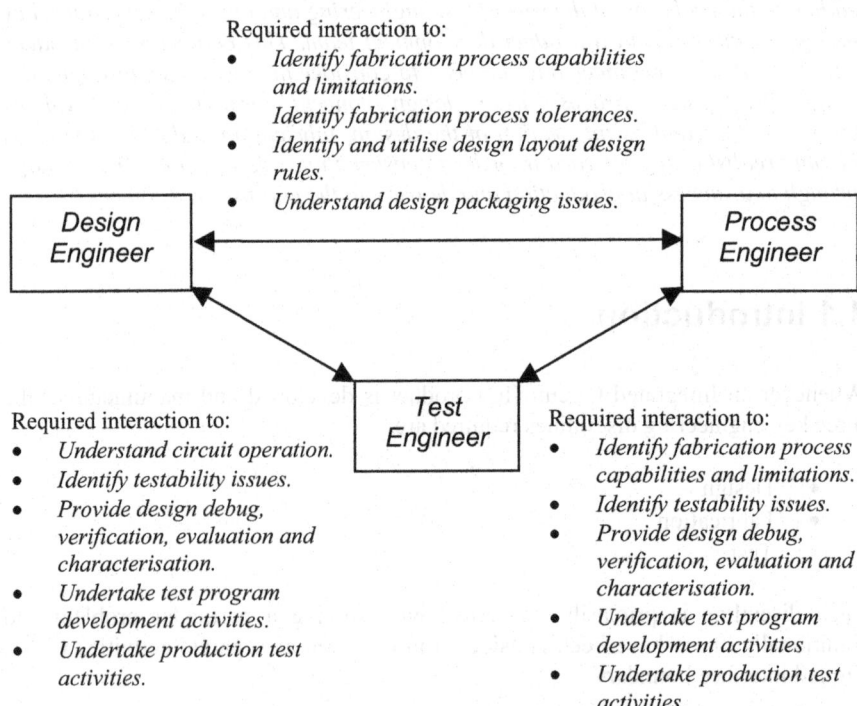

**Fig. 1.1.** Design, test and fabrication interaction

The barriers that have developed between the disciplines over the years has led to a non-optimal situation, since each activity is actually interdependent on the other, and the introduction of barriers between the disciplines has limited the ability for the disciplines to communicate effectively. For example, the fact that the disciplines have been placed within different departments within organisations, have been structured with their own sets of priorities, and have evolved their own methods and terminology (the terminology may differ between disciplines even though they may describe the same thing), has been accepted in previous years. Now however, with the ever-increasing design complexities, the newer fabrication processes, the ever increasing problems for test and the need for lower product costs, this has required a major rethink and reorganisation of activities. Fortunately, over the last few years, there has been a concerted effort on behalf of the semiconductor community to remove the barriers and ensure that better approaches to product development are undertaken. This is especially so in the closer linkages between design and test. The move towards adopting a *Design for Testability* (DfT) approach, whereby testability issues are identified at an early stage in the design development and special test circuitry may be added to address specific test access problems, has led to the fusing of design and test disciplines into a DfT discipline.

## 1.2 The Rule of Ten

A useful indicator that is quoted in order to demonstrate the importance of test and discovering faults in an electronic circuit once it has been fabricated and before it is used, is referred to as the *"Rule of Ten"* [2], see Fig. 1.2. In this, the cost multiplies by ten every time a faulty item is not detected but is used to form a large electronic circuit or system. Whilst this is a generalisation, it identifies the cost escalation of failing to detect faults whenever an electronic circuit/system is manufactured/fabricated. Here, if the cost of detecting a faulty device (IC) when it is produced is one unit, the cost to detect that faulty device when used at the board level (PCB) is ten times that. The cost to detect that faulty board when inserted in its system is ten times the cost of detecting the faulty board at production, and so on. Hence, there is a cost escalation that must be avoided.

## 1.3 The Evolution of Test Engineering

Over the last few years, the evolution of test engineering has brought it closer to design, bridging the "traditional" gap between design and test. Here, where once the design and test activities were separate and distant, the gap has been bridged by developing and adopting the *"Design for Testability"* (DfT) – or sometimes referred to as *"Design for Test"* - approach, see Fig. 1.3. Here, test is now closer to design and may influence specific details on how a design is created in order to identify testability issues and improve test access to specific circuitry within the design. In this scenario, specialist engineers in design and test are supported with a generalist DfT engineer - with the ability to bridge the gap, but not necessarily required to be a specialist in either field – the need for specialists is based on the need for in-depth knowledge of specific design and test issues, roles which a single person could not realistically be expected to undertake. The exact structure of any DfT activities would ultimately be specific to the particular company.

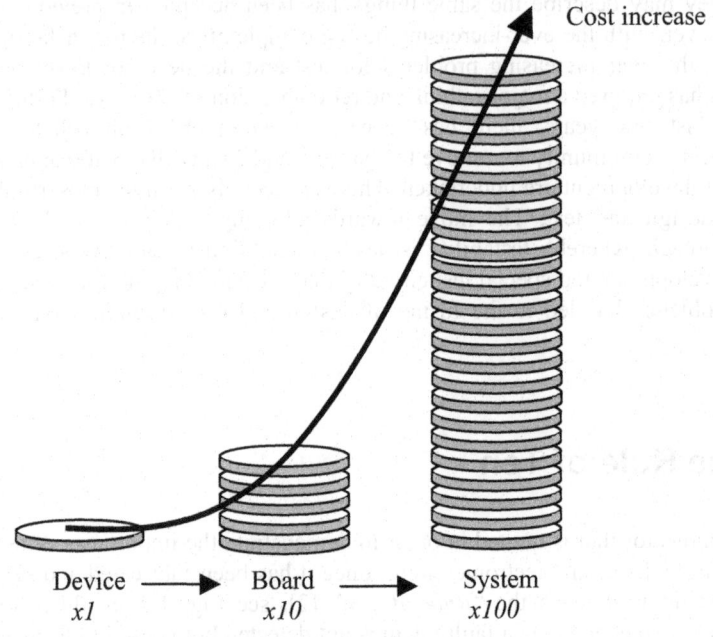

Device ⟶ Board ⟶ System
*x1*        *x10*      *x100*

**Fig. 1.2.** The Rule of Ten

# 1.4 Test Engineer Activities

It could be argued that, given a perfect design and a perfect fabrication process, there is no need for test. This would be true if the world was perfect, but given that even the most minor defect in either the design and/or fabrication process can cause a circuit fault, the importance of test cannot be underestimated.

In order for a design to be tested, there is a requirement to ensure that the design is both controllable and observable:

- **Controllability**    The ability to control specific parts of a design in order to set particular values at specific points. In a digital logic design, this would be a particular logic value. In an analogue circuit, this would be a particular voltage or current level.

- **Observability**                    The ability to observe the response of a circuit to a particular circuit stimulus.

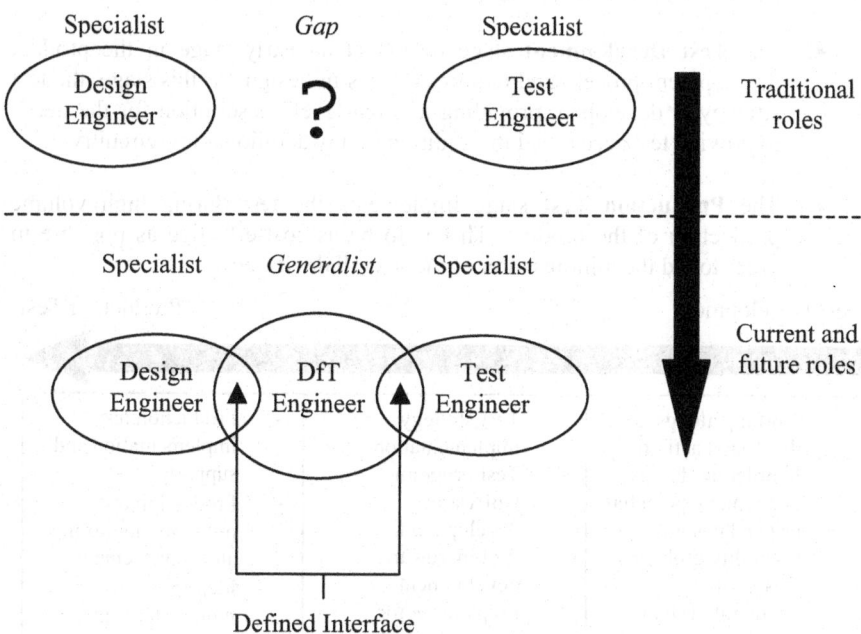

**Fig. 1.3.** Changing role for design and test

The need for access to parts of the design by addressing the controllability and observability is fundamental to test, but it must be addressed in the most cost-effective means possible. Adequate circuit test coverage must be provided at the lowest cost possible. This is a non-trivial problem to solve. For example, the number of nodes within a circuit is increasing with increases in design complexity. This is causing additional problems for test in the ability to access these nodes within the circuit that cannot have signals directly applied to them. A **traditional approach** to test is that once a design has been completed, it is given over to the test engineer to identify and implement a test procedure (the procedure will be implemented as a software program that will control the target tester hardware that then physically interface to the device under test (DUT)). At this stage, any testability issues relating to the ability to develop an adequate and minimal cost test procedure, are only identified. However, at this stage, if it is found that the design cannot be adequately tested, it may be too late to rectify the problem and a greater potential for faulty devices to be passed onto the customer exists.

A **modern approach** to test involves the proactive involvement of the test engineer alongside the design engineer in order to assess test and design testability issues at an early stage in the design process – starting at the design specification stage and prior to detailed design activities. Potential problems can be identified and, if possible, removed or their effects minimised. In this, the test engineer would be involved with design at a **Test Development** stage and fabrication (process) at the **Production Test** stage. A coherent test strategy would be developed and implemented, see Fig. 1.4. In this:

- The **Test Development** stage occurs at an early stage in the product development process and requires a basis in design. At this stage, the test strategy is developed, providing a "high-level" description for the needs to provide test access and the addition of any additional test circuitry.

- The **Production Test** stage implements the test during high volume production of the product. This is to be as cost-effective as possible in order to aid the minimisation of the overall device cost.

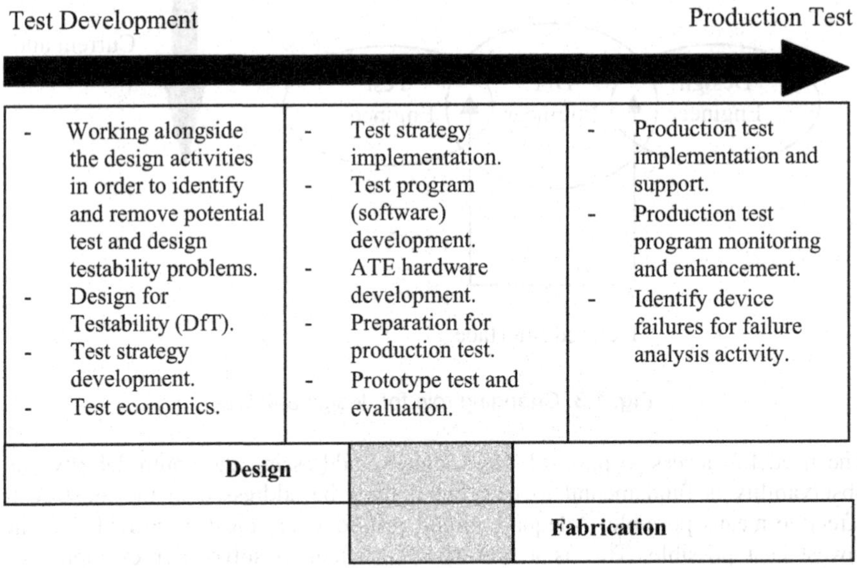

Test Development                                                    Production Test

| Working alongside the design activities in order to identify and remove potential test and design testability problems.<br>- Design for Testability (DfT).<br>- Test strategy development.<br>- Test economics. | - Test strategy implementation.<br>- Test program (software) development.<br>- ATE hardware development.<br>- Preparation for production test.<br>- Prototype test and evaluation. | - Production test implementation and support.<br>- Production test program monitoring and enhancement.<br>- Identify device failures for failure analysis activity. |
|---|---|---|
| **Design** | | |
| | | **Fabrication** |

**Fig. 1.4.** Test activities (from Test Development to Production Test)

The steps between "*Test Development*" and "*Production Test*" stages involve the translation of high-level strategies into the hardware and software infrastructure that is required to undertake design test in high volume production. The test program development, in a suitable programming language (commonly based on C), targets the required tester hardware system. In production test, a suitable **Automatic Test Equipment** (ATE) system would be identified and resourced. Such a system would be required to test the device to the required quality levels but within the shortest time as possible.

# 1.5 Device Testing

Testing is undertaken on a circuit for a number of reasons and at different points in time from design concept through to full device fabrication, see Table 1.1.

**Table 1.1.** Testing the design

| Simulation | During design development, the model of the design within the CAD tool is simulated in order to verify the correctness of the design. At this time, the test program development would commence and the program operation simulated. |
|---|---|
| Device debug/verification and characterisation | Undertaken on fabricated samples of a device prior to the production phase. This is undertaken to verify the operation of the design and also to undertake exhaustive functional and parametric (DC and AC) tests. The operation of the prototyped device can be compared to the design simulation model. At this time, it is possible for the production test program operation to be verified. |
| Production test | Once the design has been fully prototyped, the device will then go into full scale production (fabrication). The production test will need to be undertaken using **Automatic Test Equipment** (ATE) and each device must be tested in the shortest time possible without compromising the quality of the test. |
| Failure analysis | When devices fail production test, there will be the need to identify the causes of the failure so that corrective action can be undertaken. The particular test that a device fails would be extracted from the test results database within the ATE system. |

**Functional testing** is undertaken by applying specific test vectors to the design and then measuring the response of the circuit. This is undertaken to verify the correct functionality of the design. **Parametric tests** are undertaken to measure specific electrical characteristics of a device. **DC parametric** tests measure voltages and currents, along with open/short circuit tests. These are not time dependent. **AC parametric** tests measure time dependent characteristics such as delays and rise and fall times of signals. In addition to functional and parametric tests, during production testing, **structural tests** are performed. Here, the test vectors are developed to detect specific faults that may have been introduced into the circuit due to processing defects. When considering a structural test, it is taken that the device design is correct and the purpose of the test program is to identify failures due to processing defects. Any design errors should have been eliminated at the device characterisation stage.

During **production test**, there are a number of different points in the fabrication process from wafer through to packaged device that require tests to be undertaken [3]. Figure 1.5 identifies the key test points. The key point is to identify failures quickly after a major fabrication step so that these failures are not passed onto the next stage in fabrication:

- The initial stage is to test the wafer to ensure that the electrical properties (*e.g.* material resistivity, MOS transistor threshold voltage, *etc.*) are within the process spread (variance). Special "drop-in" circuits with defined circuit test structures are placed at key points on the wafer for this test. If the wafer is within limits, then each die on the circuit is tested. Parametric and functional tests would be applied and only those dies that pass the test are then used. Faulty dies are marked (either with ink dots placed on the die or marked electronically within an electronic database).

- The wafer is diced and the individual (working) dies are packaged.

- The packaged device is tested (1st level test). Parametric, functional and structural tests would commonly be undertaken.

- After packaging, an optional step will be to perform **burn-in** [26]. Here, the device is tested for a set time (*e.g.* 24 h) at an elevated temperature (*e.g.* 125°C) and at an elevated power supply voltage (*e.g.* 1.4 x $V_{DD}$) [3]. This is undertaken to accelerate circuit failure that would be caused by any circuit defects that may exist within the device. These faults may not necessarily occur during production test, but would cause the device to fail early within the final application.

- After burn-in, a final 2nd level package test is undertaken. Parametric, functional and structural tests would commonly be undertaken.

The cause of a failure will also require analysis and the test results will be fed-back to the process engineer (if the failure was process induced) or the design engineer (if the failure was due to the design).

## 1.6 Production Test: ATE Systems

Test programs will be implemented on semiconductor testers (**ATE**: Automatic Test Equipment) that require both hardware and software parts in order to set-up and control the test program execution. In general, a range of test equipment will be required, the exact nature of the equipment being dependent on the signal generation, capture and results analysis requirements. In general, testers may be categorised to fall into one of two types of system:

- **Dedicated testers:** Specially designed to measure specific parameters for a device. This will be dedicated to a particular device or small set of devices.

- **General purpose testers**: Used to test a range of devices, where the devices may have vastly different operational parameters. With this type of tester, it is temporarily customised to a particular IC to test.

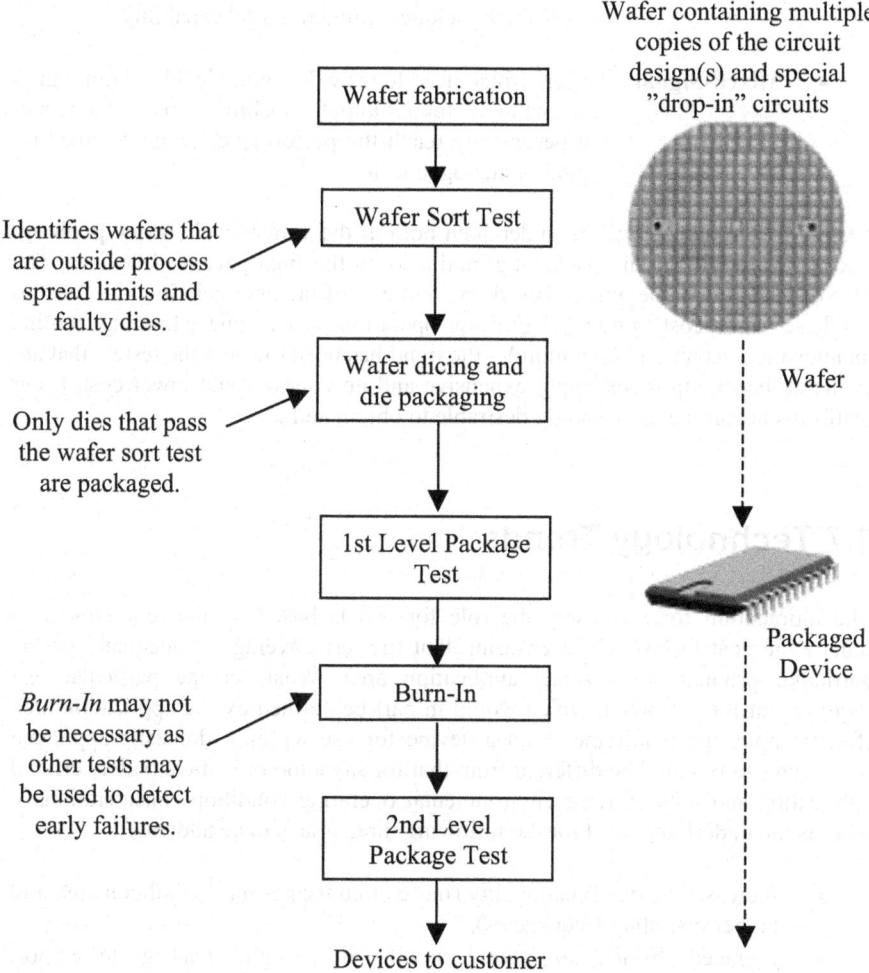

**Fig. 1.5.** Production test steps

Additionally, the type of device to test will determine the resources required by the semiconductor tester. In general, semiconductor testers will be categorised as:

- **Digital**: Optimised for digital circuits and systems with typically a large number of high-speed digital Input/Output pins and a limited analogue capability.

- **Memory**: Optimised for the testing of memory devices.

- **Analogue**: Optimised for analogue circuits with high performance analogue Input/Output current and voltage pins, with high performance data acquisition (the analogue to digital conversion) and signal analysis capabilities. These systems will include a limited digital capability.

- **Mixed-Signal**: Testers which need to provide a good level of both digital and analogue Input/Output capabilities but which may not necessarily reach the performance levels attained by digital or analogue testers.

Testing of the device will be undertaken both at the wafer level (**wafer probing**, prior to wafer dicing and packaging) and also on the final packaged devices. The type of tester available will be based on a number of factors, including the "cost to purchase" and "cost to own" (*e.g.* tester operation, test engineer training, routine maintenance, repair and upgrading) – the trend has been towards the testers that are available becoming increasingly expensive and now, newer and lower cost, tester platforms becoming increasingly desirable to obtain and use.

# 1.7 Technology Trends

The momentum for increasing the role for test is based on the requirement to reduce the cost to test whilst ensuring that the test coverage is adequate for the particular product and product application area. Whatever the particular test requirement for a product, which would in part be defined by the application area (for example, the requirements on a device for use within a domestic appliance (*e.g.* television) would be different from that for say a motor vehicle (safety critical application and with extreme environmental operating condition considerations)), there is the underlying need for the following three issues to be addressed:

- Increased device functionality (more circuitry per $mm^2$ of silicon area and higher operating frequencies).
- Reduced physical size (more circuitry in a smaller package to support miniaturisation of the product and aid portability (for mobile applications)).
- Lower cost (the need to sell products with higher performance for less cost to the customer).

The cost to test an IC is of primary concern given three main points:

- The cost of design is reducing.
- The cost of fabrication is reducing.
- The cost to test relative to the cost of design and fabrication is increasing. That is, for a particular product, the cost to test the fabricated design is

becoming significant. Whilst exact costs would be commercially sensitive and may be difficult to obtain, it has been quoted, for example, that test costs can be as high as 50% of the overall IC costs. This is a significant problem for the higher performance mixed-signal products.

# 1.8 International Technology Roadmap for Semiconductors (ITRS)

Of importance to the semiconductor industry globally is the International Technology Roadmap for Semiconductors (ITRS) [7-9]. This identifies industry trends, highlights technical obstacles, and provides companies with the information to align their product cycles with the developing technologies. The latest roadmap was published in 2003 and is an update of the previous roadmap published in 2001. The ITRS identifies the semiconductor industry technological challenges and needs over the next 15 years. This assessment (roadmapping) is a co-operative effort of the industry manufacturers and suppliers, government organisations and universities. It is sponsored by the following organisations:

- European Semiconductor Industry Association (ESIA)
- Japan Electronics and Information Technology Industries Association (JEITA)
- Korean Semiconductor Industry Association (KSIA)
- Semiconductor Industry Association (SIA)
- Taiwan Semiconductor Industry Association (TSIA)

The 2003 roadmap consists of a set of documentation categorised into the following. These are currently available for download from the ITRS homepage:

- Executive Summary
- System Drivers
- Design
- Test and Test Equipment
- Process Integration, Devices and Structures
- RF and Analogue/Mixed-Signal Technologies for Wireless Communications
- Emerging Research Devices
- Front End Processes
- Lithography
- Interconnect
- Factory Integration
- Assembly and Packaging
- Environment, Safety and Health
- Yield Enhancement
- Metrology
- Modeling and Simulation

## 1.9 Computer Aided Test

Reflecting the previous emphasis on improving the design activities, requiring ever more sophisticated software design and analysis tools, Computer Aided Design (CAD) has evolved to produce software design tools that allow for the efficiency and effectiveness needs of the designer to be met in order to meet the trends towards increasingly complex IC products. Test, on the other hand, has not traditionally been supported with the types of tools that have supported design. Now with the need for greater emphasis on test development and closer links to the design activities, software tools will now support the test engineer in the generation of the test programs. Computer Aided Test (CAT) tools provide the software tools that link into the design database and allow for **test program generation** (TPG) to be effectively undertaken.

## 1.10 Virtual Test

With the increase in device complexity and the subsequent increase in the problems associated with the creation of test programs, there is a need to reduce test program development time. **Virtual Test** [21] is undertaken using a software model of the design, a model of the ATE and the test program in order to simulate the operation of the test program prior to application on the ATE itself. This enables, through simulation, the development and debugging of the test program and has a direct link to the target ATE. It allows for much of the test program development and debug to be undertaken prior to device fabrication and without the need to undertake the tasks directly on the target ATE system.

## 1.11 Moore's Law

Back in the 1960s, in the early days of the integrated circuit revolution, it was feasible for tens of transistors to be integrated onto a single die. There could only be predictions as to the limit of future integration. Gordon Moore, later the founder of Intel® Corporation, provided the prediction in 1965 that has become a widely used benchmark to the future of IC integration. This prediction was based on the economics of device integration and the prediction of 65,000 components on a single silicon die would be possible by 1975.

*Moore's Law [10, 11] now is a prediction that the number of transistors within an integrated circuit will double every 18 to 24 months.*

The actual trend has kept broadly in line with this for the high-end digital circuits – *i.e.* microprocessors and memory. However, it is not the case for analogue and mixed-signal circuits where such levels of integration are not a fundamental requirement. An example of the substantial increase in transistors per IC is provided by Intel® Corporation in relation to their x86 and Pentium microprocessor products [11]. This shows the rapid increase in integration over the last 30 years for MPU (Microprocessor Unit) devices. Table 1.2 identifies a number of processors that have been developed and which show the device complexities now possible. In general, the precise definition of integration would need to be considered with care. For example, does the term mean the number of transistors on a single die, or within the IC? The IC may contain multiple dies, with each die containing fewer transistors than the total number for the IC itself.

**Table 1.2.** Example digital IC complexities

| Processor | Information |
|---|---|
| 4004 (Intel® Corporation) [11] | 2250 transistors, introduced 1971 |
| 486™ DX processor (Intel® Corporation) [11] | 1,180,000 transistors, introduced 1989 |
| Pentium 4 processor (Intel® Corporation) [11] | 42,000,000 transistors, introduced 2000, 130 nm and 90 nm CMOS processes |
| 100 GOPS vision processor [23] | 12,000,000 transistors, 0.35 µm CMOS process |
| Sun Microsystems "Niagara" processor [24] | 90 nm process |
| IBM, PowerPC 603™ microprocessor [25] | 1,600,000 transistors, 0.5 µm CMOS process |

# 1.12 Rent's Rule

Rather than considering the increase number of transistors within an integrated circuit versus time, Rent's Rule [12-14] determines the number of nets crossing a boundary between groups of gates.

A special case of this provides a relationship between the number of transistors *vs* the number of pins. As a result of the increase in the number of transistors within the integrated circuit, then there is a substantial increase in the number of components and interconnect within the IC. This has not been matched by the subsequent increase in the number of pins on the package. Whilst the high performance processors and digital systems may have in excess of 1000 package pins, the number of components and interconnect within the IC is substantial more. The approximate relationship between the number of I/O pins (Np) and the number of transistors (Nt) is given by:

$$Np = K.\sqrt{Nt}$$

where K is a constant.

## 1.13 Benchmark Circuits

Acknowledging the need to provide a standardised approach to the provision of metrics for the performance of Design for Testability (DfT) approaches and Automatic Test Pattern Generation (ATPG) tools, a set of benchmark circuits [15-17] from companies and academia have been developed and made freely available. These benchmark circuits are named after the conference/symposium that they are associated with:

- ITC'99          International Test Conference, 1999
- ITC'02          International Test Conference, 2002

- ISCAS'85        IEEE International Symposium on Circuits and Systems, 1985
- ISCAS'89        IEEE International Symposium on Circuits and Systems, 1989
- ISCAS'99        IEEE International Symposium on Circuits and Systems, 1999

## 1.14 DfX

Whilst much of the emphasis is placed on Design for Testability (DfT), in general the following are also considered and approaches developed:

- DfA          Design for Assembly
- DfD          Design for Debug [19, 20]
- DfM          Design for Manufacturability [18]
- DfR          Design for Reliability
- DfT          Design for Testability
- DfY          Design for Yield [22]

## 1.15 Summary

This chapter has introduced issues relating to IC product test and the role of the test engineer within a modern electronic/microelectronic product development environment. The changing role of test has led to the need to break down the traditional barriers with design and fabrication, and is leading to a fusion of these activities. Where once the test engineering activities may have been considered only once a design had been completed, they are now considered at an early stage in the design process. The emergence of Design for Testability (DfT) is one indication of the changes that have taken place, are in the process of happening, and will emerge in the future.

# 1.16 References

[1]    Smith M., "Application Specific Integrated Circuits", Addison-Wesley, 1999, ISBN 0-201-50022-1

[2]    Bushnell M. and Agrawal V., "Essentials of Electronic Testing for Digital, Memory & Mixed-Signal VLSI Circuits", Kluwer Academic Publishers, 2000, ISBN 0-7923-7991-8

[3]    Rajsuman, R., "System-on-a-Chip Design and Test", Artech House Publishers, USA, 2000, ISBN 1-58053-107-5

[4]    Hurst S., "VLSI Testing digital and mixed analogue/digital techniques", IEE, 1998, ISBN 0-85296-901-5

[5]    Needham W., "Designer's Guide to Testable ASIC Devices", Van Nostrand Reinhold, 1991, ISBN 0-442-00221-1

[6]    Burns M. and Roberts G.W., "An Introduction to Mixed-Signal IC Test and Measurement", Oxford University Press, New York, 2001, ISBN 0-19-514016-8

[7]    Edenfeld D. et al., "2003 Technology Roadmap for Semiconductors", Computer, IEEE Computer Society, January 2004, pp47-56

[8]    International Technology Roadmap for Semiconductors, 2003 Edition, "Executive Summary"

[9]    International Technology Roadmap for Semiconductors, 2003 Edition, "Test and Test Equipment"

[10]    Moore G., "Cramming more components onto integrated circuits", Electronics, Vol. 38, No. 8, April 1965

[11]    Moore's Law, Intel, http://www.intel.com

[12]    Landman B. and Russo R., "On a pin versus block relationship for partitions of logic graphs", IEEE Transactions on Computers, C-20, 1971, pp1469-1479

[13]    Christie P. and Stroobandt D., "The Interpretation and Application of Rent's Rule", IEEE Transactions on Very Large Scale Integration (VLSI) Systems, Vol. 8, No. 6, December 2000, pp639-648

[14]    Yazdani M., Ferry D.K. and Akers L.A, "Microprocessor Pin Predicting", IEEE Circuits and Devices Magazine, Vol. 13, No. 2, March 1997, pp28-31

[15]    Harlow J., "Overview of Popular Benchmark Sets", IEEE Design and Test of Computers, Vol. 17, No. 3, July-September 2000, pp15-17

[16]    ITC 1999 benchmark circuits, http://www.cerc.utexas.edu/itc99-benchmarks/bench.html

[17]    ITC 2002 benchmark circuits, http://www.extra.research.philips.com/itc02socbenchm/

[18]    Pennino T.P. and Potechin J., "Design for Manufacture", IEEE Spectrum, September 1993, pp51-53

[19]    Vermeulen B. and Goel S. K., "Design for Debug: Catching Design Errors in Digital Chips", IEEE Design and Test of Computers, Vol. 19, No. 3, May-June 2002, pp37-45

[20]    Vranken H.P.E. and Segers M.T.M., "Design-for-Debug in Hardware/Software Co-Design", Proceedings of the 5th International Workshop on Hardware/Software Co-Design, 1997, pp35-39

[21]    Hogan T. and Heffernan D., "Virtual Test reduces semiconductor product development time", IEE Electronics and Communication Engineering Journal, April 2001, pp77-83

[22]    Zorian Y. and Gizopoulos D., "Guest Editors' Introduction: Design for Yield and Reliability", IEEE Design and Test of Computers, Vol. 21, No. 3, May-June 2004, pp177-182

[23]    Raab W. et al., "A 100GOPS Programmable Processor for Vehicle Vision Systems", IEEE Design & Test of Computers, January-February 2003, pp8-16

[24]    Geppert L., "Sun's Big Splash", IEEE Spectrum magazine, January 2005, pp50-54

[25]    Hunter C. Vida-Torku E. and LeBlanc J., "Balancing Structured and Ad-hoc Design for Test: Testing of the PowerPC 603[TM] Microprocessor", Proceedings of the International Test Conference, 1994, pp76-83

[26]    United States Department of Defense, MIL-STD-883F, "Test Method Standard Microchips", 18th June 2004

# Chapter 2

# Fabrication Processes for Integrated Circuits

*Integrated Circuit designs will be realised in one of a number of possible fabrication processes. The circuit fabricated on a die will then be packaged within a suitable protective housing. The particular fabrication process to use will be dependent on a number of issues including cost, availability, experience in the use of, and circuit component capabilities.*

## 2.1 Introduction

A number of fabrication processes exist for realising microelectronic circuit designs. In the main, CMOS (Complementary Metal Oxide Semiconductor) technology is dominant, with a number of key advantages over the alternative processes. However, the available fabrication processes are classified as:

- Bipolar
- CMOS          Complementary Metal Oxide Semiconductor
- BiCMOS        Bipolar and CMOS
- SiGe BiCMOS   Silicon germanium BiCMOS
- GaAs          Gallium arsenide
- Memory

The above processes will be reviewed, with detail given to CMOS technology. These fabrication processes will be required to support the design and realisation of specific circuit components, the availability of components being dependent on the particular process. In the main, the following circuit components will be required:

- Transistor
- Diode

- Resistor
- Capacitor
- Inductor

In digital circuits, the transistor is the key circuit component used and digital processes will be optimised to support this. In analogue and mixed-signal processes, the use of high quality resistors, capacitors and potentially inductors, will be required in addition to the transistor. The basic structure of the IC die is shown in Fig. 2.1. Here, the circuit is fabricated within the die and this is mounted in a suitable package.

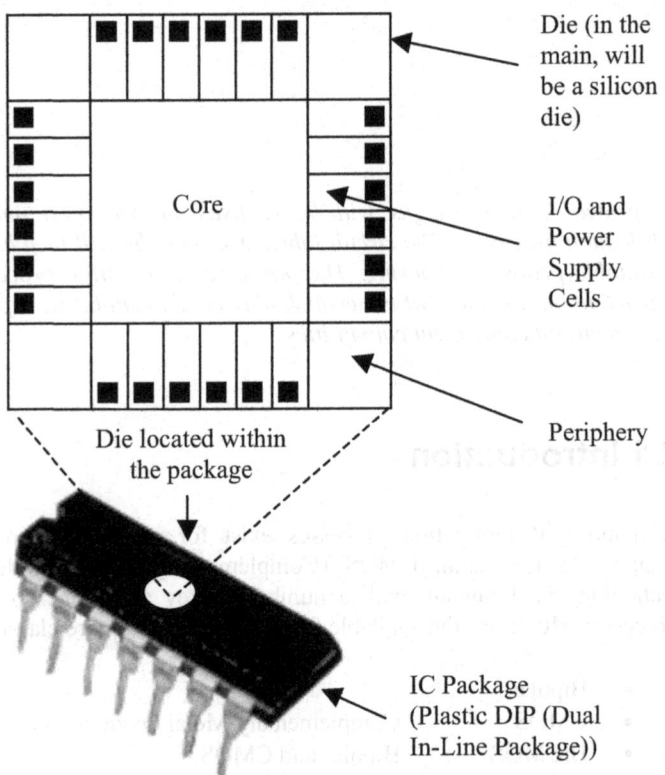

**Fig. 2.1.** Structure of the IC

The (silicon) die will be square or rectangular in shape and will be fabricated on a wafer. The die will have two identifiable areas, the periphery and the core:

- The core contains the bulk of the circuitry.
- The periphery will contain Input/Output cells and Power Supply cells. These cells are physically much larger than the core cells and will be required to bond the die to the package, as well as providing Electrical Overstress (EOS) protection.

The die is mechanically secured to the package and electrically bonded to the package pins. Electrical, mechanical and thermal considerations must be taken into account in the choice of the package used.

## 2.2 Technology Nodes

Driven by the need for higher levels of integration and higher operating frequencies for digital systems, the move is to reduce the circuit component geometries. This allows for more components per $mm^2$ of die area and a reduction in the interconnect lengths between components. There are two main figures given to represent the level of integration:

- Minimum transistor gate length
- Technology node

The **minimum transistor gate length** defines the smallest transistor gate length that can be designed (this is the **printed gate length**) by the designer. Once fabricated, this will actually be smaller due to processing issues (this is the **physical gate length**). For example, a 0.18 μm CMOS process would define a minimum printed MOSFET transistor gate length of 0.18 μm. The minimum gate length figure has been linked to microprocessor unit (MPU) and ASIC (Application Specific Integrated Circuit) devices.

The **technology node** is also used as a metric, and has been historically linked to the introduction of new generations of DRAM (Dynamic Random Access Memory). The figure quoted is the DRAM lithography half-pitch. In the International Technology Roadmap for Semiconductors (ITRS) 2003 edition [1-4], the predicted gate lengths and DRAM ½ pitch dimensions are linked. The roadmap is separated into near years (2003-2009) and far years (2010-2018). Table 2.1 provides a snapshot of the ITRS technology nodes at three points in time and an example of the Austria Mikro Systems (AMS) [20] process roadmap.

**Table 2.1.** Example available roadmaps and fabrication processes

| Source | Year | | |
|---|---|---|---|
| | **2005** | **2010** | **2015** |
| ITRS Roadmap Technology Node (DRAM ½ pitch (nm)) | 80 | 45 | 25 |
| ITRS Roadmap Technology Node (printed gate length (nm)) | 45 | 25 | 14 |

| Source | Roadmap Fabrication Processes | | |
|---|---|---|---|
| Austria Mikro Systems (AMS) (roadmap 2006 onwards – for the 0.18 μm process) [20] | Mixed-Signal/RF CMOS | SiGe BiCMOS | High Voltage CMOS |
| | High Speed Opto-CMOS | One Time Programmable / Non-Volatile Memory | |

## 2.3 Wafer Size

The wafer size (wafer diameter) has been increasing over recent years so that now the wafer diameters have progressed from 4" through 6" and 8" to the current 12" (300 mm) diameter. It is not predicted that the wafer diameter will increase beyond 300 mm for the foreseeable future due to cost issues.

The circuit dies are initially fabricated on the wafer, these being circular slices of semiconductor material (*e.g.* silicon). The wafer contains multiple copies of a single die, or may contain multiple copies of several die designs, along with process monitor die, see Fig. 2.2. The number and types of monitor die design depend on the maturity of the process – a new fabrication process will have different process monitoring requirements than a mature fabrication process.

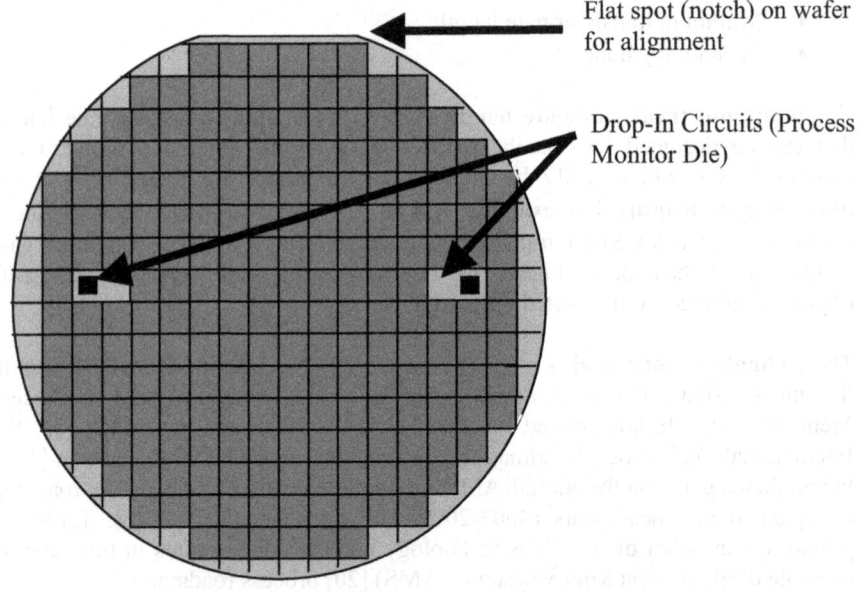

Flat spot (notch) on wafer for alignment

Drop-In Circuits (Process Monitor Die)

**Fig. 2.2.** Silicon wafer

Where several die designs are placed on the silicon wafer, this is referred to as a **Multi-Project Wafer** (MPW). These contain multiple "project" designs and would be used for design prototyping in that only a small number of dies are typically required (much less than a wafer can readily contain). The wafer fabrication costs can be spread across multiple projects and so reduce the design prototyping costs. With the introduction of 300 mm wafers, the ability for each wafer to contain more copies of a die can aid in the reduction of the fabrication costs of the die. However, the necessary fabrication equipment purchase and operational costs for the lower process geometries are much higher than those for the coarser process geometries using smaller diameter wafers. The cost implications and potential cost benefits need careful consideration.

## 2.4 Bipolar Technology

Within the bipolar process [5, 6], the fabrication of the bipolar transistor (BJT – Bipolar Junction Transistor) is of primary concern, although a fabrication process may also support other forms of component (in particular the inclusion of resistors and capacitors). The bipolar process was the only available technology available in the early years of IC fabrication and today, bipolar transistors are used for their advantages in high-speed operation (*e.g.* communication systems) and high-current drive capability for power applications (*e.g.* automotive). However, they do not allow for the high levels of integration and low-power operation that can be achieved with CMOS (see next section).

The bipolar transistor is based around a sandwich of n and p-type silicon, see Fig. 2.3.

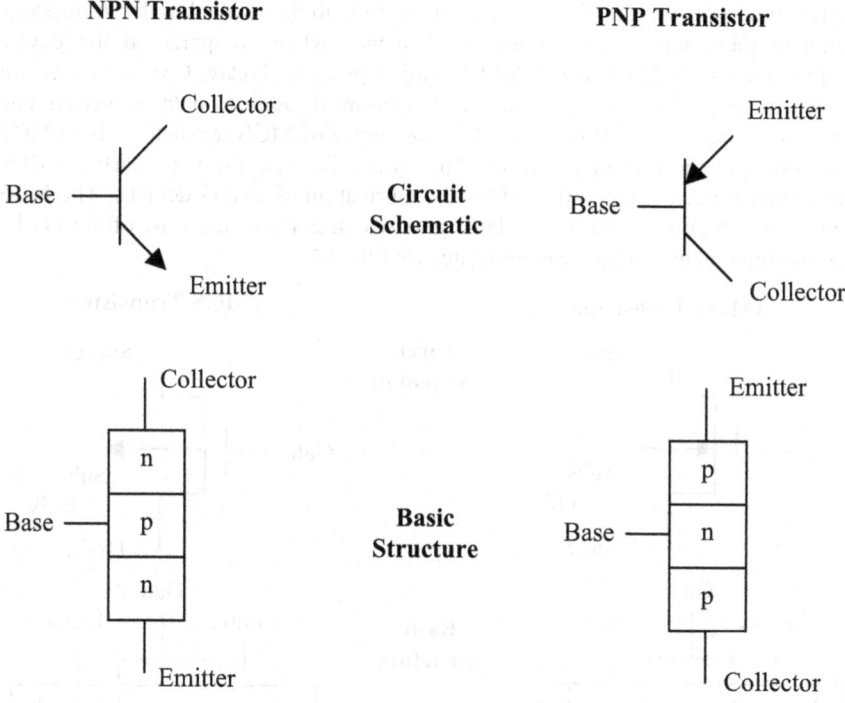

**Fig. 2.3.** Bipolar transistor forms

The basic operation, with reference to the NPN transistor, is that a current flowing into the base of the transistor (and exiting out of the emitter) allows for a larger collector current to flow from the collector through to the emitter. This current gain can be used to implement analogue (*e.g.* amplification) and digital (*e.g.* digital logic) functions. In the diagrams shown in Fig. 2.3, a 4th connection (substrate) connection has been omitted, although this would need to be accounted for in an IC design.

## 2.5 Complementary Metal Oxide Semiconductor (CMOS) Technology

Following on from the early days of the bipolar fabrication process, the Complementary Metal Oxide Semiconductor (CMOS) [7-9] fabrication process has been the mainstay of the microelectronics industry since the early 1970s. In this, the MOSFET (Metal Oxide Semiconductor Field Effect Transistor) is of primary concern, although a fabrication process may also support other forms of component (in particular the inclusion of resistors and capacitors). Additionally, a CMOS fabrication process can also include the ability to fabricate bipolar transistors, although these tend to be of low performance when compared to the device performance available in a bipolar fabrication process. Today, CMOS is used for many circuit applications from digital processors through to data converters and analogue amplifiers. CMOS is based on two types of MOS transistor – the nMOS transistor and the pMOS transistor. This is a follow-on from the earlier nMOS fabrication process which allowed for the fabrication of nMOS devices. The MOS transistor is based around an insulated gate which controls the flow of current by varying the voltage on the insulated gate, see Fig. 2.4.

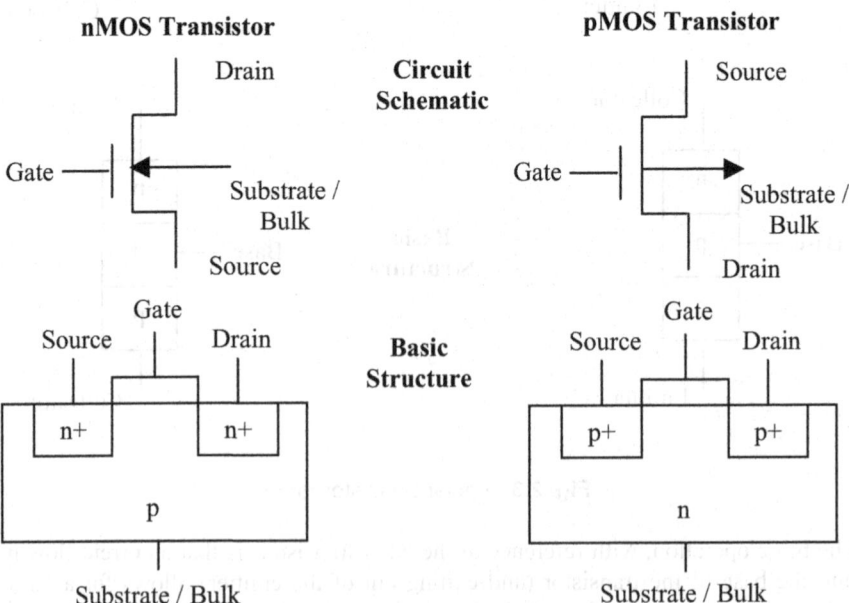

**Fig. 2.4.** MOS transistor forms

Here, the transistor has four connections (gate, drain, source and substrate (or bulk)). With reference to the nMOS transistor, a positive voltage between the gate and source allows for a current to flow between the drain and source. By varying the gate-source voltage, the drain current will vary. This operation can be used to implement analogue (*e.g.* amplification) and digital (*e.g.* digital logic) functions.

A 3D view of the nMOS transistor is shown in Fig. 2.5.

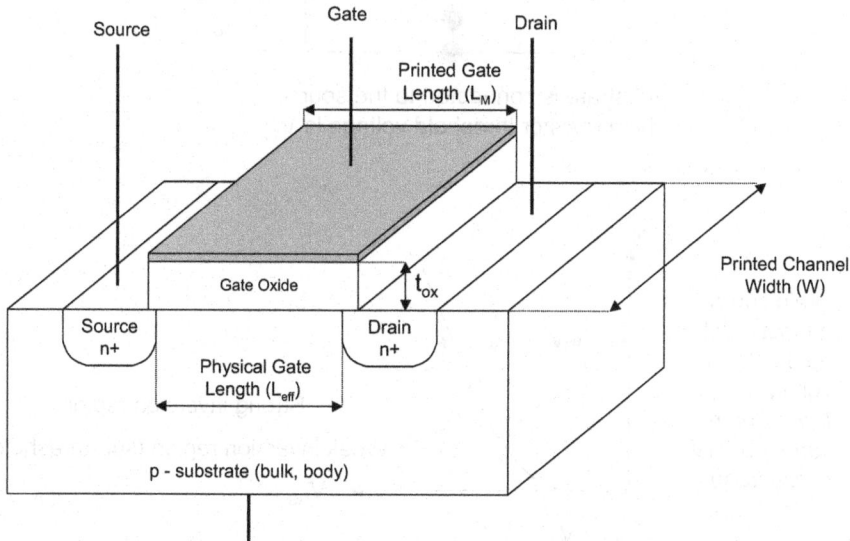

**Fig. 2.5.** 3D view of the nMOS transistor (simplified)

The printed gate length ($L_M$), the physical gate length ($L_{EFF}$ – effective gate length) and the insulating gate oxide thickness ($t_{ox}$) are shown.

The operation of the MOSFET is considered with respect to:

- Gate-source voltage ($V_{GS}$) vs drain current ($I_D$).
- Drain-source voltage ($V_{DS}$) vs drain current ($I_D$) for various gate-source voltage values.

The transistor can operate in both a linear region (the transistor acts as a voltage controlled resistor) and a saturation region (the transistor acts as a voltage controlled current source). Figure 2.6 shows the basic relationship plots for the nMOS transistor. When the gate-source voltage is below a threshold voltage ($V_{TO}$), the drain current is, to a first approximation, zero. It is however small but finite – this is the sub-threshold current. It is this small sub-threshold current that gives CMOS its low-power operation capability and the effect is used in $I_{DDQ}$ testing for digital logic. However, with the smaller transistor gate lengths found in deep submicron (DSM) processes, this sub-threshold current is significantly larger.

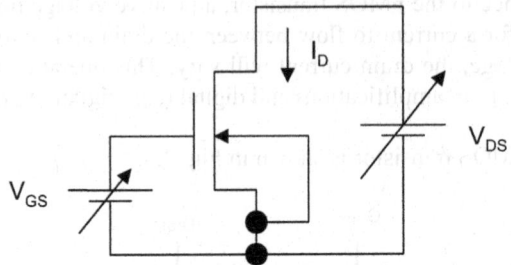

Substrate is connected to the source
The transistor threshold voltage is $V_{TO}$

Drain current as a function of gate-source voltage: *transistor in saturation: ($V_{DS}$ is constant)*

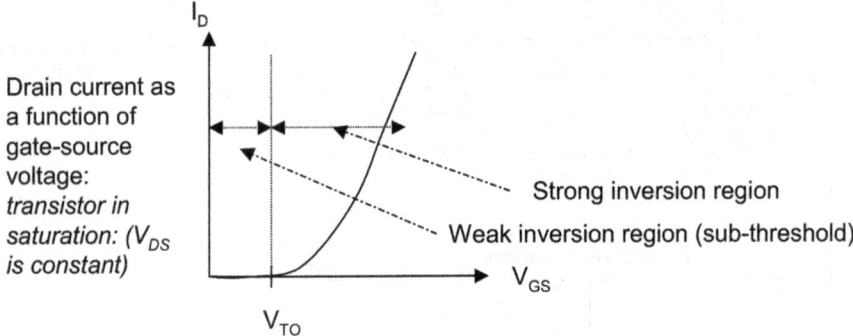

Strong inversion region

Weak inversion region (sub-threshold)

$V_{TO}$

Pinch-off:
$V_{DS} = (V_{GS} - V_{TO})$

Current-voltage relationship including channel length modulation effect ($\lambda$)

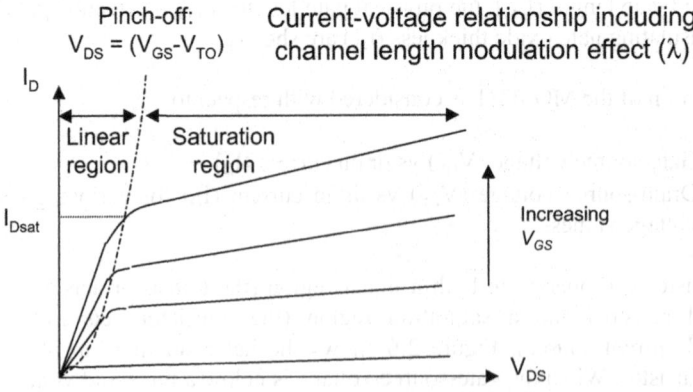

Linear region

Saturation region

$I_{Dsat}$

Increasing $V_{GS}$

**Fig. 2.6.** $V_{GS}$ / $V_{DS}$ vs $I_D$ plots

In addition to the MOSFET, the CMOS fabrication process will also be capable of implementing passive components (resistors, capacitors and, to a very limited extent, inductors), along with a basic bipolar transistor. With the resistor and capacitor, there are a number of variants on the basic R and C, see Figs. 2.7 and 2.8. In Fig. 2.7, a layer within the CMOS process will have a certain resistivity ($\rho$). By defining a suitable shape of material (and assuming that the thickness of the layer will remain constant), a resistor can be fabricated. Resistors are however physically large (when compared to transistors) and it is difficult to fabricate accurate values.

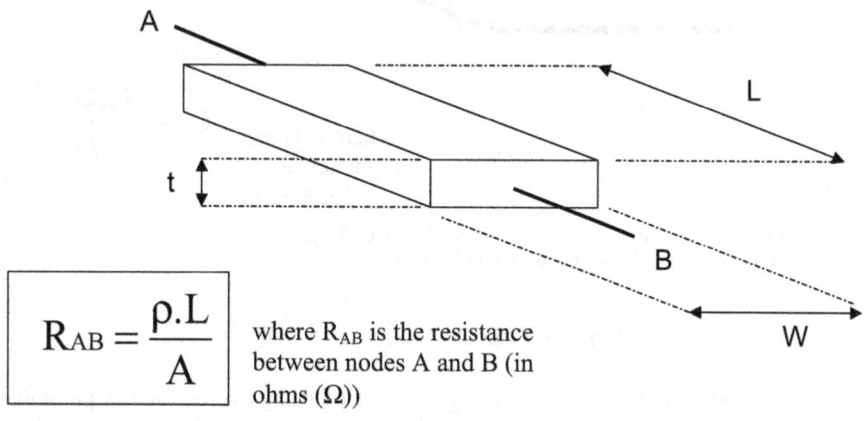

$$R_{AB} = \frac{\rho.L}{A}$$

where $R_{AB}$ is the resistance between nodes A and B (in ohms ($\Omega$))

Where A is the cross-sectional area (W.t)

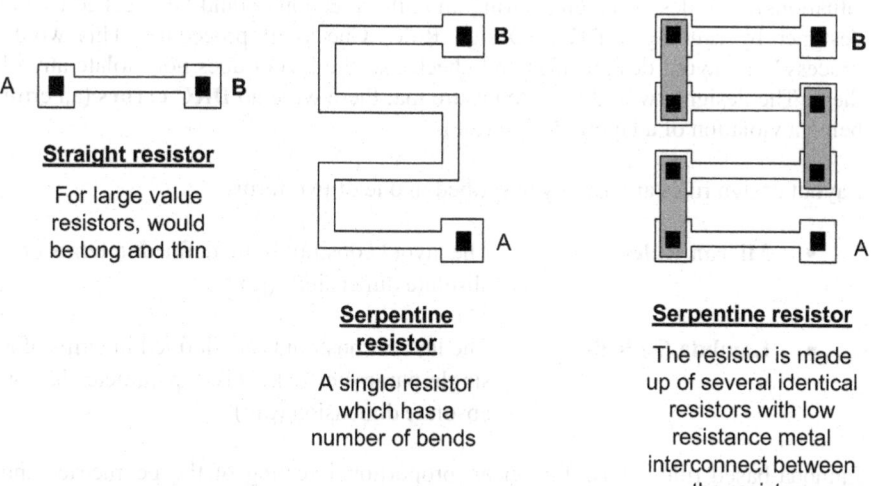

**Straight resistor**

For large value resistors, would be long and thin

**Serpentine resistor**

A single resistor which has a number of bends

**Serpentine resistor**

The resistor is made up of several identical resistors with low resistance metal interconnect between the resistors

**Fig. 2.7.** Resistor structures in CMOS

In Fig. 2.8, the capacitor is fabricated with two conductors separated by an insulator (dielectric). This forms a capacitor structure. A range of capacitor structures can be formed in the CMOS process.

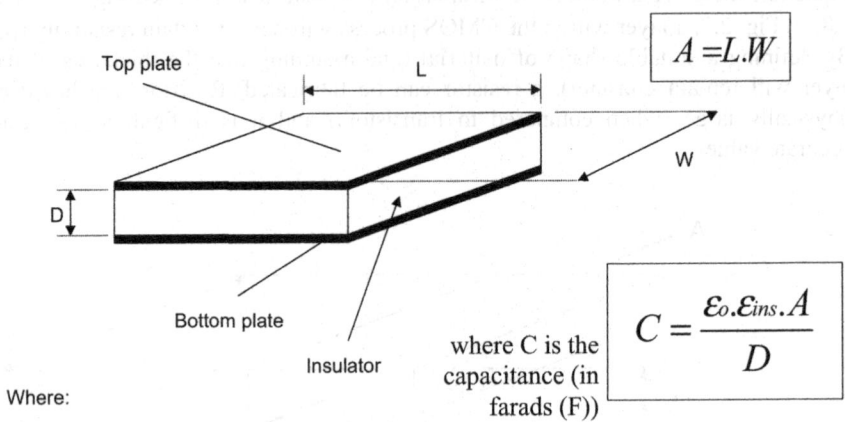

$$A = L.W$$

$$C = \frac{\varepsilon_o . \varepsilon_{ins} . A}{D}$$

where C is the capacitance (in farads (F))

Where:
- $\varepsilon_{ins}$ is the relative permittivity of the insulator ($\approx 4.0$ for $S_iO_2$)
- $\varepsilon_o$ is the permittivity of free space ($8.85 \times 10^{-14}$ F/cm)

**Fig. 2.8.** Capacitor structure in CMOS

The layout will be designed according to a set of design rules for the particular fabrication process. These design rules will have been generated to allow for a high overall yield (working circuits after fabrication) and reliability of the circuits, whilst using the smallest possible area. For a design to be accepted for fabrication, the layout must conform to the layout design rules (except in special agreed situations). The design layout, during and after creation, would be checked by the designer by running a **DRC** (Design Rules Checking) procedure. This would "access" the layout design rules and check that the layout does not violate any of these. The designer would need to ensure that there were no **DRC errors** (an error being a violation of a layout design rule).

Layout design rules are usually described in one of two forms:

- **Micron Rules**      The layout constraints are defined in terms of absolute dimensions (μm).

- **Lambda (λ) Rules**      The layout constraints are defined in terms of a single parameter (λ). This λ parameter has an absolute dimension (μm).

Lambda-based rules allow for linear, proportional scaling of the geometries and were originally devised to simplify industry-based micron rules, also allowing for scaling of various processes.

For fabrication, the layout will be exported from the CAD (Computer Aided Design) tool to a file that will describe the circuit layout in a suitable format. The two file formats used are:

- GDSII          Stream File Format
- CIF            Caltech Intermediate Format

CIF is a low-level graphics language for defining layout geometries. The CIF file consists of a number of statements (and where appropriate, comments). This is a machine-independent ASCII text file. Consider the CMOS inverter shown in Fig. 2.9. Here, the design is a static CMOS logic inverter shown as a schematic (left) and a layout (right).

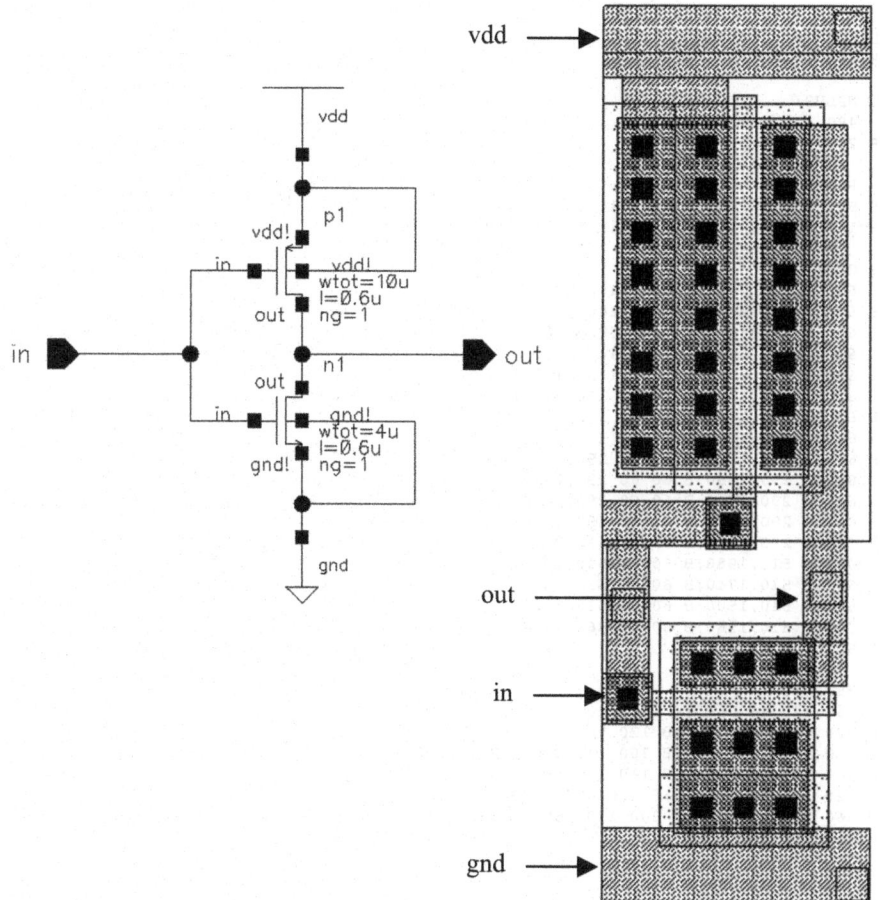

**Fig. 2.9.** Schematic diagram and layout for a CMOS inverter

An example of a CIF file describing this inverter layout is shown in Fig. 2.10. This file contains information on the geometries on each layer that are used to create the inverter design (the transistors and metal interconnect). All dimensions are in centimicrons.

```
(CIF file for Inverter cell);

DS 1 1 1;

9 Inverter;

L NWELL;
B 760 1360 380,1680;

L PPLUS;
B 480 200 400,250;
B 420 1080 410,1680;

L DIFF;
B 400 540 400,460;
B 540 1000 310,1680;

L NPLUS;
B 480 420 400,560;
B 200 1080 100,1680;

L POLY1;
B 520 60 400,550;B 140 140 70,560;
B 140 140 360,1050;B 60 1120 400,1680;

L CONTACT;
B 60 60 280,260;B 60 60 400,260;
B 60 60 520,260;B 60 60 280,440;
B 60 60 400,440;B 60 60 520,440;
B 60 60 280,660;B 60 60 400,660;
B 60 60 520,660;B 60 60 110,2100;
B 60 60 110,1980;B 60 60 110,1860;
B 60 60 110,1740;B 60 60 110,1620;
B 60 60 110,1500;B 60 60 110,1380;
B 60 60 110,1260;B 60 60 290,2100;
B 60 60 290,1980;B 60 60 290,1860;
B 60 60 290,1740;B 60 60 290,1620;
B 60 60 290,1500;B 60 60 290,1380;
B 60 60 290,1260;B 60 60 510,2100;
B 60 60 510,1980;B 60 60 510,1860;
B 60 60 510,1740;B 60 60 510,1620;
B 60 60 510,1500;B 60 60 510,1380;
B 60 60 510,1260;B 60 60 360,1050;
B 60 60 70,560;

L METAL1;
B 360 120 400,660;B 120 120 70,560;
B 120 120 360,1050;B 120 960 510,1680;
B 300 960 200,1680;B 300 130 200,2225;
B 760 200 380,2390;B 110 120 635,660;
B 120 1440 630,1440;B 120 370 70,805;
B 360 300 400,350;B 300 120 150,1050;
B 760 200 380,100;

DF;
C 1;
E
```

**Fig. 2.10.** Example CIF file for a CMOS inverter

## 2.6 BiCMOS Technology

By combining the bipolar and CMOS processes, a BiCMOS process is developed. Here, the low-power and high levels of integration of CMOS and the power drive capabilities of bipolar are integrated. This is desirable for applications that require a large amount of digital signal processing (in CMOS) and actuator drive (in bipolar). An example application would be the automotive industry for the intelligent control of actuators such as DC motors. There is a trade-off required in the performance of each type of transistor required.

## 2.7 SiGe BiCMOS Technology

Silicon-Germanium (SiGe) [10-12] is a technology that combines the integration and cost benefits of silicon with the speed of more expensive technologies such as gallium-arsenide. In this process, germanium is introduced into the base layer of an otherwise all-silicon bipolar transistor. This creates significant improvements in operating frequency, current, noise, and power capabilities of the resulting transistor when compared to an all-silicon bipolar transistor. The result is a SiGe **Heterojunction Bipolar Transistor** (HBT). The application areas are mainly high frequency (GHz) RF (Radio Frequency) communications circuits and systems. Examples include wireless communications and high-speed data converters.

## 2.8 Gallium Arsenide (GaAs) Technology

Gallium Arsenide (GaAs) was developed and initially used for IC fabrication to overcome the operating frquency limitations with silicon devices in high frequency RF (Radio Frequency) communications systems applications. The GaAs material electron mobility is much higher than that of silicon and it was put forward as a complement to silicon for high-speed circuits and systems. However, the high fabrication costs have limited its uptake and have mainly been superseded by both improvements in silicon processing and the introduction of SiGe processes.

## 2.9 Memory Processes

Memory processes are optimised for the high device density required for the integration of high capacity memories and low memory access times:

- RAM                SRAM and DRAM

- ROM                PROM, EPROM, EEPROM, FLASH

In particular, the process advances are being driven by the DRAM and FLASH memory market areas requiring increasing memory capacity and faster memories (reducing memory access times for both writing data to the memory and reading data from the memory).

# 2.10 Packaging

At some point, the die will need to be packaged for protection and further use. Unless the bare die is to be directly mounted onto a substrate (MCM, SiP or PCB), it will be packaged in either a Through-Hole or Surface Mount package type. The package material will be either plastic or ceramic. The following lists the package names referring to the main available types:

## Through-Hole Device Package Types

**CERQUAD**       Ceramic Quadruple Side

**DIP**           Dual In-line Package:

|  |  |
|---|---|
| **CERDIP** | Ceramic DIP |
| **HDIP** | Hermetic DIP |
| **PDIP** | Plastic DIP |

**SIP**           Single In-line Package

**ZIP**           Dual In-line Zig-Zag Package

**PGA**           Pin Grid Array:

|  |  |
|---|---|
| **CPGA** | Ceramic Pin Grid Array |
| **PPGA** | Plastic Pin Grid Array |
| **SPGA** | Staggered Pin Grid Array |

## Surface Mount Device Package Types

**BGA**           Ball Grid Array:

|  |  |
|---|---|
| **CBGA** | Ceramic Ball Grid Array |
| **FBGA** | Fine Pitch Ball Grid Array |
| **PBGA** | Plastic Ball Grid Array |

**CQFP**          Ceramic Quad Flat Pack

| **LCC** | Leadless Chip Carrier*: | |
| --- | --- | --- |
| | **CLCC** | Ceramic Leadless Chip Carrier |
| | **PLCC** | Plastic Leadless Chip Carrier |

| **LCC** | Leaded Chip Carrier*: | |
| --- | --- | --- |
| | **JLCC** | J-Leaded Chip Carrier |
| | **CLCC** | Ceramic Leaded Chip Carrier |
| | **PLCC** | Plastic Leaded Chip Carrier |

| **QFP** | Quad Flat Pack: | |
| --- | --- | --- |
| | **CQFP** | Ceramic Quad Flat Pack |
| | **PQFP** | Plastic Quad Flat Pack |
| | **QFJ** | Quad Flat Pack (J-lead) |
| | **TQFP** | Thin Quad Flat Pack |
| | **VQFB** | Very-thin Quad Flat Pack |

| **SOIC** | Small Outline IC: | |
| --- | --- | --- |
| | **CSOIC**   Ceramic Small Outline Integrated Circuit | |

| **SOP** | Small Outline Package: | |
| --- | --- | --- |
| | **PSOP** | Plastic Small-Outline Package |
| | **QSOP** | Quarter Size Outline Package |
| | **SOJ** | Small Outline (J-lead Package) |
| | **SSOP** | Shrink Small-Outline Package |
| | **TSOP** | Thin Small-Outline Package |
| | **TSSOP** | Thin Shrink Small-Outline Package |
| | **TVSOP** | Thin Very Small-Outline Package |

Note: * - the leaded and leadless chip carriers can be identified by the same acronyms and can be easily confused.

Packages are defined by the following Military Standards:

- **MIL-STD-1835D**      Electronic Component Case Outlines.
- **MIL-HDBL-6100**      Case details for Discrete Semiconductor Devices.
- **MIL-STD-2073-1D**    Packaging of Microcircuits (Military Packaging).
- **MIL-STD-1285D**      Marking of Electrical and Electronic Parts.

# 2.11 Die Bonding

The bonding of the die to the package is primarily required to allow for electrical connections (signal I/O and power) between the package pins and the die core circuitry. Once the die has been secured to the package (allowing for thermal bonding for heat removal (using materials with similar Thermal Coefficient of Expansion (TCE) and sometimes electrical connection from the back of the die to the package), the die bonding is required. If the materials have different Thermal Coefficients of Expansion, during device heating and cooling under normal and extremes of operation, thermal stresses may be introduced which can lead to problems in the interconnect such as broken joints.

Wire bonding (using fine gold or aluminum wires), Tape Automated Bonding (TAB) or flip-chip solder bonding allow for the electrical connections to be made.

# 2.12 Multi-Chip Modules

Many packaged devices will contain a single die. However, there are situations where a package will contain multiple dies, *e.g.* where sensors and circuits are to be housed in a single package but cannot be fabricated on a single die. These types of devices are referred to as **Multi-Chip Modules** (MCMs) [13] and were originally referred to as **Hybrid Circuits**. The MCM is a structure consisting of two or more Integrated Circuits (ICs) and passive components on a common circuit base (substrate), and interconnected by conductors fabricated within that base. The ICs may either be packaged dies or bare dies. This addresses a number of issues relating to the physical size reduction problem and the degradation of signals passing through the packaging and interconnect on a PCB. The MCM may provide advantages in certain electronic applications over a conventional PCB implementation. The advantages include:

- Increased system speed
- Reduced overall size
- Ability to handle ICs with a large number of Inputs and Outputs (I/O)
- Increased number of interconnections in a given area
- Reduced number of external connections for a given functionality

In addition, an MCM may contain dies produced with different fabrication processes within a single packaged solution (e.g. mixing low-power CMOS with high-power Bipolar technologies). There are a number of types of MCMs that can be realised:

**MCM-D**
Modules whose interconnections are formed by thin film deposition of metals on deposited dielectrics. The dielectrics may be polymers or inorganic dielectrics.

**MCM-L**
Modules using advanced forms of PCB technologies, forming copper conductors on laminate-based dielectrics.

## MCM-C

Modules constructed on co-fired ceramic substrates using thick film (screen printing) technologies to form conductor patterns. The term co-fired relates to the fact that the ceramic and conductors are heated in the oven at one time.

## MCM-D/C

Deposited dielectric on co-fired ceramic.

## MCM-Si

Silicon based substrate similar to conventional silicon ICs.

It is essential to obtain good quality dies that are known to be fully functional. Obtaining a **Known Good Die** (KGD) is a requirement for MCM design. During the production test of an IC, rather than progressing the die into a packaging stage, it may provided as a KGD to the customer. In this case, it may be necessary to undertake a more rigorous test on the die than would have been undertaken if the die was then to be later packaged and re-tested by the producer. With the ability to provide bare silicon dies, then an electronic system may be produced in one of a number of ways – from bare silicon die through to PCB. There are four package levels:

**Die level**                    Bare silicon die.

**Single IC level**              Packaged silicon die (a single packaged die).

**Intermediate level**           An intermediate level where silicon dies (die level) and/or packaged dies (single IC level) are placed onto a suitable substrate that may or may not then be further packaged.

**PCB level**                    Printed Circuit Board.

Combining these four levels, four types of packaged electronics can be identified:

**Type 1**                       Packaged silicon die mounted on a PCB.

**Type 2**                       Packaged silicon die mounted on an intermediate substrate that is then mounted onto a PCB.

**Type 3**                       A bare silicon die mounted on an intermediate substrate that is then mounted onto a PCB.

**Type 4**                       A bare die mounted directly onto a PCB.

Figure 2.11 shows the relationship between package levels and the types of packaged electronics. Package types 2 and 3 are commonly referred to as MCMs.

| Package Level | Package Type | | | |
|---|---|---|---|---|
| | 1 | 2 | 3 | 4 |
| Die level | ✸ | ✸ | ✸ | ✸ |
| Single IC level | ✸ | ✸ | | |
| Intermediate level | | ✸ | ✸ | |
| PCB level | ✸ | ✸ | ✸ | ✸ |

**Fig. 2.11.** Package levels and types

The MCM structure is in-line with the ITRS definition for a **System in Package** (SiP) device. The ITRS definition for the SiP is:

*"any combination of semiconductors, passives, and interconnects integrated into a single package"*

However, SiP designs [14-18] extend the concept of the MCM from devices placed horizontally side-by-side and bonded to a substrate (as in a PCB), to vertically stacked devices with bonding to the substrate. Wire bonding to the substrate is common.

# 2.13 Foundry Services

A number of companies provide their own foundry services, either for in-house use or as a service to other organisations. Table 2.2 identifies a number of the key foundry services available.

**Table 2.2.** Foundry services

| Foundry | URL |
|---|---|
| AMI Semiconductor (AMIS) | http://www.amis.com |
| Austria Mikro Systems (AMS) | http://www.austriamicrosystems.com |
| Chartered Semiconductor | http://www.charteredsemi.com |
| International Business Machines (IBM) | http://www.ibm.com |
| Hewlett Packard (HP) | http://www.hp.com |
| Intel Corporation | http://www.intel.com |
| Maxim | http://www.maxim-ic.com |
| ST Microelectronics | http://www.st.com |
| Taiwan Semiconductor Manufacturing Company (TSMC) | http://www.tsmc.com |
| United Microelectronics Corporation (UMC) | http://www.umc.com |

In addition, organisations provide access to tools and foundry services for groups of users (e.g. universities). The two key organisations are:

- **Europractice**    http://www.europractice.com
- **Mosis**    http://www.mosis.org

## 2.14 Process Variations

Whenever a wafer is fabricated, each wafer will have variations from each other as to the electrical properties of the circuit components fabricated on the wafer. Additionally, each circuit die on the wafer will vary from each other as to the electrical properties of the circuit components fabricated. It is necessary during the fabrication process to ensure that the process remains within a specific tolerance band. Additionally, it is a common approach for the circuit design to be created so that it will operate according to the required specifications over the guaranteed process variations. Circuit simulation, using a range of component models, will be used by the designer in order to verify the circuit operation (through simulation). With the finer process geometries, the spread in the electrical properties of the individual transistors can be significant and cause for concern.

## 2.15 Electromigration

Electromigration [5, 6, 19] occurs in the interconnect metal due to high current densities. This is a significant problem when the track widths reduce to narrow dimensions. The current density in the interconnect (amps/m$^2$) flowing through the small cross-section area of metal can be high enough to cause bulk movement of the metal due to the momentum of the current carrying electrons. This can cause the narrowing of the track width and increased current density. The result will be an increase in the electrical resistance of the track and eventual fusing.

## 2.16 Future Directions

Future advances [1-4] in the processing and packaging are based on the required end user applications. A summary is provided in Table 2.3.

**Table 2.3.** Future process issues

| Technology driver | Description |
|---|---|
| Faster operating speeds | Higher signal frequencies<br>Driven by communications applications<br>Reduced device geometries (gate length and gate oxide thickness)<br>Moving onto the next technology node to improve performance of components and reduce interconnect delays<br>Lower k (dielectric constant) dielectrics to |

|  | reduce interconnect delay. A move away from silicon dioxide ($SiO_2$) to insulators with a lower dielectric constant (k)<br>Low resistivity metal for interconnect. Move away from aluminum interconnect to copper<br>New packaging materials<br>Move into the nanotechnology domain |
|---|---|
| Higher levels of integration (Higher density) | Reduced device geometries (gate length and gate oxide thickness)<br>Higher k dielectrics for DRAM capacitors to reduce capacitor area<br>Thinner dielectrics for memory devices<br>Reduced package dimensions<br>More pins on a package<br>Move into the nanotechnology domain |
| Lower operating voltages | A need to improve device reliability by reducing electric field strength in dielectrics<br>Aim to lower device power consumption<br>Portable, battery operated circuits |
| SoC and SiP devices | Multiple dies and passive components within packages<br>Integration of MEMs (Micro-Electromechanical) devices (e.g. sensor integration)<br>Process integration – analogue, digital and memory on the same die |

## 2.17 Summary

This chapter has provided an introduction to the basic fabrication processes and packaging requirements for Integrated Circuits. Whenever a circuit design is to be realised, the right fabrication process and packaging technology must be chosen. This will be dependent on a number of requirements including the availability of a fabrication process and the ability for the process to allow for the circuit functionality to be designed at the right cost. CMOS is by far the most prevalent fabrication process in use today, with bipolar, BiCMOS, SiGe BiCMOS and GaAs also utilised. This chapter has introduced the available processes as an overview with references provided for further reading. The purpose of this was to introduce the basic fabrication processes in order to identify a number of the process issues that will relate to test program development.

# 2.18 References

[1]     International Technology Roadmap for Semiconductors, 2003 Edition, "Executive Summary"

[2]     International Technology Roadmap for Semiconductors, 2003 Edition, "Test and Test Equipment"

[3]     International Technology Roadmap for Semiconductors, 2003 Edition, "Assembly and Packaging"

[4]     International Technology Roadmap for Semiconductors, 2003 Edition, "Lithography"

[5]     Chang C.Y. and Sze S.M., "ULSI Technology", McGraw-Hill International Editions, Singapore, 1996, ISBN 0-07-114105-7

[6]     Sze S.M., "Semiconductor devices Physics and Technology", Wiley, New York, 1985, ISBN 0-471-83704-0

[7]     Laker K.R. and Sansen W.M.C. "Design of Analog Integrated Circuits and Systems", McGraw-Hill International Editions, Singapore, 1994, ISBN 0-07-113458-1

[8]     Bellaouar A. and Elmasry M., "Low-Power Digital VLSI Design Circuits and Systems", Kluwer Academic Publishers, The Netherlands, 1995, ISBN 0-7923-9587-5

[9]     Kang S. and Leblebici Y., "CMOS Digital Integrated Circuits Analysis and Design", McGraw-Hill International Editions, Singapore, 1996, ISBN 0-07-114423-4

[10]    Meyerson B.S., "High speed silicon-germanium electronics," Scientific American, vol. 270, no. 3, pp. 42-47, 1994.

[11]    IBM Research, http://www.research.ibm.com/

[12]    Singh R., Modest M.M. and Oprysko D.H., "Silicon Germanium Technology, Modeling and Design", Wiley-IEEE Press, 2003, ISBN 0-471-44653-X

[13]    Doane D. A. and Franzon P.D., "Multichip Module Technologies and Alternatives, The Basics", Van Nostrand Reinhold, New York, 1993, ISBN 0-442-01236-5

[14]   Evans-Pughe C., "Got to get a packet or two", IEE Review, December 2004, pp40-43

[15]   Edwards C., "Questions hover over the package path to Integration", IEE Electronics Systems and Software, August-September 2004, pp30-31

[16]   Miettinen, J., Mantysalo, M., Kaija, K. and Ristolainen, E.O. "System design issues for 3D system-in-package (SiP)", Proceedings of the Electronic Components and Technology Conference (ECTC), 2004, Vol. 1, pp610-614

[17]   Tai K.L., "System-In-Package (SIP): challenges and opportunities", Proceedings of the Asia and South Pacific Design Automation Conference, 2000, pp191-196

[18]   Song Y., et al., "The reliability issues on ASIC/memory integration by SiP (system-in-package) technology", Proceedings of the IEEE International SOC Conference, 2003, pp7-10

[19]   O'Connor P., "Test Engineering, A Concise Guide to Cost-effective Design, Development and Manufacture", John Wiley & Sons Ltd., England, 2001, ISBN 0-471-49882-3

[20]   Austria Mikro Systems (AMS), Austria, "Process Roadmap", http://www.austriamicrosystems.com/05foundry/roadmap.htm

# Exercises

## Question 1

From the device datasheets, identify the following for the current range of microprocessors used in <u>desktop</u> PCs:

- The device type and manufacturer
- Package type and number of pins
- Number of pins dedicated to the following:
  - Power Supply
  - I/O
  - Test
- Power consumption and cooling requirements

## Question 2

Repeat Question 1, except now identify the following for the current range of microprocessors used in <u>notebook</u> PCs.

## Question 3

Repeat Question 1, except now identify the following for the current range of microprocessors used in <u>Personal Digital Assistants (PDAs)</u>.

## Question 4

Identify the applications that use bipolar technology. What is the range of signal frequency that they work with?

## Question 5

Identify the applications that use CMOS technology. What is the range of signal frequency that they work with?

## Question 6

Identify the applications that use BiCMOS technology. What is the range of signal frequency that they work with?

## Question 7

Identify the applications that use SiGe technology. What is the range of signal frequency that they work with?

## Question 8

Identify the applications that use GaAs technology. What is the range of signal frequency that they work with?

# Chapter 3

# Digital Logic Test

*Digital logic forms the basis of many electronic circuits and systems from simple decoding logic through to complex microprocessor based systems. Whatever the application and complexity of the design, digital logic testing is based on a number of core principles and, provided that the design can be suitably accessed, particular test stimuli applied and the results observed, the device test problem can be addressed.*

## 3.1 Introduction

Despite the world being analogue in nature, the role that is being demanded by the end user for digital electronic circuits and systems [1-4] is driving many of the advances in integrated circuit design, fabrication and test. Electronic circuit functions can therefore be undertaken in either the analogue or digital domains:

- *The **analogue domain** is a representation of a signal that varies continuously over a range of values.*

- *The **digital domain** represents a signal that varies in discrete levels over a range of values.*

Given that electronic circuit functions can be undertaken in either the analogue or digital domains, integrated circuits will be classified as one of three types of circuit:

- **Analogue**: this type of circuit manipulates continuously varying signals (voltages and currents) over a range of values.

- **Digital**: this type of circuit manipulates signals which are in the form of discrete (usually **binary** – a logical 0 or 1 value) values that change at discrete points in time.

- **Mixed-Signal**: this type of circuit combines the functionality of analogue and digital circuits, usually for interfacing analogue signals to a digital processor system, see Fig. 3.1.

The revolution in the development and use of processor based electronics, in applications ranging from the home PC to the automobile, has led to the need for faster and (physically) smaller packaged digital electronic circuits and systems. In many cases, the digital electronic circuit is based around a processor IC. The main types of processors used are the:

- Microprocessor (μP)
- Microcontroller (μC)
- Digital Signal Processor (DSP)

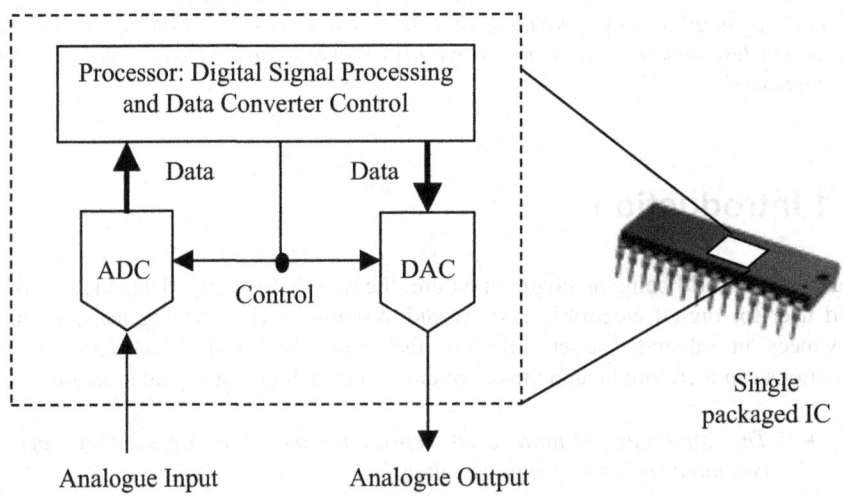

**Fig. 3.1.** Digital processor with on-board ADC and DAC

In the view shown in Fig. 3.1, a digital processor is used to undertake the main signal processing operations on an analogue input signal. This signal is sampled using an Analogue to Digital Converter (ADC) and converted from the analogue to digital domains. The digital processor manipulates the digitised signal in a suitable manner. The digital result is then applied to a Digital to Analogue Converter (DAC) that accepts the result of the digital signal processing and converts this back to the analogue domain. The digital electronic circuit uses discrete logic levels (logic levels 0 and 1) to represent information at discrete points in time. The combination of 0s and 1s into groups provides the means to store and transfer a large amount of information quickly and for the digital signals to represent accurately an analogue equivalent:

- A single binary digital value (0 or 1) is referred to as a **bit** (Binary Digit).
- A group of four bits is referred to as a **nibble**.
- A group of eight bits is referred to as a **byte**.

The early microprocessor systems were based on data bytes – that is, the data stored and manipulated within the microprocessor system was in the form of bytes. Today, 8-bit microprocessor systems are used in simpler embedded systems, whilst the high-end microprocessors manipulate 16, 32, 64 or 128 bit data structures. Digital systems have a number of advantages over their analogue equivalents, in:

- **Ease of design**: the design is based around the manipulation and storage of discrete level (usually binary) values rather than continuously varying analogue voltages and currents. The design process is aided by the use of **Hardware Description Languages** (HDLs) [5], circuit **synthesis**, and design automation (**EDA** – Electronic Design Automation) tools. Analogue circuit design on the other hand tends to be application specific and relies more on the experience of the designer. This is, in the main, a manual design process with limited design automation support.

- **Information storage**: binary values are easily stored in special memory circuits (see Chap. 4) and bistable circuit elements. Analogue signal storage is more difficult to design and implement, and is based around charge storage on a capacitor structure.

- **Higher levels of integration**: in digital logic, the emphasis is to make the digital logic gates operate at increasingly higher operating frequencies. This requires minimising the size of the individual transistors that make up the logic gates and minimising the interconnect between logic gates. This is achieved by the use of smaller transistor gate lengths (and moving to the next technology node). As such, the move is towards physically smaller logic gates with more logic gates within the IC itself.

- **Programmability**: certain digital ICs lend themselves more to the ability for reprogramming the IC functionality after the device has been fabricated. This can be achieved in a number of ways through the use of digital memory storing a processor program code and/or the use of Programmable Logic Devices (PLDs) such as the Field Programmable Gate Array (FPGA) and the Complex Programmable Logic Device (CPLD).

- **Accuracy and precision**: the functionality of the digital circuit is not distorted by effects such as process variations, temperature and age that analogue circuits components are subject to.

The digital circuit consists of a number of logic gates, which will be combinational or sequential circuit elements. In terms of the number of logic gates within a digital logic IC, the number will range from a few logic gates on the device through to hundreds of thousands (and ultimately into millions of gates per IC). In previous times, when the potential for higher levels of integration was far less than is now possible, the digital IC was classified by the level of integration – that is, the number of logic gate equivalents per IC, see Table 3.1. With the increases in the levels of integration, the following levels were identified as follow-on descriptions from VLSI, but these are not in common usage:

- **ULSI**    Ultra-Large Scale Integration
- **WSI**    Wafer Scale Integration

**Table 3.1.** Levels of integration

| Level of integration | Acronymn | Number of gate equivalents per IC |
|---|---|---|
| Small-Scale Integration | SSI | <10 |
| Medium-Scale Integration | MSI | 10-100 |
| Large-Scale Integration | LSI | 100-10,000 |
| Very-Large-Scale Integration | VLSI | >10,000 |

The equivalent logic gate is defined to consist of four transistors. In static CMOS logic, the 2-input NAND and 2-input NOR, see Fig. 3.2, are four transistor logic gate structures (2 nMOS and 2 pMOS transistors).

An important aspect of test is the controllability and observability of nodes within a design. For test access, this will be limited to the **primary inputs** and **primary outputs** of a design. Access to internal nodes within a design will need to be made via the **primary I/O**. In the above logic gates, the primary I/O are the logic gate connections. At the packaged IC level, then the primary I/O will be the package pins.

## 3.2 Logic Families

Digital logic may be implemented using different fabrication processes and different circuit architectures:

- TTL              Transistor-Transistor Logic (Bipolar)
- ECL              Emitter Coupled Logic (Bipolar)
- CMOS           Complementary Metal Oxide Semiconductor
- BiCMOS        Bipolar and CMOS

These are complemented with the digital logic capabilities of GaAs and SiGe technologies. Today, CMOS is by far the dominant process used for digital logic. The logic gates may take one of a number of different circuit architectures at the transistor level:

- Static CMOS
- Dynamic CMOS
- Pass Transistor Logic CMOS

Today, static CMOS logic is by far the dominant logic cell design structure used.

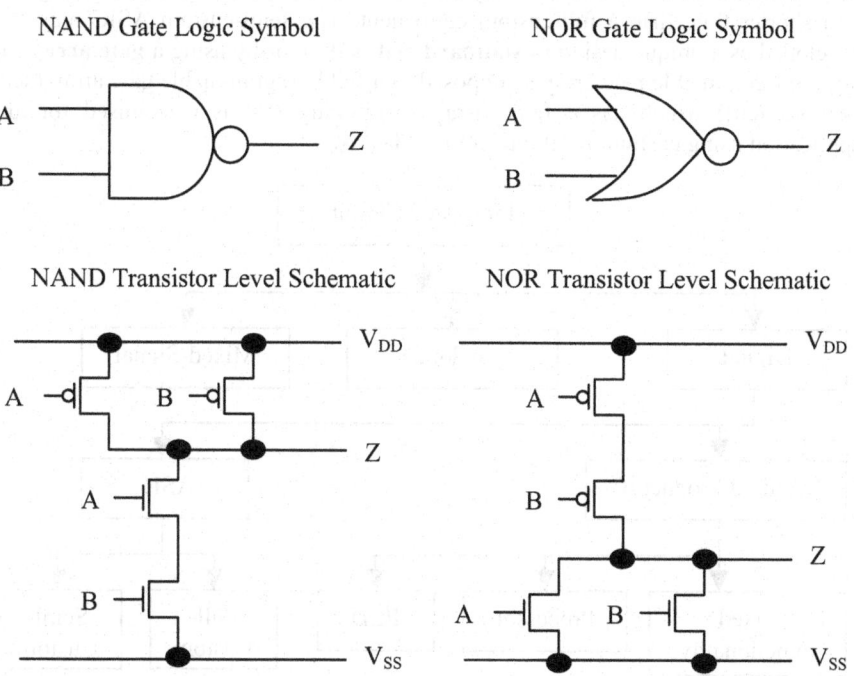

**Fig. 3.2.** 2-Input NAND and NOR circuits in static CMOS

## 3.3 Digital Logic ICs

Digital logic ICs may be obtained for use as either a pre-designed IC (a COTS (Commercial Off The Shelf) **standard product** device), or as a custom IC design (**ASIC** – Application Specific Integrated Circuit - either a **full-custom** or **semi-custom** design), see Fig. 3.3. The standard product ICs may be fixed functionality, processor based, or Programmable Logic Device (PLD).

A **full-custom** ASIC is designed using the basic available circuit components (transistors, etc) and is a highly manual design process. This is however time consuming, but can produce the most optimum solution. However, with the time and cost penalties associated with full-custom design, this is limited to analogue circuit design requiring specific circuit performance and the development of library cells to be used in a semi-custom design.

For larger designs, in particular digital system ASIC design, **semi-custom** ICs are developed. These designs use a library of pre-designed cells that are used to build the larger circuit/system. The designer then creates a design based on the performance availability of the library cells. A semi-custom design can be designed in a shorter time than a full-custom equivalent. The semi-custom ASIC may be developed as a unique design (a **standard cell** ASIC), or by using a **gate array** (a mask programmable gate array as opposed to a field programmable gate array (see next section)) which has a fixed array of circuitry that is customised for an application using custom metal interconnect layers.

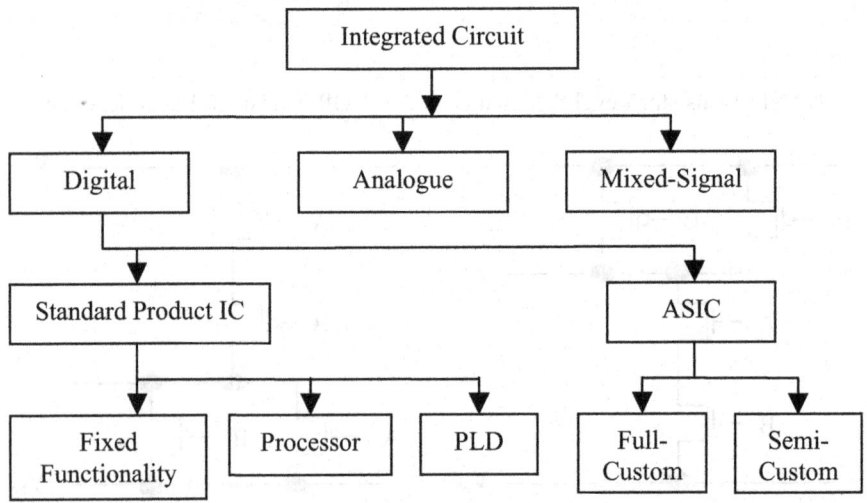

**Fig. 3.3.** Availability of digital ICs

# 3.4 Programmable Logic

Today, the use of the Programmable Logic Device (PLD) is widespread in industrial applications. The ability to configure and readily modify complex digital circuits and systems provides many time and cost advantages. The basic idea is for the PLD to consist of an array of logic with programmable interconnect. The design that has been entered in the CAD tool is then downloaded to the PLD in order to commit the device. In the simplest terms, a single PLD can replace a PCB containing multiple COTS (Commercial Off The Shelf) discrete digital ICs, providing a potentially smaller and faster solution with lower power consumption. However, the devices have more use for their ability to prototype rapidly complex ASIC (Application Specific Integrated Circuit) designs. A range of PLDs exist, see Fig. 3.4.

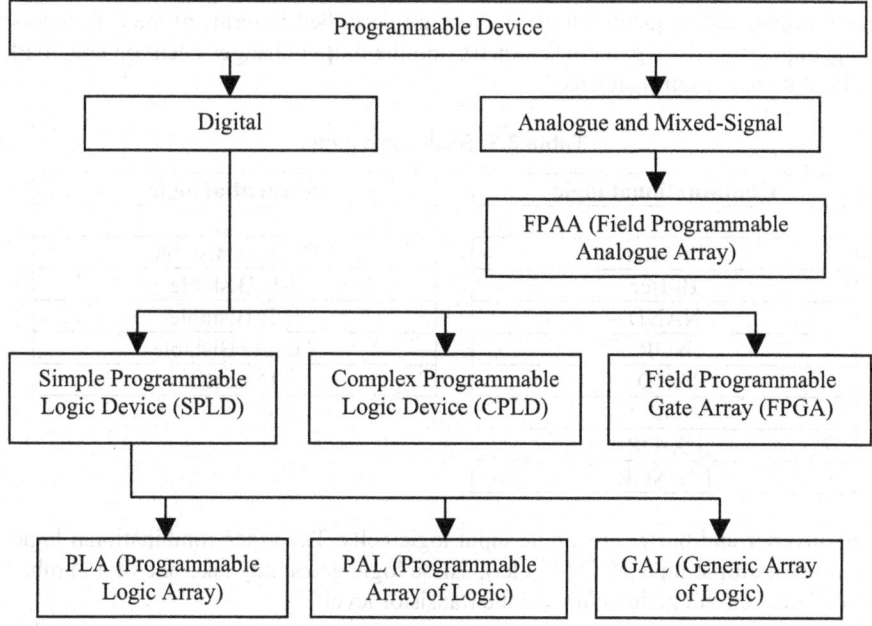

**Fig. 3.4.** Programmable Devices

The current types of programmable logic available today includes devices capable of implementing only a handful of logic equations through to large FPGA *systems* that can implement a complete processor system, including peripherals. Table 3.2 identifies the key programmable logic vendors.

**Table 3.2.** Programmable logic vendors

| Vendor | Internet resource |
| --- | --- |
| Altera Corporation | http://www.altera.com |
| Xilinx Inc. | http://www.xilinx.com |
| Lattice Semiconductor Corporation | http://www.latticesemi.com |
| Cypress Semiconductor | http://www.cypress.com |
| Atmel Corporation | http://www.atmel.com |

## 3.5 Basic Logic Gates

The basic digital logic gates used to implement digital logic functions are listed in Table 3.3. These are commonly combined to produce additional macro cells with a unique logical behaviour. The combinational logic circuits are described in terms of **Boolean Logic** expressions and **truth-tables** defining an input-output relationship. The sequential logic circuits are described in terms of the data inputs and control signals (*e.g.* a clock signal), and the output changes occur on specified values of these control signals.

**Table 3.3.** Basic logic gates

| Combinational logic | Sequential logic |
|---|---|
| Inverter | D-Type Bistable |
| Buffer | J-K Bistable |
| NAND | S-R Bistable |
| NOR | Toggle Bistable |
| AND | D-Latch |
| OR | |
| EX-OR | |
| EX-NOR | |

The inverter and buffer are single input logic cells. The other combinational logic cells are multiple input (2, 3, 4 , etc.). These logic gates may take one of a number of different circuit architectures at the transistor level:

- Static CMOS
- Dynamic CMOS
- Pass Transistor Logic CMOS

Today, **static CMOS** logic is by far the dominant logic cell design used. The basic structure of a static CMOS logic gate is based around a network of pMOS and nMOS transistors connected between the power supplies, see Fig. 3.5. The input signals (one or more) will be connected to the gates of the transistors and the output is taken from the common connection between the transistor networks. The transistors will act as switches with the switch connections between the drain and source of the transistor, and the switch control via the gate voltage:

- An **nMOS** transistor will be switched ON when a high voltage (logic 1) is applied to the transistor gate. A low voltage (logic 0) will turn the switch OFF.

- A **pMOS** transistor will be switched ON when a low voltage (logic 0) is applied to the transistor gate. A high voltage (logic 1) will turn the switch OFF.

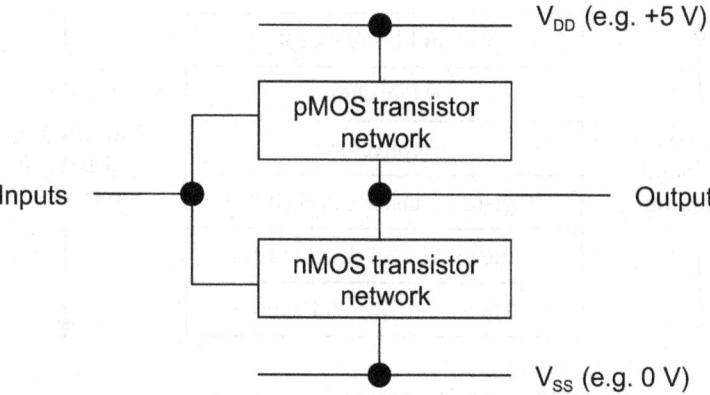

**Fig. 3.5.** Static CMOS logic structure

In the network, a series connection of nMOS transistors will produce an AND effect (*i.e.* both transistors need to be switched ON for the combined effect to be ON). A parallel connection of nMOS transistors will produce an OR effect (*i.e.* any one transistor will need to be switched ON for the combined effect to be ON). For the pMOS transistor network, a series connection of nMOS transistors will require a parallel connection of pMOS transistors. A parallel connection of nMOS transistors will require a series connection of pMOS transistors. With this arrangement, the full range of combinational logic circuit types can be made and significant current is only drawn from the power supply during input logic level changes.

# 3.6 Hardware Description Languages

With the simpler circuit designs, the design can be created using a circuit schematic, graphically creating and viewing the circuit in terms of the logic gates, the interconnections between the gates and the primary inputs and outputs. Within the Computer Aided Design (CAD) tools used, the design is saved in a suitable format within the design database. When the design increases in complexity, then schematic capture method becomes difficult to undertake and any modifications to a design can be time consuming to perform.

In modern digital circuit and system design, the alternative to the schematic uses a suitable **Hardware Description Language** (HDL) [5-6]. The HDL is a software programming language used to model digital logic circuit functionality and physical logic gate connections. Here, the design is entered as a textural based description of the design and also allows for the design description to be simulated. An important feature of the HDL is that it allows for a design to be described at various levels of abstraction, see Fig. 3.6.

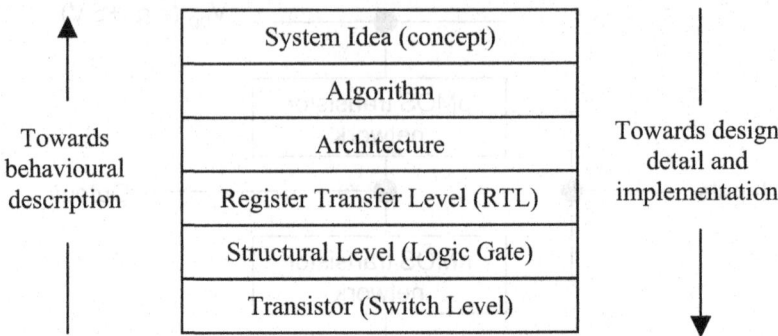

**Fig. 3.6.** HDL description abstraction

Starting at the highest level, the **System Idea** or **Concept** is the initial high-level description of the design that provides the design specification. The **Algorithm** describes a high-level **behavioural** description of the design operation at a high-level, mathematical level. This algorithm structure in hardware is described by the **Architecture**, identifying high level functional blocks (*e.g.* an ALU (Arithmetic and Logic Unit) may be a functional block). The Algorithm and Architecture levels would describe the behaviour of the design that would be verified in simulation. The next level down from the Architecture is the **RTL** (Register Transfer Level). This level describes the storage and movement of data around the design. It is this level that is usually used by **synthesis** tools to turn the design description into a **Structural Level** (the netlist of the design in terms of logic gates and interconnections between the logic gates). The logic gates are themselves implemented using transistors. The HDL may also support **Switch Level** descriptions that model the transistor operation as a switch (ON/OFF).

There are a range of HDLs available for use, although the two IEEE standards are VHDL [24] and Verilog®-HDL [25] are the most commonly used.

## 3.7 Digital Circuit and System Design Flow

A typical design flow for a digital IC would involve a combination of schematic capture and HDL design entry, sometimes utilising a combination of HDLs in describing different parts of a design. Increasingly, the design process is undertaken by multiple design groups that are based globally in design centres around the world, but which share a common design database. An example design flow for a design developed in VHDL is shown in Fig. 3.7. Here, the starting point is the design concept. This is initially developed as a design specification document before detailed design work starts. It is at this point that the design test issues are identified and design testability concepts are developed.

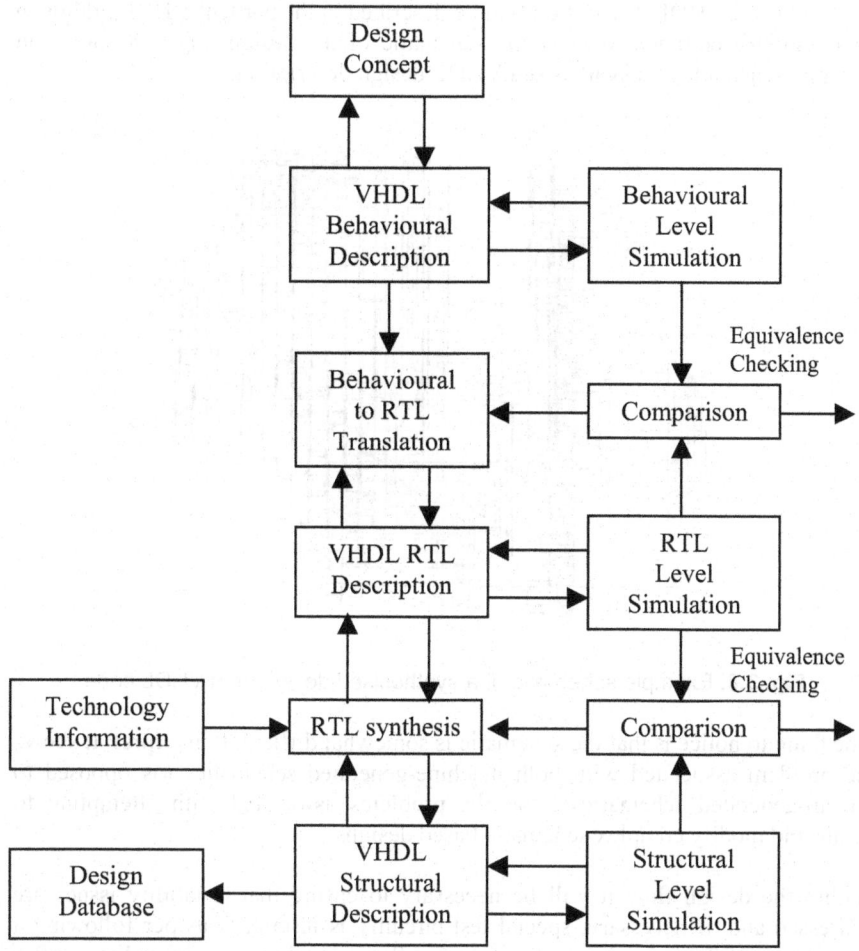

**Fig. 3.7.** VHDL based design flow

A behavioural level model of the design is initially developed and verified through simulation. Once complete, the behavioural level model is translated to an RTL level description, usually by a manual process. The operation of the RTL model is verified (equivalence checking) through simulation and compared to the behavioural level model in order to ensure translation correctness. The RTL level model can then be synthesised to a structural level model for a particular technology (ASIC or PLD) using a suitable synthesis tool and target technology information. Up to the synthesis, the design is **technology independent** and this provides an advantage that the design can be targeted to alternative technologies if required. The structural level description is simulated and compared to the RTL level model (equivalence checking) in order to ensure that the design was synthesised correctly. The design is then taken for implementation either as an ASIC or PLD solution.

The synthesised HDL description will be described in the particular HDL, although it is also still common to view the schematic of the design. Fig. 3.8 shows an example schematic of a synthesised VHDL design description.

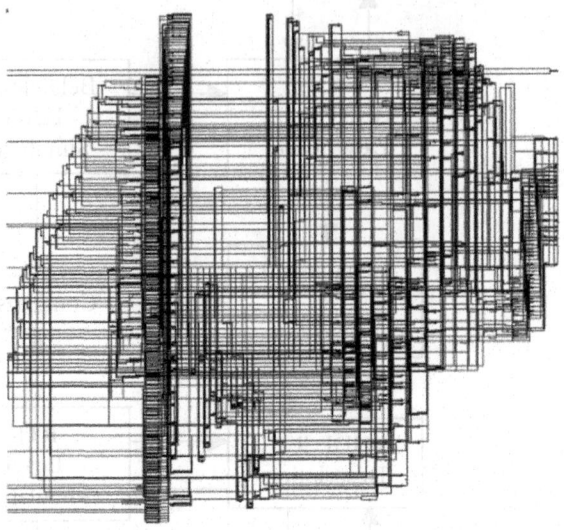

**Fig. 3.8.** Example schematic of a synthesised design from VHDL code

One thing to notice is that the schematic is somewhat difficult to interpret. It shows the problem associated with both machine-generated schematics (as opposed to human-generated schematics), and the problems associated with attempting to create and modify complex schematic based designs.

Within the design flow it will be necessary to ensure that testability issues are addressed and, if necessary, special test circuitry is inserted – as per following a DfT approach. Test circuit insertion is normally undertaken at or immediately after the design synthesis stage – at this stage the structural level description in terms of logic gates and interconnections is set and the test circuitry inserted into this netlist description. Test insertion software tools would normally be used to facilitate this.

## 3.8 Number Systems

A range of number systems [7] are used to represent values in the digital domain:

- Binary
- Decimal
- Octal
- Hexadecimal

Within the digital electronic circuit/system, the values are represented as binary (base 2) values. For positive numbers only, unsigned binary is used. Where the binary number is to represent both positive and negative numbers, signed binary (usually 2s complement notation) is used. Table 3.4 provides a summary of the different number notations for decimal numbers ranging from $0_{10}$ to $+15_{10}$.

**Table 3.4.** Number System Conversion ($0_{10}$ to $+15_{10}$)

| Decimal | Unsigned binary | Octal | Hexadecimal |
|---------|-----------------|-------|-------------|
| 0 | 0000 | 0 | 0 |
| 1 | 0001 | 1 | 1 |
| 2 | 0010 | 2 | 2 |
| 3 | 0011 | 3 | 3 |
| 4 | 0100 | 4 | 4 |
| 5 | 0101 | 5 | 5 |
| 6 | 0110 | 6 | 6 |
| 7 | 0111 | 7 | 7 |
| 8 | 1000 | 10 | 8 |
| 9 | 1001 | 11 | 9 |
| 10 | 1010 | 12 | A |
| 11 | 1011 | 13 | B |
| 12 | 1100 | 14 | C |
| 13 | 1101 | 15 | D |
| 14 | 1110 | 16 | E |
| 15 | 1111 | 17 | F |

The octal (base 8) and hexadecimal (base 16) number systems are used for representing large binary numbers. The octal number system has eight possible digits (0, 1, 2, 3, 4, 5, 6, 7), where each digit represents a decimal number between 0 and 7 (decimal). The hexadecimal number system has sixteen possible digits (0-9, followed by A, B, C, D, E and F), where each digit represents a decimal number between 0 and 15 (decimal). Table 3.4 shows integer numbers only. It is common for fractional numbers to be represented within the computations undertaken by the digital logic.

## 3.9 CMOS Inverter

The CMOS inverter is used to identify important characteristics of combinational logic. Fig. 3.9 shows a static CMOS inverter as:

- Logic symbol (1)
- Boolean expression (2)

- Transistor level schematic (3)
- Spice simulation file (netlist and simulation commands) (4)
- Truth-table (5)
- Layout (6)

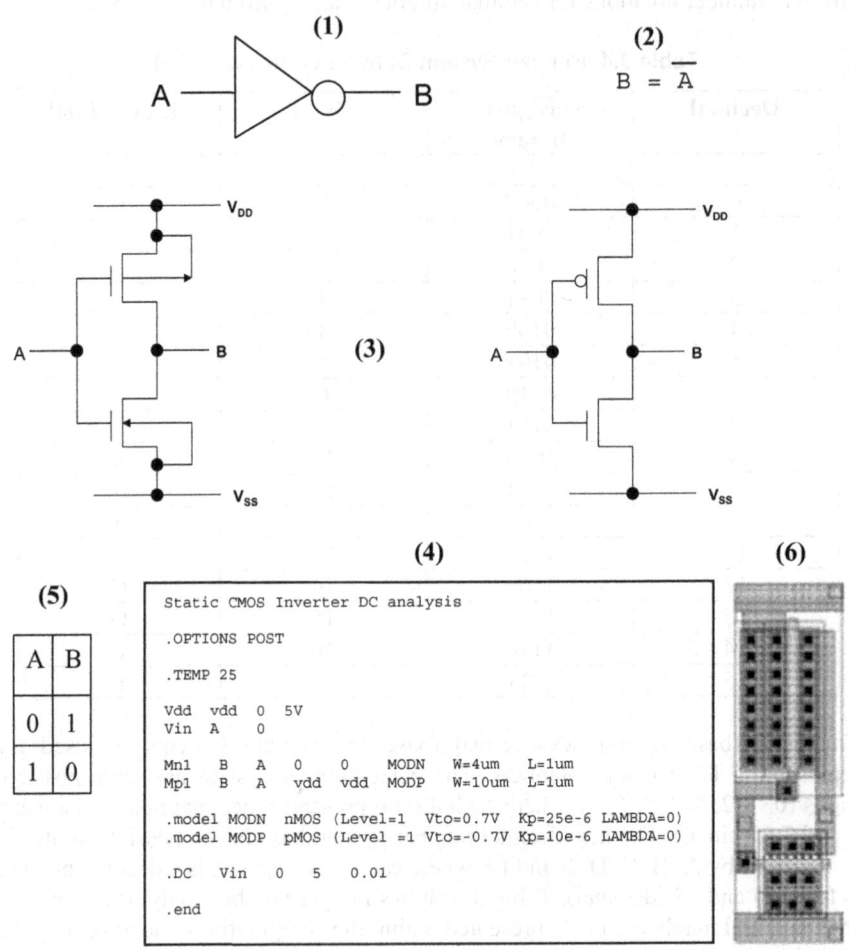

**Fig. 3.9.** CMOS inverter views

Two views of the transistor level schematic are shown. First (left), the full 4-terminal transistor schematic symbol is used. Second (right), the 3-terminal transistor schematic is used. In this case, the transistor substrate/bulk connection is assumed to be connected to $V_{DD}$ (pMOS) and $V_{SS}$ (nMOS).

An important issue for the use of a logic gate is the amount of circuitry that is connected to the gate output. The circuitry connected to the gate output produces an electrical load. In CMOS this is a capacitive load representing the gate connections of the individual transistors within the following logic gates. A cell will be designed to have a certain **fan-out**. This is the maximum number of logic inputs (of following logic gates) that can be driven reliably by a logic gate output.

The CMOS inverter is now considered for analysis with **static** and **dynamic** operation. With **static** operation, the input-output relationship is considered by applying a DC voltage at the gate input and observing the output voltage and power supply current, see Fig. 3.10. There are five regions of operation (A to E) identifying the operation of each transistor from cut-off through the linear region to the saturation region.

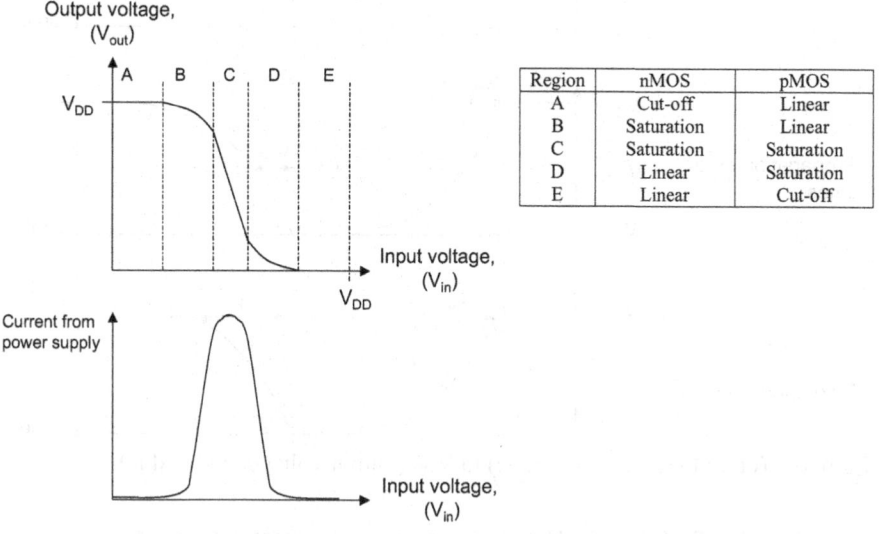

| Region | nMOS | pMOS |
|--------|-----------|------------|
| A | Cut-off | Linear |
| B | Saturation | Linear |
| C | Saturation | Saturation |
| D | Linear | Saturation |
| E | Linear | Cut-off |

**Fig. 3.10.** Static CMOS inverter – static characteristics

With the static CMOS inverter, an interesting characteristic is that when the input is static at a logic level, then the current drawn from the power supply is minimal and, to a first approximation, is zero. There is however a small but finite **sub-threshold current** drawn in the static condition. For the coarser process geometries, the sub-threshold current is low and this current characteristic could be used for test purposes. $I_{DDQ}$ testing measures the power supply current in static conditions and certain faults within the circuit have been seen to increase significantly this current level. By applying a pass-fail threshold current level, then faulty devices can be detected. For the finer process geometries, there is a rise in the sub-threshold current for the transistors within the logic gate and, coupled with more logic gates per IC, the static current has been seen to increase significantly for fault-free devices. This effect has resulted in questioning of the effectiveness of $I_{DDQ}$ testing in modern digital circuits.

In addition to the static characteristics, the CMOS inverter exhibits **dynamic** characteristics. Here, the output voltage changes due to a dynamic change in the input voltage. The dynamic characteristics will determine the maximum operating frequency of the logic gate and will be dependent on the output load conditions. The dynamic characteristics for the inverter is shown in Fig. 3.11. Here, a step change at the input (with zero rise and fall times) is applied. The voltages vary

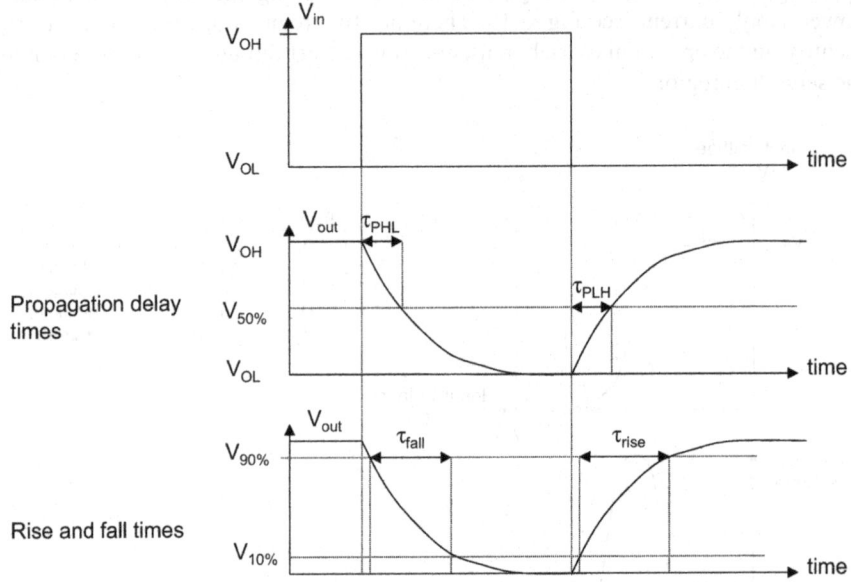

from $V_{OL}$ (output voltage, low value) to $V_{OH}$ (output voltage, high value).

**Fig. 3.11.** Static CMOS inverter – dynamic characteristics (1)

**The propagation delay** is defined as the time taken for the output signal voltage to change by 50% between its initial and final values. High-to-Low ($\tau_{PHL}$) and Low-to-High ($\tau_{PLH}$) delays are defined.

The **rise and fall times** are defined as the time taken for the output signal voltage to change between 10% and 90% of the difference between the initial and final values. High-to-Low (fall - $\tau_{fall}$) and Low-to-High (rise - $\tau_{rise}$) times are defined.

For a non-zero time change at the gate input, the propagation delay times are taken from the 50% change point in the input signal, see Fig. 3.12.

In the digital logic circuit, a logic level (0 or 1) will be created by a voltage with variance. Therefore a logic level will be represented by a range of voltages from a minimum value to a maximum value.

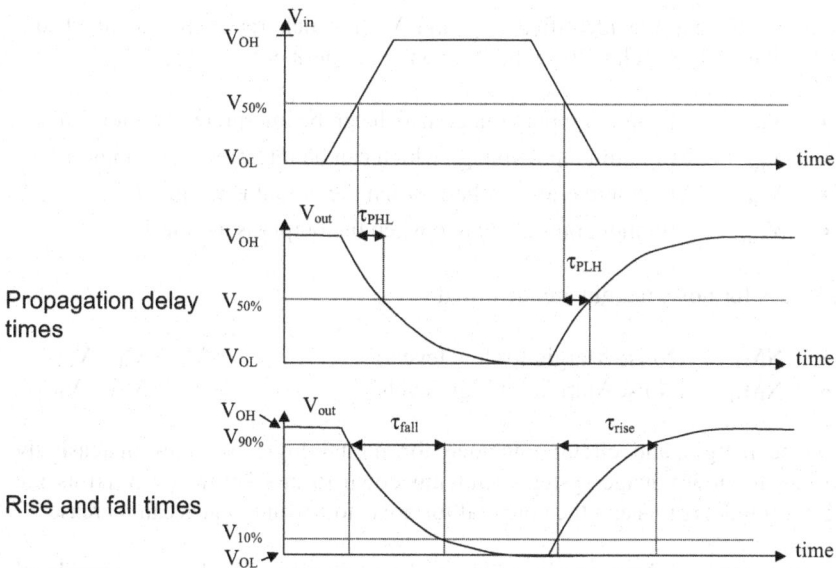

**Fig. 3.12.** Static CMOS inverter – dynamic characteristics (2)

The **noise margin** for a logic gate, see Fig. 3.13, will provide an indicator as to how tolerant a logic gate is to variations in the signal voltages creating the logic value. These variations may be due to noise that can be added to the signal from either neighbouring signal lines through capacitive or inductive coupling, or from outside the system. This has the potential to corrupt the signal (*i.e.* change its logic value). This potential corruption of the signal is increasingly important for logic operating at lower power supply voltages.

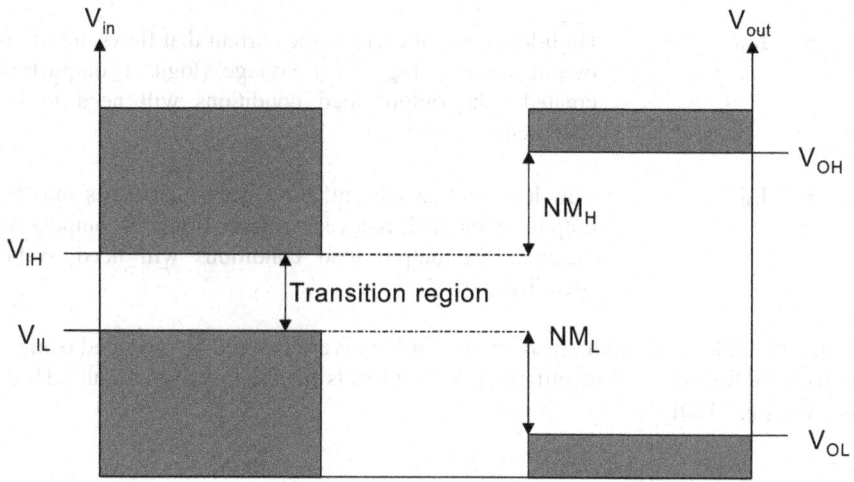

**Fig. 3.13.** Noise margin

Here, two voltages are identified ($V_{in}$ and $V_{out}$) which represent the input and output voltages. For each voltage, the following are defined:

- $V_{IL}$      Maximum input voltage which can be interpreted as a logic 0
- $V_{IH}$      Minimum input voltage which can be interpreted as a logic 1
- $V_{OL}$      Maximum output voltage when the output is a logic 0
- $V_{OH}$      Minimum output voltage when the output is a logic 1

Two values for noise margin are identified:

- $NM_L$    (Noise Margin for low levels)      $NM_L = V_{IL} - V_{OL}$
- $NM_H$    (Noise Margin for high levels)      $NM_H = V_{OH} - V_{IH}$

The noise margin and circuit tolerance for digital logic becomes increasingly important for low-voltage systems (moving down to and below 1V $V_{DD}$) as the noise margin decreases and the potential for noise to corrupt values can increase.

In addition to the voltages defined above, the logic gate will also have low-level and high-level input and output currents:

- $I_{IH}$                High-level input current: the current that flows into an input when a high-level voltage (value to be specified) is applied.

- $I_{IL}$                Low-level input current: the current that flows out of an input when a low-level voltage (value to be specified) is applied.

- $I_{OH}$                High-level output current: the current that flows out of an output when a high-level voltage (logic 1 output) is created. The output load conditions will need to be specified.

- $I_{OL}$                Low-level output current: the current that flows into an output when a low-level voltage (logic 0 output) is created. The output load conditions will need to be specified.

In the digital logic domain, the operation of the inverter would be modelled using a suitable HDL. Fig. 3.14 identifies example models for the inverter in both VHDL and Verilog®-HDL.

---

**Example VHDL model for the Inverter**

```
--//////////////////////////////////////////
-- Entity/Architecture description of an
-- Inverter logic gate
--//////////////////////////////////////////
-- Input:-  A
-- Output:- B
--//////////////////////////////////////////


------------------------------------
-- Entity Definition
------------------------------------

ENTITY inverter_model IS
    PORT ( A    :   IN     STD_LOGIC;
           B    :   OUT    STD_LOGIC);
end ENTITY inverter_model;


------------------------------------
-- Architecture Definition
------------------------------------

ARCHITECTURE simple OF inverter_model IS

BEGIN

    B <= not (A);

END ARCHITECTURE simple;
```

---

**Example Verilog®-HDL model for the Inverter**

```
//////////////////////////////////////////
// Module description of an Inverter
// logic gate
//////////////////////////////////////////
// Input:-  A
// Output:- B
//////////////////////////////////////////

module inverter_model(B, A);

output B;
input A;

inv (B, A);

endmodule
```

**Fig. 3.14.** Example VHDL and Verilog®-HDL models for the inverter

A more complex logic gate is shown in Fig. 3.15. Here, the circuit schematic and Spice simulation file are shown with the simulation set to perform a transient analysis.

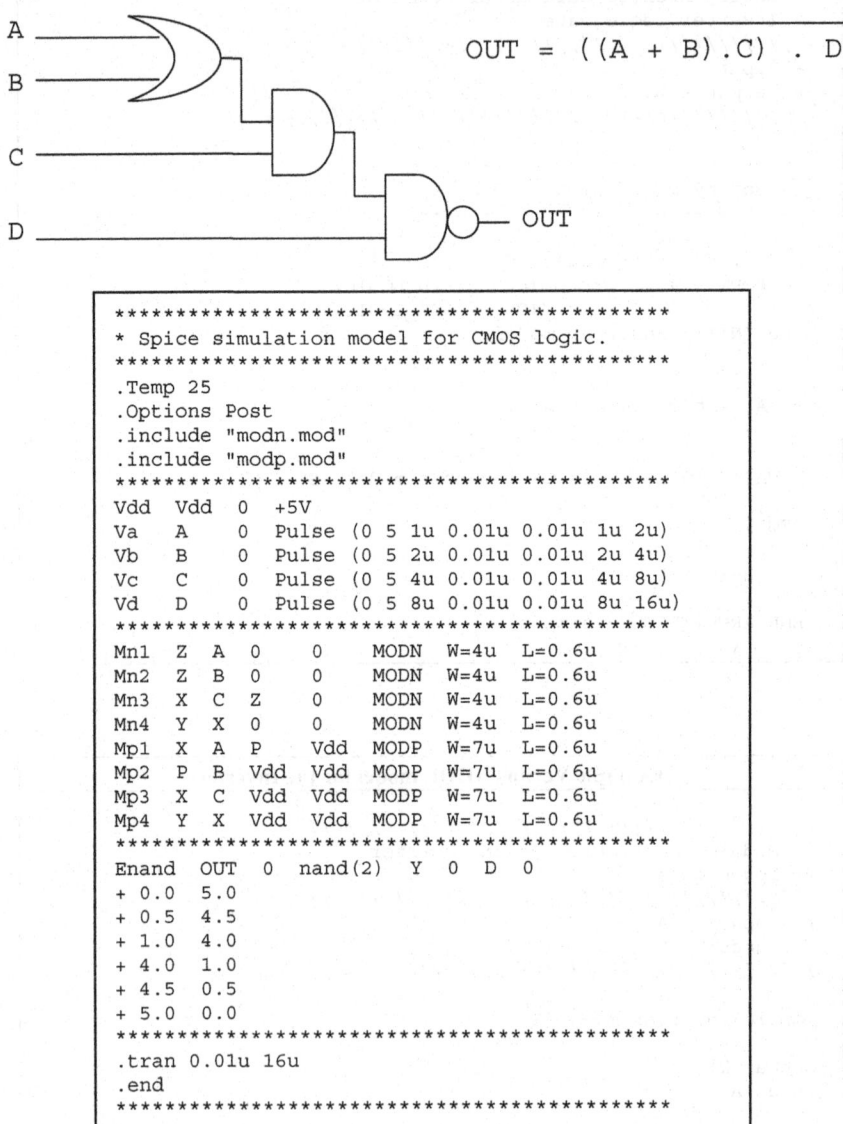

$$\overline{OUT = ((A + B).C) . D}$$

```
*************************************************
* Spice simulation model for CMOS logic.
*************************************************
.Temp 25
.Options Post
.include "modn.mod"
.include "modp.mod"
*************************************************
Vdd  Vdd  0    +5V
Va   A    0    Pulse (0 5 1u 0.01u 0.01u 1u 2u)
Vb   B    0    Pulse (0 5 2u 0.01u 0.01u 2u 4u)
Vc   C    0    Pulse (0 5 4u 0.01u 0.01u 4u 8u)
Vd   D    0    Pulse (0 5 8u 0.01u 0.01u 8u 16u)
*************************************************
Mn1  Z  A  0    0    MODN  W=4u  L=0.6u
Mn2  Z  B  0    0    MODN  W=4u  L=0.6u
Mn3  X  C  Z    0    MODN  W=4u  L=0.6u
Mn4  Y  X  0    0    MODN  W=4u  L=0.6u
Mp1  X  A  P    Vdd  MODP  W=7u  L=0.6u
Mp2  P  B  Vdd  Vdd  MODP  W=7u  L=0.6u
Mp3  X  C  Vdd  Vdd  MODP  W=7u  L=0.6u
Mp4  Y  X  Vdd  Vdd  MODP  W=7u  L=0.6u
*************************************************
Enand  OUT  0  nand(2)  Y  0  D  0
+ 0.0  5.0
+ 0.5  4.5
+ 1.0  4.0
+ 4.0  1.0
+ 4.5  0.5
+ 5.0  0.0
*************************************************
.tran 0.01u 16u
.end
*************************************************
```

**Fig. 3.15.** Combinational logic circuit

This Spice model incorporates both transistor level (nMOS and pMOS) and macro level (Voltage Controlled Voltage Source (VCVS)) models for simulation purposes. Here, the transistor models (MODN and MODP) are stored in a separate file and "included"

## 3.10 Latch-Up

A problem that may occur in IC designs is due to **latch-up**. Here, a low resistance path between the power supply rails ($V_{DD}$ and $V_{SS}$) is created by a parasitic thyristor action created by the combination of p-n junctions in the circuit layout, see Fig. 3.16. The top view shows a simplified cross-section of the fabricated circuit and the circuit that is produced. The latch-up condition allows for a high current to flow through the device. This may cause heating of the device and eventual failure.

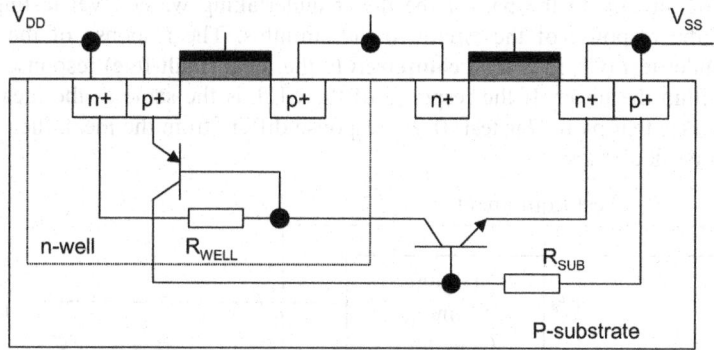

*Simplified view*

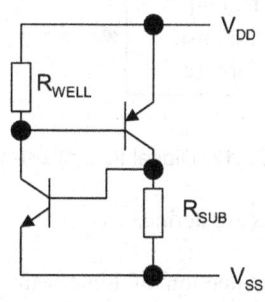

*Circuit model*

**Fig. 3.16.** Latch-up in CMOS circuit

In this view, a CMOS inverter is fabricated using an n-well process (and p-substrate). Parasitic resistance and bipolar transistors form a thyristor structure between the power supplies. A small base current in the pnp transistor connected to $V_{DD}$ allows for a larger collector current to flow. Some of this current is used to create a base current for the npn transistor connected to $V_{SS}$. This in turn allows for the base current in the pnp transistor to be sustained. A small current in the well or substrate material can cause the initial operation of this parasitic device to start. Layout design rules will need to be followed to minimise the risk of latch-up occurring.

# 3.11 Introduction to Digital Logic Test

When it comes to testing digital logic [8-11], the main purpose is to undertake the main test effort in terms of the logical input-output relationship of the IC. However, there will also be the need to undertake specific parametric (analogue) tests.

The basic arrangement for testing a digital IC [12] is shown in Fig. 3.17. Here, the role of the test equipment is to apply (**drive**) specific digital stimulus to the pins of the IC (or directly to the pad of the die if undertaking wafer level testing) and **capture** the response of the circuit to the stimulus. The response of the actual circuit under test (CUT) is then **compared** to the ideal (fault-free) response that is stored within the tester. If the response of the CUT is the same as the ideal, then the IC passes that particular test. If the response differs from the ideal, then the IC fails that particular test.

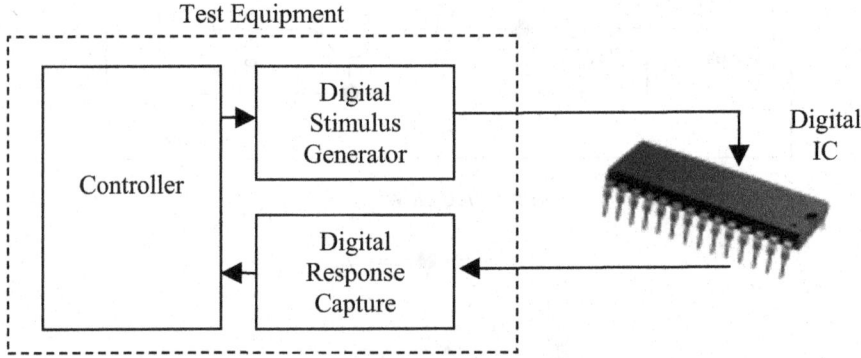

**Fig. 3.17.** Digital IC test set-up

Test **vectors**, **patterns** and **sets** are defined as:

- A **test vector** is a collection of logic values (0s and 1s) that will be applied to the circuit under test.
- A **test pattern** is a test vector with the addition of the expected output values from the circuit under test when the circuit is fault-free.
- A **test set** is a series of digital patterns.

The above definitions may vary, so it is important to identify the exact meaning of the terms in the particular scenario. The digital vectors would need to be applied at specific points in time. This may be fixed within the software test program controlling the tester electronics, or may be undertaken with the timing changing on a vector-by-vector basis. When the timing is changed between vectors, this is referred to as **timing on the fly**. Additionally, parametric testing of the IC I/O and power supply current would be undertaken. In this arrangement, the generator and capture electronics are under the control of the controller circuitry with the tester.

When it comes to the testing of the logical correctness of the design, the aim will be to test adequately the design, but in a minimal time. This is particularly important for production testing in that the test procedure must be as cost effective and thorough as possible. This requires a suitably comprehensive test procedure to be undertaken in as minimal a time as possible. For the logical testing of the IC, then either functional or structural testing, or a combination of both is undertaken:

- A **functional test** will exercise the design in such a way as to exercise the operation of the design through the various functional operations that it would be designed to undertake. For complex digital circuits and systems, this can be extremely time consuming and hence costly.

- A **structural test** will exercise the design in such a way as to stimulate faults that may exist in the design due to fabrication defects. The idea is to apply suitable digital vectors that will sensitise the fault such that the faulty circuit will produce a different result at the primary output from a fault-free circuit. This requires suitable fault models to be created to model fabrication defects and for these models to be simulated in the design in order to identify the right set of digital vectors to apply to the actual fabricated circuit.

- In addition, **parametric testing** of the IC I/O will be undertaken (see Chap. 7 for I/O test).

The purpose of the structural test is to reduce the number of vectors required to test the circuit when compared to an **exhaustive functional test**. In production test, the design itself is considered to be correct and the role of the test process is to identify faulty circuits due to fabrication defects within a minimum time possible. The structural test will aim to detect 100% of the considered faults. However, it may not necessarily be possible to detect 100% of faults in the circuit. Typically, **the fault coverage** would be required to be in excess of 95% and those parts of the circuit that may contain faults not covered by purely structural testing would require additional functional testing to be undertaken. The fault coverage is calculated as:

$$FC\ (\%)\ =\ \frac{Number\ of\ faults\ detected}{Total\ number\ of\ faults\ considered}\ \ x\ 100\ (\%)$$

It should be noted that the fault coverage figure would need to be used carefully. The figure will relate only to those faults considered. Faults that are not considered may have importance, but their effect will not be included. If care is not taken in the faults considered, then a fault coverage figure of 100% may be attained for specific faults, but faults of more relevance to the particular design and fabrication process might not be detected. The above definition of fault coverage may also be referred to as the **fault model coverage** (FMC) [10].

For analysis and test development purposes, digital logic is considered as both **combinational** and **sequential** logic. Each type of logic circuit has associated test issues that require to be addressed. The need for **structural test** is considered by considering the number of patterns required to exhaustively functionally test a combinational logic circuit. For an n-input circuit, $2^n$ combinations of input would be required to exhaustively test the circuit, see Fig. 3.18. It is a requirement that the test program is to effectively test a circuit in a minimal time. For large combinational logic circuits (*i.e.* with a large number of inputs), the required test time may become excessive and so this approach would be impractical. When $n$ is small, exhaustive test may be a practical solution. As $n$ increases, then it becomes impractical to undertake exhaustive functional tests.

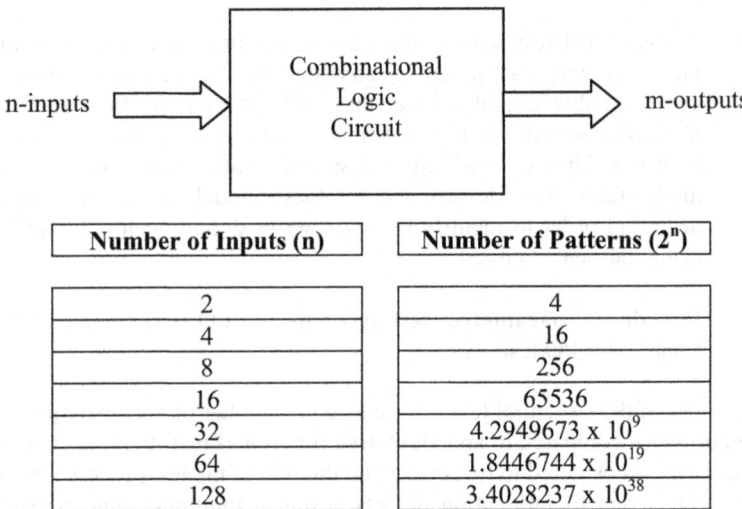

| Number of Inputs (n) | Number of Patterns ($2^n$) |
| --- | --- |
| 2 | 4 |
| 4 | 16 |
| 8 | 256 |
| 16 | 65536 |
| 32 | $4.2949673 \times 10^9$ |
| 64 | $1.8446744 \times 10^{19}$ |
| 128 | $3.4028237 \times 10^{38}$ |

**Fig. 3.18.** Input pattern complexity for combinational logic

A suitable structural test would reduce the number of patterns required and hence the test time required. Structural tests will be based on the detection of specific faults represented by their own set of fault models. The fault models will be considered to be either **logical fault models** or **defect oriented fault models**. Fault models will however be based on electrical faults caused by process defects within a circuit, and:

- **Logical fault models** are translations of electrical faults into logic level models. The models are based on logic levels and potentially timing.

- **Defect oriented fault models** are electrical faults based on the properties of the defect that created the fault. These are not simple digital (logical and potentially timing) models, but consider the electrical operation (voltage and current) of the fault in terms of analogue circuit primitives (resistance, capacitance, *etc.*).

Examples of electrical faults within a circuit include:

- Process variations outside the normal process spread (*e.g.* an excessive change in transistor threshold (MOS) voltage value).
- Open circuits in metal interconnect. These will be resistive open circuits, the value of the resistance dependent on the physical nature of the open.
- Short-circuits (bridges) between metal interconnect tracks. These will be resistive short circuits, the value of the resistance dependent on the physical nature of the short.
- Transistor stuck-open and stuck-short faults where the transistor is considered as a switch.
- Excessive steady-state (quiescent) power supply current.
- Transistor (MOS) open/short circuits (defects in the gate oxide and resistive opens/shorts between nodes in the transistor)

A number of fault models have been developed for IC test purposes. The key models used for digital circuits are:

**Logical fault models** include:
- Stuck-At-Fault
- Bridging fault (Wired-AND and Wired-OR)
- Delay fault (the delay fault may be considered as a defect-oriented fault rather than a logic fault)
- Memory fault (logical faults considered)

**Defect-Oriented fault models** include:
- Bridging fault (resistive)
- Memory fault (non-logical faults considered)
- Open-circuit fault in interconnect metal
- Stuck-open and stuck-short faults
- Transistor (MOS) open/short faults (used for analogue circuit analysis)
- $I_{DDQ}$ Fault [13-18]

Usually a **single fault assumption** is made in that a faulty circuit is to contain only one fault. However, **multiple faults** may exist and can be considered. Additionally, the relevance of the model to the fabrication process and how accurately the model reflects the physical defect needs to be questioned. However, obtaining up-to-date and accurate process defect information can be difficult, and models may need to be developed based on public-domain (published) material unless access to specific foundry information can be obtained. This however may not be possible due to commercial requirements and certain company specific information may not be accessible outside the company. Additionally, the rapid progress in fabrication processes and moving along the technology roadmap introduce additional problems given the introduction of new defect mechanisms that may not have existed in the *coarser* fabrication processes.

## 3.12 Fault Models

### 3.12.1 Stuck-At-Fault

The **Stuck-At-Fault** (SAF) is based around the fault model that considers a fault to create nodes within the design to be stuck at a logic level, no matter what logic level the circuit is trying to set, see Fig. 3.19. A node is considered to be either stuck-at-logic 0 (**SA0**) or stuck-at-logic 1 (**SA1**). The belief here is that a defect within the circuit will cause this type of logical fault. The circuit is considered to have one of three scenarios:

- **Fault-free** operation: no fault is considered to exist in the circuit.

- **Single-Stuck-at-Fault** (SSAF) operation: the circuit is considered to contain a single stuck-at-fault (this is a **single fault assumption**).

- **Multiple-Stuck-at-Fault** (MSAF) operation: the circuit is considered to contain multiple stuck-at-faults.

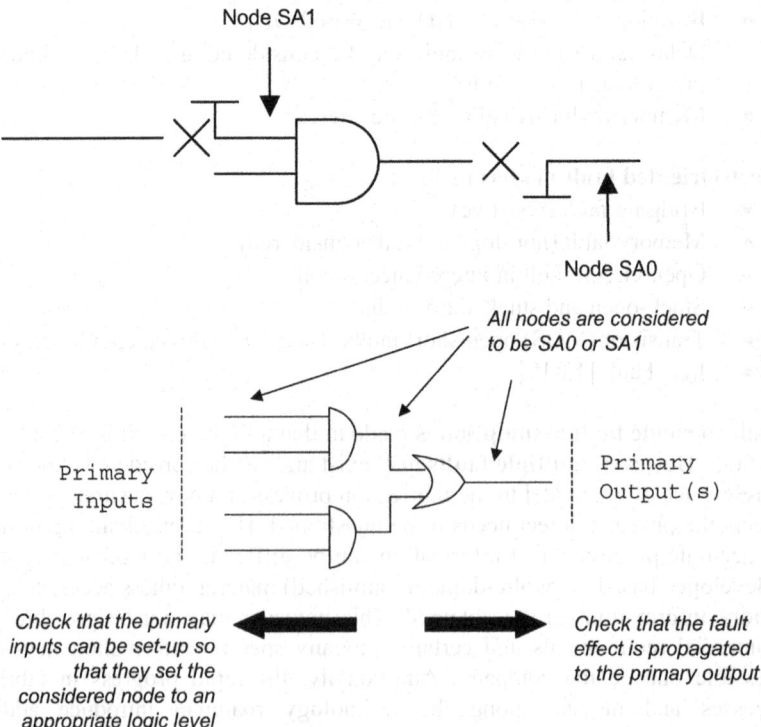

**Fig. 3.19.** Stuck-At-Fault

The fault-free operation will define the expected response. When considering a faulty circuit, it is common to use a single fault assumption rather than to consider multiple faults. This:

- Provides for a fault model that is independent of the fabrication process used and the design style adopted.

- Makes the test pattern generation (TPG) process manageable – simplifies the test pattern generation process. If multiple faults are considered to exist, then the decision has to be made as to how many faults are to be considered (two, three or more) and the locations of the faults. This is not trivial.

However, the single fault assumption requires the belief that the circuit will contain only one fault at a time and should multiple faults exist, then these will be detected by the patterns that are created to detect the single stuck-at-faults. Additionally, the SAF is considered to detect unmodelled fabrication defects. In order to detect the stuck-at-fault, each node is considered in turn to be SA0 and then SA1. The effect of the fault needs to be propagated to the primary output in order to set an output that logically differs between the fault-free and faulty circuits. The primary inputs must then be established in order to sensitise the fault. This is achieved by setting the primary inputs in order to set the opposite logic level at the node to the value of the stuck-at-fault.

## 3.12.2 Bridging Fault

The **bridging fault** [29-30] considers two (or more) nodes to be unintentionally connected. An example of this would be resistive material in the metal interconnect layer connecting tracks where tracks are placed close to each other on the physical circuit layout. It is common to consider low resistance value bridges. To identify the nets (interconnect) that may be bridged, this can be done using:

- The **circuit schematic**: the nodes within the circuit netlist are used. All nodes may be considered, but this can lead to a large number of bridging faults (the fault list) and does not reflect the physical positioning of the faults on the layout.

- The **circuit layout**: the nodes within the circuit that are physically close on the layout can be considered. This can reduce the number of bridges considered and potential bridging faults can be weighted (probability of occurrence) – that is, the nets more likely to be bridged are those that are physically close on the layout and run close to each other for a substantial distance.

The bridging faults can be determined from the layout using **Inductive Fault Analysis** (IFA) techniques [19-23] – either using **critical area** (geometry based) or **Monte Carlo** (randomly scattering of defects on a layout and determining the faults created) techniques. IFA has been used in the creation of fault lists based on a **defect oriented approach** for testing and has been used for the extraction of a range of component (*e.g.* transistor) and interconnect faults. In considering the creation of bridging faults, if the fault is considered to be logical in nature, then digital fault simulation of the circuit can be undertaken and there are two fault models considered, see Fig. 3.20:

- **Wired-AND**:
  In the Wired-AND model, two nodes are considered and the fault is modelled to be logic 0 dominant.

- **Wired-OR**:
  In the Wired-OR model, two nodes are considered and the fault is modelled to be logic 1 dominant.

In considering the creation of bridging faults, if the fault is considered to be analogue in nature (*e.g.* resistive), then analogue fault simulation (at the transistor level) is undertaken in order to determine the stimulus required to detect the fault. However, analogue fault simulation is time consuming (when compared to digital fault simulation) and requires extensive computing facilities in order to be undertaken. The value of the resistive connection between the nodes also needs to be ascertained.

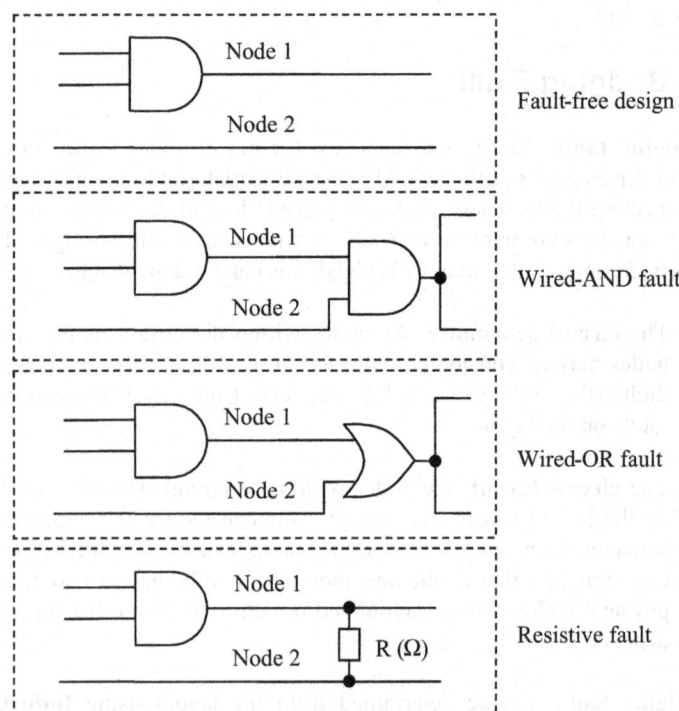

**Fig. 3.20.** Circuit schematic representation of bridging fault models

## 3.12.3 Delay Fault

The **delay fault** considers that the fault will not cause a logical error in the circuit output, rather it will cause an error in the timing. The output will then reach its final logical value at a later time than expected. For a system under the timing control of a clock signal, the delay may set-up the situation where a signal has not stabilised to its final logic value before the next clock edge.

Five types of delay faults [4] are noted:

- **Gate-delay fault**:
  This fault will consider the input to output delay of a single logic gate in the design whilst all other gates retain their expected delays.
- **Line-delay fault**:
  Delays in a signal rising and falling times in a given signal line are modelled. The effect of the fault is modelled through the longest path that can propagate the fault effect.
- **Transition fault**:
  This fault will consider the input to output delay of a single logic gate in the design whilst all other gates retain their expected delays. The delay is long enough to set-up the situation where a signal has not stabilised to its' final logic value before the next clock edge. This will occur even when the shortest signal path is considered. This fault can be considered as a temporary stuck-at-fault.
- **Path-delay fault**:
  This fault will consider a combined delay in a path consisting of combinational logic to exceed a maximum time.
- **Segment-delay fault**:
  This fault will consider delays in a signal propagating through a segment of circuitry containing a chain of combinational logic gates.

A common technique used for testing for delay faults in ICs is to vary (*i.e.* increase) the IC clock frequency [26] until the IC fails. It is then possible to rate the speed of operation of a device and to provide for a device family that operates at different speeds.

## 3.12.4 Memory Fault

With the high density of circuitry and interconnect found in memory circuits when compared to the densities found in logic circuits, and the array structure of the memory, faults relating only to memories can be encountered. Memory faults will be discussed in Chap. 4.

## 3.12.5 Stuck-Open and Stuck-Short Faults

This fault considers the transistor as an ideal switch. The switch may be stuck-open or closed (stuck-short) independent of the gate controlling signal.

## 3.12.6 $I_{DDQ}$ Fault

$I_{DDQ}$ [13-18] testing is based around the measurement of the power supply current ($I_{DD}$) drawn from the circuit power supply when the input signals and internal circuitry are stable (static – $_Q$). The premise is that for certain circuit designs, in a static condition, then the power supply current is low, see Fig. 3.21. A process defect causing a circuit fault will create a larger than expected current. If a limit on the current value is set, then circuits with lower current than the threshold pass the $I_{DDQ}$ test. Circuits with higher current than the limit will fail the test. In this test, the primary inputs are set, but the primary outputs are not monitored.

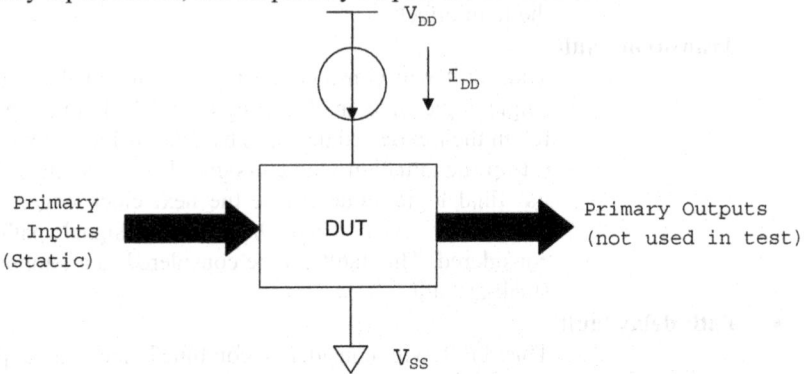

**Fig. 3.21.** $I_{DDQ}$ measurement principle

$I_{DDQ}$ testing will detect devices that have a higher than expected static power supply current but which may otherwise pass logical based testing. That is, in a digital IC, the circuit might pass logical tests, but fail the $I_{DDQ}$ test. Despite this logic test pass, then an $I_{DDQ}$ failure:

- May cause a reliability problem – the device will fail in its final application earlier than expected.

- For portable (battery powered) applications, the power consumption of the IC will be higher than necessary, so reducing the time that the circuit will operate on a single battery pack.

Whilst the device works logically, it may fail earlier in the final application than expected. As such, it is a reliability hazard and these potential problem devices need to be identified during production test. Burn-In (which will electrically and thermally stress ICs in order to accelerate the failure of reliability hazard devices) screening of devices can be reduced, or replaced by $I_{DDQ}$ testing. The types of defects that can be detected by $I_{DDQ}$ testing include:

- Resistive bridges between metal interconnect
- Resistive opens in metal interconnect
- Transistor (MOS) gate oxide defects

Both $I_{DDQ}$ (measurement in the $V_{DD}$ line) and $I_{SSQ}$ (measurement in the $V_{SS}$ line) have been considered in the past, although any $I_{SSQ}$ measurement would interfere with the common node potential (voltage) during measurement. Measuring $I_{DDQ}$ can be undertaken using one of the following methods:

- By using the **PMU** (Precision Measurement Unit) within an external tester – that is, using the tester measurement resources.

- By placing a discrete current monitor circuit close to the device under test on the tester **DIB** (Device Interface Board).

- By building into the IC a current monitor circuit – referred to as a **BICS** (Built-In Current Sensor).

However, care has to be taken as to the impact of the measurement circuit on the power supply voltage value at the DUT power supply pins. Additionally, the need to apply inputs and let the circuit stabilise before taking a current measurement means that $I_{DDQ}$ testing is a relatively slow (and time consuming) test method and hence has costs associated with the test times. The $I_{DDQ}$ pass/fail threshold will be set by the sampling of a statistically significant number of fabricated ICs and identifying the spread. This spread will be a Gaussian for both fault-free and faulty-devices, see Fig. 3.22. In this view, a single pass/fail threshold ($I_{th}$) is identified and the fault-free and faulty distributions are separated.

Frequency of occurrence

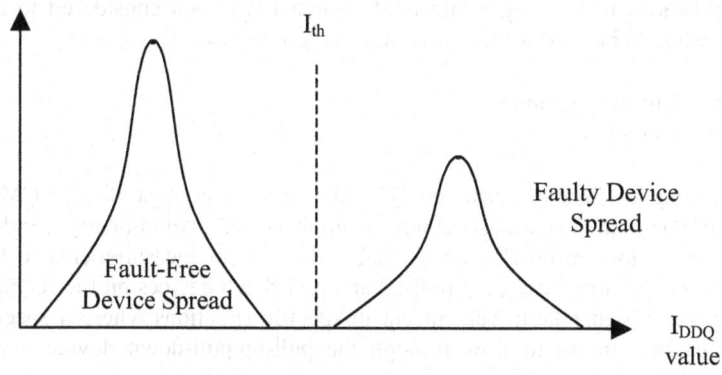

**Fig. 3.22.** $I_{DDQ}$ variations (1)

Here, the fault-free and faulty distributions are separated, and the threshold current is readily set. However, the distributions may overlap, see Fig. 3.23, and this leads to test escapes (faulty devices that have passed the test) and yield losses (good devices that have failed the test).

Frequency of occurrence

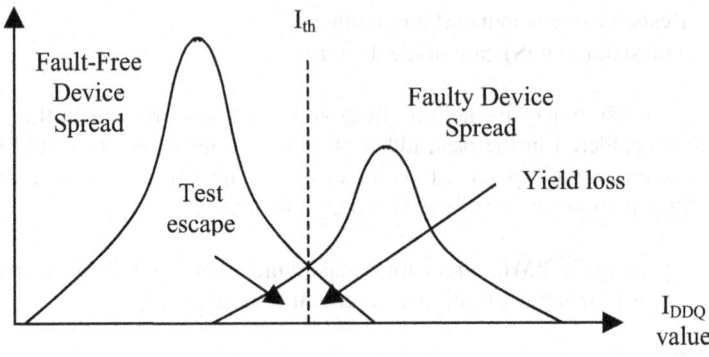

**Fig. 3.23.** $I_{DDQ}$ variations (2)

The above figures identify the use of a single threshold value ($I_{th}$). In the finer process geometries (moving along the technology roadmap), then higher fault-free $I_{DDQ}$ levels are encountered. In these processes, the use of lower power supply voltage levels are required for reliability aspects – the traditional 5V TTL level power supply voltage is replaced with sub-2V logic moving down to and beyond 1V operation. A consequence of the lower power supply voltage is that, unless the (MOS) transistor threshold voltage is reduced, then the speed of operation of the logic reduces. In order to avoid a speed loss, then the transistor threshold voltage is reduced. A consequence of the lower threshold voltage is a higher sub-threshold current in the transistor and so a higher $I_{DDQ}$ in the logic gate. This, coupled with the large numbers of logic gates within an IC, gives rise to a higher $I_{DDQ}$ for the device and difficulties in setting the $I_{DDQ}$ threshold value – the high current caused by a defect may be masked by the high fault-free $I_{DDQ}$ level.

Modifications to the single threshold current have been considered to allow for $I_{DDQ}$ testing to be used for the finer process geometries:

- Current signatures
- Delta $I_{DDQ}$

$I_{DDQ}$ testing has been primarily considered here in the context of static CMOS logic designs due to the low quiescent power supply current. Additionally, the design has to be in a "low current" state to enable the current measurements to be taken. Certain circuit structures (*e.g.* pull-up and pull-down devices on the IC input pads) can cause a higher fault-free current in specific situations where a specific logic level causes current to flow through the pull-up/pull-down device. Faults may occur either within a logic gate, or in the interconnect between the gates (signals) and also the power supply. Faults sited within the interconnect may be:

- Between signal lines.
- A signal line and a power supply line ($V_{DD}$ or $V_{SS}$).
- Power supply lines ($V_{DD}$ and $V_{SS}$).

Faults to a signal line will be **pattern dependent** – that is, the detection of the fault will be dependent on the value on the node. The faults between the power supply lines will be **pattern independent**.

As an example, Fig. 3.24 shows a resistive bridge within a static CMOS inverter that produces an $I_{DDQ}$ fault but not a logical fault. The detection of the fault is sensitive to the pattern applied.

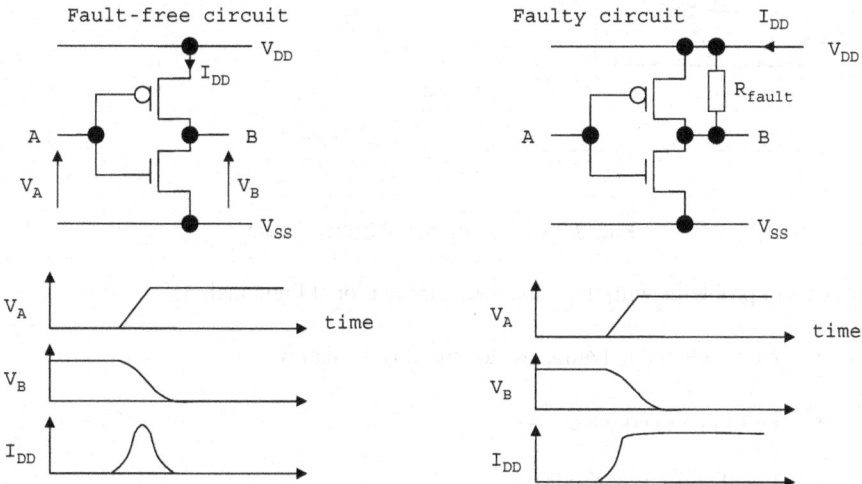

**Fig. 3.24.** Static CMOS inverter fault

When the input is a logic 0, then the output is a logic 1 (the voltage being $V_{DD}$). With the resistive bridge connected to $V_{DD}$, there will be no potential difference across the resistor and so no fault current flowing through the resistor. When, however, the input is a logic 1, the output is a logic 0 (the voltage being $V_{SS}$) and a fault current ($I_{fault}$) flows through the resistor:

$$I_{fault} = \frac{(V_{DD} - V_{SS})}{R_{fault}} \, amps$$

A significant current flows only in the fault-free circuit during an input signal change. However, current flows in the faulty circuit during the input signal change and when the input is at a logic 1 level.

## 3.13 Combinational Logic Test

### 3.13.1 Introduction

In a combinational logic circuit, the operation of the circuit can be described in terms of a Boolean Logic expression. On the application of a digital vector at the circuit input, after a short time delay (the propagation delay due to the gates and interconnect in the design), the output becomes a valid logic value. Fig. 3.25 is an example of a combinational logic circuit.

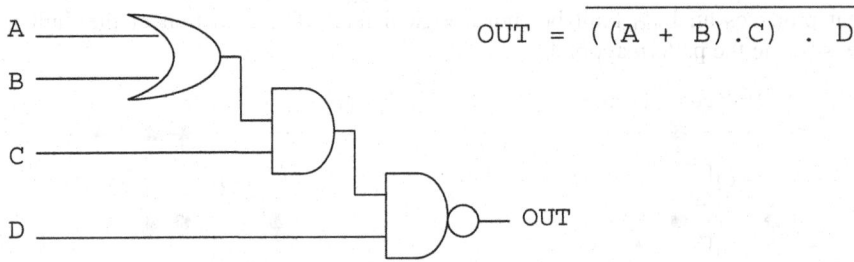

$$OUT = \overline{((A + B).C) . D}$$

**Fig. 3.25.** Combinational Logic Circuit

In general, problems with the testing of combinational logic include:

- Non-Detectable Faults due to circuit redundancy

- Fan-out and reconvergence

- Local and global feedback

- Multiple faults and fault masking

- Limitations of the fault models used in the development of the structural test programs

### 3.13.2 Test Pattern Generation

Fault models are used in the development of structural test programs and the aim is to detect as many of the faults considered as possible. The process of creating the patterns to apply is referred to as **Test Pattern Generation** (TPG). The patterns can be created manually (time consuming and expensive), by generating pseudorandom patterns, or by using a special software tool that automates the process – this is referred to as **Automatic Test Pattern Generation** (ATPG).

Automatic Test Pattern Generation is the only practical way in which to generate a set of test patterns for any large circuit design. For combinational logic circuits, these can be dealt with automatically, but when it comes to sequential logic circuits, additional problems exist due to the nature of sequential logic. The ATPG program to be used requires information on:

- The circuit design: a model of the design is created and used.
- The faults to be considered (fault list).
- Information on the individual components used, their fault-free behaviour and the way in which fault effects are propagated from component input to output.
- The means of assessing the fault coverage.

Test pattern generation is discussed in more detail in Chap. 10.

### 3.13.3 Non-Detectable Faults due to Circuit Redundancy

A problem with testing combinational logic lies in circuit redundancy that may or may not be intentionally built into the design. For example, the addition of logic gates to avoid glitches occurring at the output during input changes. As an example, consider the following three-input combinational logic circuit, see Fig. 3.26, that is represented as (i) a Boolean expression, (ii) a truth-table, (iii) a circuit schematic and (iv) a Karnaugh Map.

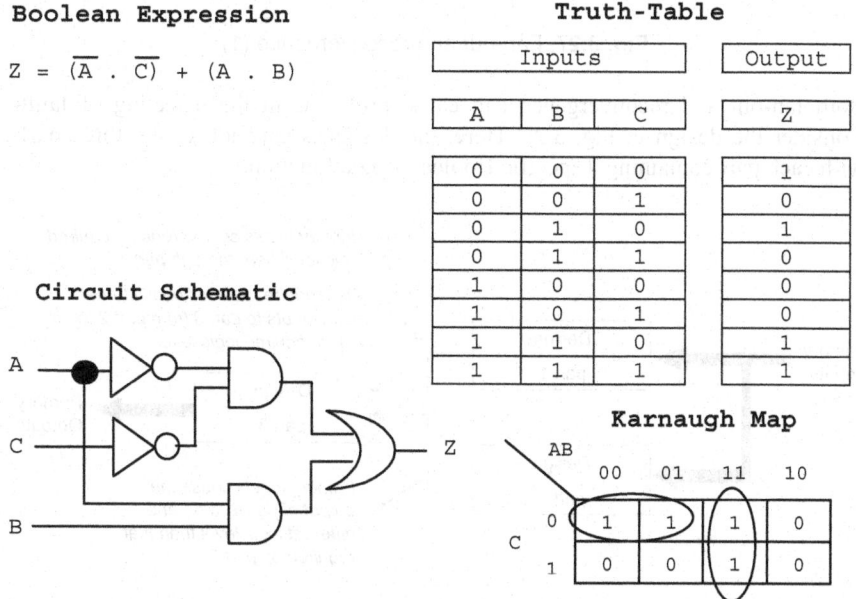

**Boolean Expression**

$$Z = (\overline{A} . \overline{C}) + (A . B)$$

**Truth-Table**

| Inputs | | | Output |
|---|---|---|---|
| A | B | C | Z |
| 0 | 0 | 0 | 1 |
| 0 | 0 | 1 | 0 |
| 0 | 1 | 0 | 1 |
| 0 | 1 | 1 | 0 |
| 1 | 0 | 0 | 0 |
| 1 | 0 | 1 | 0 |
| 1 | 1 | 0 | 1 |
| 1 | 1 | 1 | 1 |

**Circuit Schematic**

**Karnaugh Map**

**Fig. 3.26.** Combinational logic circuit no. 1

A non-minimal design which is designed to avoid glitches links the two groups in the Karnaugh Map and produces the following Boolean expression:

$$Z = (\overline{A} \cdot \overline{C}) + (A \cdot B) + (B \cdot \overline{C})$$

Both of the above Boolean expressions produce the same logic function, but using a different number of logic gates.

### 3.13.4 Fan-out and Reconvergence

When a single signal feeds the inputs of several logic gates, then this is referred to as **fan-out**. When a fanned-out signal recombines with other signals, then this is **reconvergence**, see Fig. 3.27.

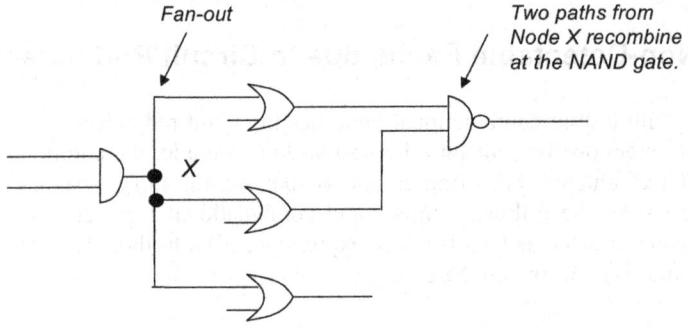

**Fig. 3.27.** Fan-out and reconvergence (1)

Both fan-out and reconvergence can cause problems in the detecting of faults. Consider the design in Fig. 3.28. Here, the design is partitioned into three parts, with each part containing a specific Boolean Logic function.

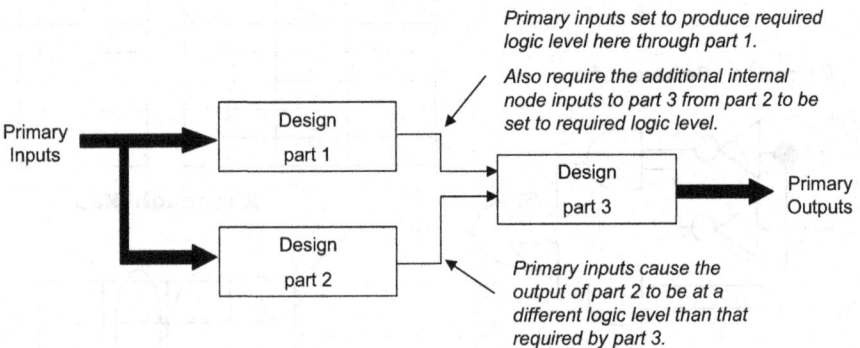

**Fig. 3.28.** Fan-out and reconvergence (2)

For the fault in part 3, to sensitise the fault and observe the fault effect at the primary output, the outputs of parts 1 and 2 need to be set to specific logic values. These blocks are supplied with the same inputs from the primary inputs. However, in this case, the output of part 2 differs from that required to propagate the fault effect to the primary output.

## 3.13.5 Local and Global Feedback

In combinational logic circuits, there may be the need to provide for feedback of the output signals of the complete design or part of the design into the logic circuit itself. Feedback can be considered as either **local** or **global**:

- **Local feedback**:       Applied around a small part of a larger circuit.

- **Global feedback**:      Applied around a whole circuit, or a major part of a circuit.

## 3.13.6 Multiple Faults and Fault Masking

It is possible that when more than one fault exists in the circuit, any one fault could be detected if that one fault occurred, but the overall effect of the multiple faults could mask the single fault effect, and so a test result would incorrectly identify a fault-free circuit.

## 3.13.7 Limitations of Fault Models

Fault models are simplifications of the process defects that are suited for use in circuit analysis, either digital or analogue circuit analysis. Fault models may not necessarily reflect the process defect accurately. The Stuck-At-Fault is a good example of this simplification. The question would be asked *"why would a process defect cause a perfect logical stuck-at-fault?"* … it probably wouldn't. However, the stuck-at-fault has been widely adopted as a model for creating structural test programs and is well adapted to Automatic Test Pattern Generation (ATPG). The benefits in using this fault model can in many circumstances outweigh the limitations, and the patterns generated to detect stuck-at-faults may also detect other circuit faults.

## 3.13.8 The Fault Matrix

A **fault matrix** can be developed to identify the minimal set of test vectors (the test set) to apply to a combinational logic circuit in order to detect the maximum number of faults. This can be identified with reference to the detection of stuck-at-faults. The fault matrix is however only suited for small circuits with a limited number of nodes. It is time consuming to develop manually, prone to errors in its creation and can be difficult to interpret. It does however serve as a useful means in which to introduce the concepts of structural test and test pattern minimisation.

The basic idea is to identify, for a combinational logic circuit and the stuck-at-fault, the input vectors that will detect a particular stuck-at-fault. These will be identified in the fault matrix – a table identifying the possible input vectors, the faults and the vectors that detect the faults.

When the fault matrix is completed, it is analysed and a minimal test set is extracted. The basic structure of the fault matrix is shown in Fig. 3.29.

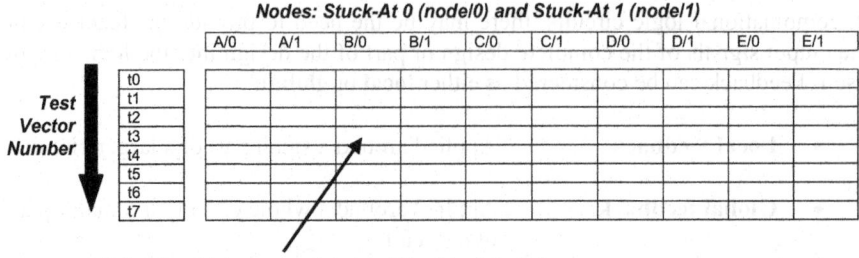

**Fig. 3.29.** Structure of the fault matrix

In this example, there are eight test vectors (t0 – t7) which determine all possible inputs to the circuit – the circuit has three primary inputs (A, B, C). The input vectors are 000, 001, 010, 011, 100, 101, 110, 111. The circuit has one primary output (E) and one internal node (D).

Each node can be considered to be stuck-at-0 (SA0 – denoted by <node>/0) or to be stuck-at-1 (SA1 – denoted by <node>/1). When a vector can be used to detect a fault, then a "tick" is placed in the box at the intersection of the fault (column) and vector number (row).

When a fault matrix has been completed, it is analysed in order to determine a minimal test-set. This is be done in two stages:

- **Stage 1**:
  Any column with only one tick identifies a fault detected by only one test vector. This is an **essential test** and must be included in the test set.

- **Stage 2**:
  With all essential tests covered, the **additional tests** required to cover all the remaining faults are to be determined. Where a column contains multiple ticks, any one of the identified test vectors can be used. The idea would be to identify and use the test vectors that detect the maximum number of faults and minimise the number of vectors to use.

A minimal set of test patterns can be derived by considering each fault in turn. As an example, consider the circuit schematic identified in fig. 3.26. In this circuit, additional node names are required:

- Node D is on the output of the inverter connected to node A
- Node E is on the output of the inverter connected to node C
- Node F is on the output of the and gate connected to nodes D and E
- Node G is on the output of the and gate connected to nodes A and B

The completed fault matrix is shown in Table 3.5.

**Table 3.5.** Completed fault matrix

| Test Vector | A/0 | A/1 | B/0 | B/1 | C/0 | C/1 | D/0 | D/1 | E/0 | E/1 | F/0 | F/1 | G/0 | G/1 | Z/0 | Z/1 |
|---|---|---|---|---|---|---|---|---|---|---|---|---|---|---|---|---|
| t0 |  | √ |  |  | √ | √ | √ |  | √ |  | √ |  |  | √ |  |  |
| t1 |  |  |  | √ |  |  |  |  |  | √ |  | √ |  | √ |  | √ |
| t2 |  |  |  |  | √ | √ |  |  | √ |  | √ |  |  | √ |  |  |
| t3 |  | √ |  | √ |  |  |  |  | √ | √ |  | √ |  | √ |  | √ |
| t4 | √ |  |  | √ |  |  |  | √ |  |  |  | √ |  | √ |  | √ |
| t5 |  |  |  | √ |  |  |  |  |  |  |  | √ |  | √ |  | √ |
| t6 |  |  | √ |  |  |  |  |  |  |  |  |  | √ | √ | √ |  |
| t7 | √ |  |  | √ |  |  |  |  |  |  |  |  | √ | √ |  |  |

All stuck-at-faults are detectable and only one fault (D/1) has an essential test (vector t4). All other faults can be detected by using one of a number of vectors. At the primary input, then the input would need to be set to the opposite logic level of the stuck-at fault considered and this reduces the number of possible vectors available to detect this fault. For the primary output, the vectors to detect these SA0 and SA1 faults can be seen directly from the truth-table of the fault-free design.

**Vector t4** can also be used to detect:

- A/0
- B/1
- D/1 (the essential test!)
- F/1
- G/1
- Z/1

| Test Vector | A/0 | A/1 | B/0 | B/1 | C/0 | C/1 | D/0 | D/1 | E/0 | E/1 | F/0 | F/1 | G/0 | G/1 | Z/0 | Z/1 |
|---|---|---|---|---|---|---|---|---|---|---|---|---|---|---|---|---|
| t4 | √ |  |  | √ |  |  |  | √ |  |  |  | √ |  | √ |  | √ |

Of the remaining faults, **vector t7** can be used to detect:

- B/0
- G/0
- Z/0

| Test Vector | A/0 | A/1 | B/0 | B/1 | C/0 | C/1 | D/0 | D/1 | E/0 | E/1 | F/0 | F/1 | G/0 | G/1 | Z/0 | Z/1 |
|---|---|---|---|---|---|---|---|---|---|---|---|---|---|---|---|---|
| t7 | √ |  | √ |  |  |  |  |  |  |  |  |  | √ |  | √ |  |

**Vector t0** can be used to detect:

- A/1
- C/1
- D/0
- E/0
- F/0

| Test Vector | A/0 | A/1 | B/0 | B/1 | C/0 | C/1 | D/0 | D/1 | E/0 | E/1 | F/0 | F/1 | G/0 | G/1 | Z/0 | Z/1 |
|---|---|---|---|---|---|---|---|---|---|---|---|---|---|---|---|---|
| t0 | | √ | | | | √ | √ | | √ | | √ | | | | √ | |

**Vector t1** can be used to detect:

- C/0
- E/1

| Test Vector | A/0 | A/1 | B/0 | B/1 | C/0 | C/1 | D/0 | D/1 | E/0 | E/1 | F/0 | F/1 | G/0 | G/1 | Z/0 | Z/1 |
|---|---|---|---|---|---|---|---|---|---|---|---|---|---|---|---|---|
| t1 | | | | | √ | | | | | √ | | √ | | √ | | √ |

All 16 stuck-at-faults can be detected using only 4 vectors (t0, t1, t4 and t7). This compares to the eight vectors that would need to be applied if an exhaustive functional test were to be undertaken. The creation of the fault matrix can be undertaken as either a paper exercise or with the aid of a suitable CAD tool and simulator. Fig. 3.30 shows a schematic of the design for modeling the circuit (i) fault-free (top) and (ii) with node F containing stuck-at-1 fault (bottom).

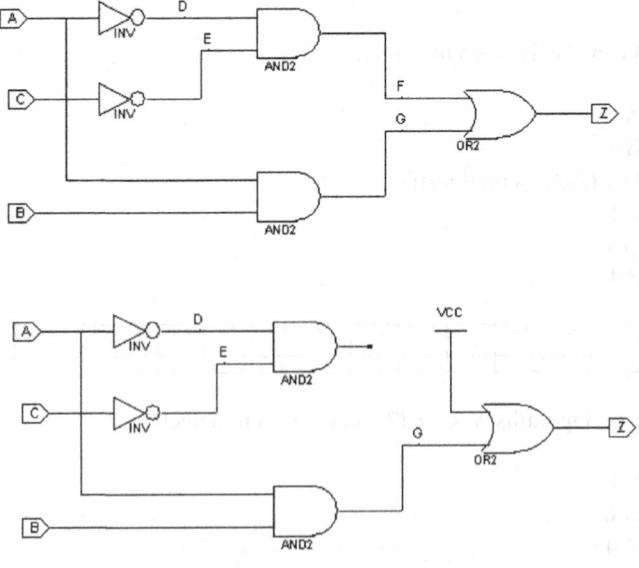

**Fig. 3.30.** Circuit schematics

## 3.14 Sequential Logic Test

### 3.14.1 Introduction

Unlike the combinational logic circuit, the sequential logic circuit input-output relationship is not simply described by a single Boolean expression. The sequential logic circuit consists of combinational logic circuit elements and memory (bistable) elements. As well as data (primary) inputs to the design, there are clock and potentially reset (and set) control inputs, see Fig. 3.31.

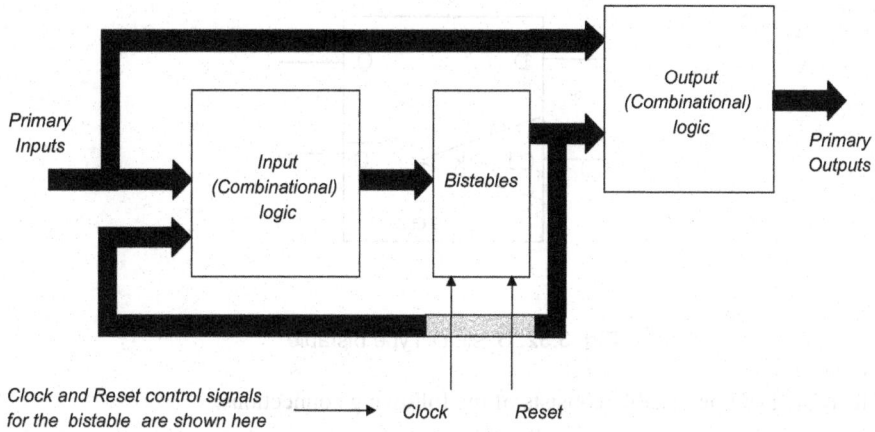

**Fig. 3.31.** General sequential logic circuit

In this view, the three main blocks consist of input and output combinational logic circuits and the bistable block. The actual structure of the sequential logic circuit will be dependent on the type of circuit. The circuits will one of the following types:

- Counter

- Finite State Machine (FSM):

  - Moore Machine
  - Mealy Machine

The output values of the bistables will define the **current state** of the circuit. On the edge of a clock signal, the circuit will move on to the **next state** in the sequence. The circuit can either change state on the positive or negative edge of the clock signal.

With the state machine designs, they differ as follows:

- With the **Moore** machine, the outputs are a function of the current state only

- With the **Mealy** machine, the outputs are a function of the current state and the circuit inputs

## 3.14.2 D-Type Bistable

The D-Type bistable lends itself to the use in testable logic circuits and will be considered further here. The basic D-Type bistable is shown in Fig. 3.32.

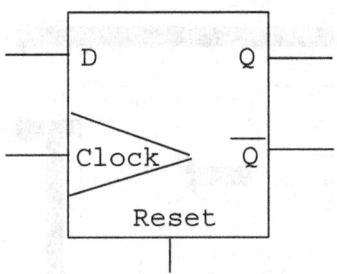

**Fig. 3.32.** Basic D-Type bistable

The basic D-Type bistable consists of the following connections:

Inputs:
- **D**              Data input.
- **Clock**          On the edge of the clock, data on the D-input is transferred to the Q output.
- **Reset**          Resets the Q-output back to '0'.

Outputs:
- **Q**              Data on D is transferred to the Q-output.
- **Q̄**              Opposite logic level to Q.

Additionally, the bistable may also have a **Set** input which sets the Q-output to '1'. The reset and set inputs may be active high or active low and may be synchronous (activates on the edge of a clock signal) or asynchronous (independent of the clock). The trigger of the bistable may be the positive or negative edge of the clock. Fig. 3.33 shows the operation of a positive edge triggered bistable. In this circuit, it is necessary for the D input to remain constant for a specific time before the clock edge (set-up time, $t_{setup}$) and for a specific time after the clock edge (hold time, $t_{hold}$) for reliable operation. The datasheet for the particular bistable will identify the required timing for the circuit under a range of output (capacitive) load conditions.

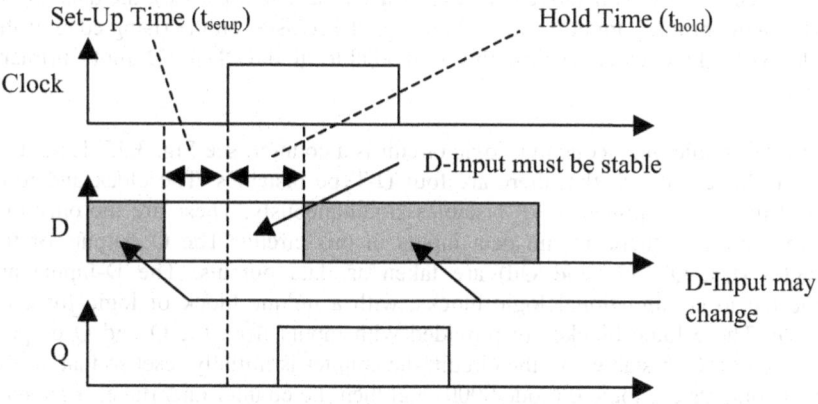

**Fig. 3.33.** D-Type bistable timing

## 3.14.3 Example Circuits

An example of a sequential logic circuit that only uses D-Type bistables is the serial-input serial-output shift register, see Fig. 3.34. This circuit consists of bistables connected with a common clock and reset signal.

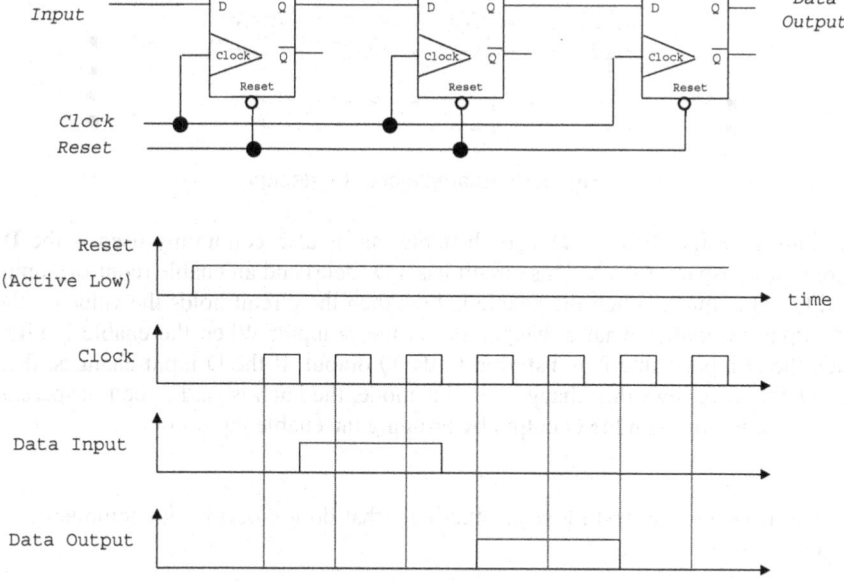

**Fig. 3.34.** Serial-input serial-output shift register

Here, when the reset is inactive (the reset is active low in this case), the data on the "Data Input" primary input is clocked through the register on the rising edge of the clock. After three clock cycles, this is available at the "Data Output" primary output.

Another example of a sequential logic circuit is a counter, see Fig. 3.35. Here, this is a four-bit counter in that there are four D-Type bistables. The clock and reset control inputs are applied to all bistables simultaneously. These are the only two primary inputs – there are no data inputs in this circuit. The Q outputs of the bistables (Qa, Qb, Qc and Qd) are taken as data outputs. The D-inputs are connected to combinational logic blocks, with a unique block of logic for each bistable. These logic blocks are provided with inputs from the Q and $\overline{Q}$ outputs from each of the bistables. In the circuit, the counter is initially reset so that all the bistable outputs are logic 0 (code 0000) and then the counter undertakes a 16-state count, repeating at code 0000, with the changes occurring on the positive edge of each clock pulse.

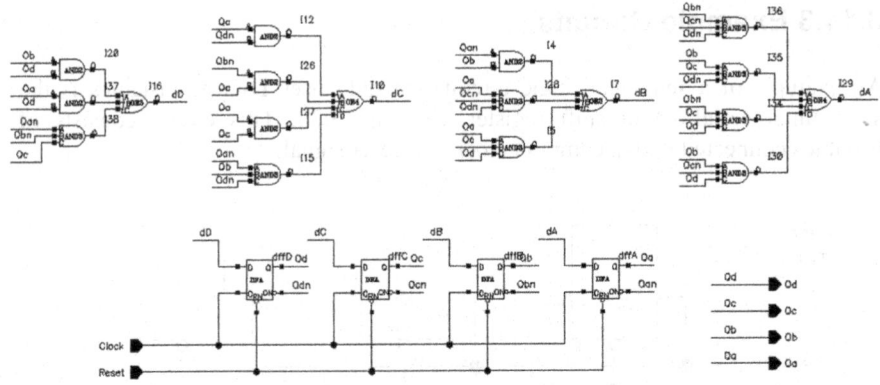

**Fig. 3.35.** Example counter design

A simpler design than the D-Type bistable that is also commonly used is the **D-latch**, or transparent latch. This circuit has a D (data) and an enable (control) input, and Q / $\overline{Q}$ outputs. When the enable is low, then the circuit holds the value on the Q output, no matter what is happening on the D input. When the enable is high, then the D input value is transferred to the Q output. If the D input changes, then the Q output follows this change – in this mode, the latch is said to be transparent. The value is stored on the Q output by bringing the enable input low.

Problems occur with testing sequential logic that do not occur with combinational logic:

- ATPG is more difficult to undertake for the sequential logic circuit as there is the need to account for the circuit states and control inputs (clock, reset, *etc.*) as well as any data inputs. The logical input-output relationship is not a simple Boolean expression as would be found in a combinational logic circuit.

- With the sequential circuit consisting of a number of states, and the circuit moving between states on the edge of a clock pulse, it may be necessary to preset the circuit into a particular state in order to undertake a test. This may be difficult to achieve when the circuit is embedded in a larger circuit, see Fig. 3.36, and may require a large number of clock cycles and data input changes to achieve this.

- There will be the need to test both the combinational and sequential circuit elements for correct operation.

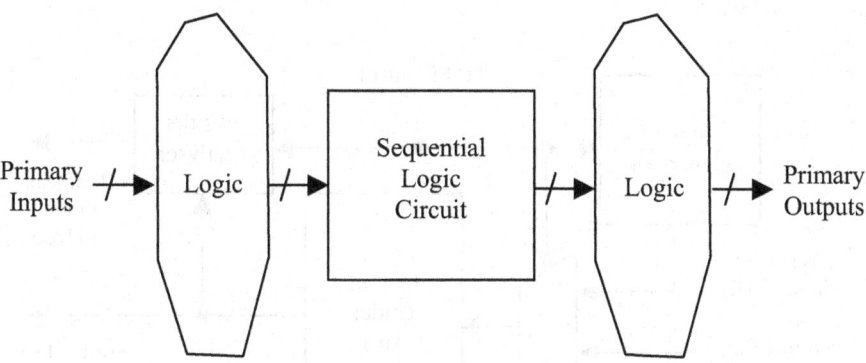

**Fig. 3.36.** Sequential logic circuit embedded within a larger design

# 3.15 DfT and BIST Overview

In order to address the test problem, a **Design for Testability** (DfT) approach is taken to addressing the controllability and observability problems associated with digital logic. Here, a suitable test strategy is developed during the design development stage and testability built into the design where required. A number of techniques have become standard within digital logic:

- The adoption of DfT guidelines during the design creation. Here, the guidelines would be provided as a document that describes all available DfT techniques. Those techniques of relevance would be adopted.

- The avoidance of certain design structures which are inherently untestable or difficult to test.

- The provision for design partitioning to aid controllability and observability – a "divide and conquer" approach to break down a large problem into smaller, more manageable tasks.

- The utilisation of scan-path test.

- The utilisation of boundary scan (IEEE standard 1149.1).

- The utilisation of Built-In Self-Test (BIST). The **BIST** approach uses **on-chip signal generation** and **analysis**, see Fig. 3.37, in order to provide local (on-chip) tester resources that would otherwise be required by external Automatic Test Equipment (ATE). Here, the circuit under test is provided with either the normal system inputs, or test inputs generated on-chip. The circuit under test output is then monitored by an on-chip results analyser.

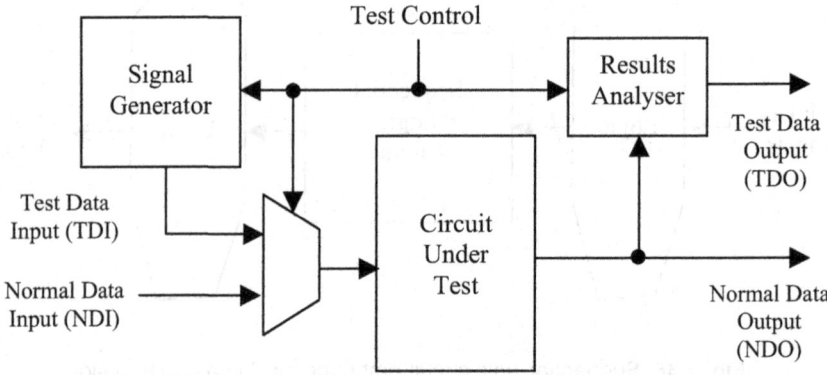

**Fig. 3.37**. Principle of operation of BIST

The DfT and BIST aspects are considered in further detail in Chap. 8.

## 3.16 Future Directions

The future advances in digital test are based on the demands of the end-user. A summary is provided in Table 3.6.

**Table 3.6.** Future digital test issues

| Technology driver | Description |
|---|---|
| Design styles | Increasing use of hardware description languages and synthesis techniques |

| | Increasing complexity of designs<br>Behavioural level synthesis and issues for test insertion<br>Increased integration – SoC and SiP test issues<br>The use of multiple power supply voltage rails on a single IC<br>Use of designs with multiple threshold voltage ($V_T$) MOS transistors<br>Move away from parallel transmission on and off-chip of data to high-speed serial bus architectures |
|---|---|
| Faster operating speeds | Higher signal frequencies<br>Driven by communications applications<br>Reduced device geometries (gate length and gate oxide thickness)<br>Reduced separation of interconnect tracks (track pitch) – capacitive and inductive coupling problems and signal cross-talk – signal integrity issues<br>Moving onto the next technology node to improve performance of components and reduce interconnect delays<br>Move into the nanotechnology domain |
| Higher levels of integration (Higher density) | Moving onto the next technology node to improve performance of components and reduce interconnect delays<br>Newer failure mechanisms - limitations of existing fault models[27-28] and need for models more representative of process defects<br>Higher fault-free $I_{DDQ}$ levels and limitations of $I_{DDQ}$ testing with the newer fabrication processes<br>Higher logic to pin ratio making it increasingly difficult to provide adequate controllability and observability<br>New test issues emerging for SoC and SiP devices<br>Move into the nanotechnology domain |
| Lower operating voltages | A need to improve device reliability by reducing electric field strength in dielectrics<br>Aim to lower device power consumption<br>Portable, battery operated circuits<br>Multi-voltage power supply device operation and interfacing |

| Test equipment and costs | Need for reduced test time and test costs<br>Need for reduced test pattern generation time and maintenance<br>More formal approaches to test program generation – test re-use<br>Problem of increasingly long test pattern generation and application times<br>Increasing requirements for test data storage in ATE<br>Increasingly difficult to perform at-speed testing of devices using current ATE<br>Need for reduced external ATE requirements and costs – simpler and less expensive ATE<br>Multi-site test<br>Increased use of BIST |
|---|---|

## 3.17 Summary

This chapter has provided an introduction to the testing of digital logic. The logic gate designs and design methods were identified as these now have an impact on the test program development requirements. The increase in digital IC complexity and speed of operation is leading to the need to use finer geometry processes,. This in turn is leading to limitations in existing test techniques. Digital IC test is based around functional, structural and parametric testing of the circuit. Structural test, based on the detection of specific faults due to processing defects is used extensively to reduce production test time as structural test programs require fewer test vectors when compared to an exhaustive functional test. Combinational and sequential logic test issues were identified and discussed.

## 3.18 References

[1]    Smith M., "Application Specific Integrated Circuits", Addison-Wesley, 1999, ISBN 0-201-50022-1

[2]    Bellaouar A. and Elmasry M., "Low-Power Digital VLSI Design Circuits and Systems", Kluwer Academic Publishers, The Netherlands, 1995, ISBN 0-7923-9587-5

[3]    Kang S. and Leblebici Y., "CMOS Digital Integrated Circuits Analysis and Design", McGraw-Hill International Editions, Singapore, 1996, ISBN 0-07-114423-4

[4]     International Technology Roadmap for Semiconductors (ITRS), 2003 Edition, "Design"

[5]     Zwolinski M., "Digital System Design with VHDL", Pearson Education Limited, 2000, England, ISBN 0-201-36063-2

[6]     Smith D., "HDL Chip Design", Doone Publications, USA, 1996, ISBN 0-9651934-3-8

[7]     Tocci R.J., Widmer N.S. and Moss G.LK., "Digital Systems 9$^{th}$ Edition", Pearson Education International, USA, 2004, ISBN 0-13-121931-6

[8]     Bushnell M. and Agrawal V., "Essentials of Electronic Testing for Digital, Memory & Mixed-Signal VLSI Circuits", Kluwer Academic Publishers, 2000, ISBN 0-7923-7991-8

[9]     Rajsuman, R., "System-on-a-Chip Design and Test", Artech House Publishers, USA, 2000, ISBN 1-58053-107-5

[10]    Hurst S., "VLSI Testing digital and mixed analogue/digital techniques", IEE, 1998, ISBN 0-85296-901-5

[11]    Needham W., "Designer's Guide to Testable ASIC Devices", Van Nostrand Reinhold, 1991, ISBN 0-442-00221-1

[12]    Burns M. and Roberts G.W., "An Introduction to Mixed-Signal IC Test and Measurement", Oxford University Press, New York, 2001, ISBN 0-19-514016-8

[13]    Sabade S. and Walker D., "$I_{DDQ}$ Test: Will It Survive the DSM Challenge?", IEEE D & T of Computers, Sept-Oct 2002, pp8-16

[14]    Mallarapu S. and Hoffman A., "IDDQ Testing on a Custom Automotive IC", IEEE Journal of Solid-State Circuits, Vol. 30, No. 3, March 1995, pp295-299

[15]    Champac V. Rubio A. and Figueras J., "Electrical Model of the Floating Gate Defect in CMOS ICs: Implications on IDDQ Testing", IEEE Transactions on Computer-Aided Design of Integrated Circuits and Systems, Vol. 13, No. 3, March 1994

[16]    Baker K. et al., "Development of a Class 1 QTAG Monitor", Proceedings of the International Test Conference, 1994, pp213-222

[17]    Wallquist K., Richter A. and Hawkins C., "A General Purpose IDDQ Measurement Circuit", Proceedings of the International Test Conference, 1993, pp642-651

[18]    Soden J.M., Hawkins C.F., Gulati R.K. and Weiwei M., "$I_{DDQ}$ Testing: A Review", Journal of Electronic Testing, Theory and Applications, No. 3, 1992, pp291-303

[19]    Shen J., Maly W. and Ferguson F., "Inductive Fault Analysis of MOS Integrated Circuits", IEEE Design and Test of Computers, Vol. 2, No. 12, December 1985, pp13-26

[20]    Jee A. and Ferguson F.J., "Carafe: An Inductive Fault Analysis Tool for CMOS VLSI Circuits", IEEE VLSI Test Symposium, 1993, pp92-98

[21]    Ferguson F.J. and Shen J.P., "A CMOS Fault Extractor for Inductive Fault Analysis", IEEE Transactions on Computer Aided Design, Vol, 7, No. 11, November 1988, pp1181-1194

[22]    Sachdev M. and Atzema B., "Industrial Relevance of Analog IFA: A Fact or a Fiction", Proceedings of the International Test Conference, 1995, pp61-70

[23]    Montanes R.R., Bruls E.M. and Figueras J., "Bridging Defects Resistance Measurements in a CMOS Process", Proceedings of the International Test Conference, 1992, pp892-899

[24]    IEEE Std 1076-2002, IEEE Standard VHDL Language Reference Manual, IEEE, USA

[25]    IEEE 1364-1995, IEEE Standard Verilog® Hardware Description Language, IEEE, USA

[26]    Datta R., Sebastine A. and Abraham J., "Delay Fault Testing and Silicon Debug Using Scan Chains", Proceedings of the 9th IEEE European Test Symposium, May 2004, pp111-116

[27]    Garcia R., "Rethink fault models for submicron-IC test", Test & Measurement World, October 2001

[28]    Aitken R., "Finding defects with fault models", Proceedings of the International Test Conference, 1995, pp498-505

[29]    Bradford J. et al., "Simulating Realistic Bridging and Crosstalk Faults in an Industrial Setting", Proceedings of the 7th IEEE European Test Symposium, 2002, pp75-80

[30]    Chess B. et al., "Logic Testing of Bridging Faults in CMOS Integrated Circuits", IEEE Transactions on Computers, Vol. 47, No. 3, March 1998, pp338-345

# Exercises

## Question 1

From the device datasheets, identify the following for the current range of microprocessors used in <u>desktop</u> PCs:

- The device type and manufacturer
- Package type and number of pins
- Number of pins dedicated to the following:
    - Power Supply
    - I/O.
    - Test
- Power consumption and cooling requirements
- Power supply voltage and $I_{DDQ}$ levels

## Question 2

Repeat question 1, except now identify the following for the current range of microprocessors used in <u>notebook</u> PCs.

## Question 3

For the following Boolean expression:

$$Z = (A . B) + ((\overline{A + B}) . (A + C))$$

- Draw the circuit schematic for the design exactly as described.
- Develop a fault matrix for the design (stuck-at-faults considered) and identify whether all faults can be detected. What is the fault coverage for this design?
- Derive a minimised Boolean expression for the design and develop a fault matrix (stuck-at-faults again considered) and identify whether all faults can now be detected. What is the fault coverage for this design?

## Question 4

For the two designs identified in Question 3, develop a structural VHDL model for the design. Simulate the design in its' fault-free state and for each of the stuck-at-faults possible in the designs.

## Question 5

For the two designs identified in Question 3, develop a structural Verilog®-HDL model for the design. Simulate the design in its' fault-free state and for each of the stuck-at-faults possible in the designs.

## Question 6

For the two designs identified in Question 3, develop a Spice model for the design. Simulate the design in its' fault-free state and for each of the stuck-at-faults possible in the designs.

## Question 7

Design a 4-bit gray code counter using D-Type bistables that are positive edge triggered and have an active low reset signal. The reset state sets the counter in count state $0000_2$. For each of the combinational logic blocks feeding the D-inputs, develop a fault matrix (stuck-at-faults considered) for the block of logic and identify whether the faults can be detected. What is the fault coverage for the complete design?

## Question 8

For the design identified in Question 7, develop a structural VHDL model for the design. Simulate the design in its' fault-free state and for each of the stuck-at-faults possible in the designs.

## Question 9

For the design identified in Question 7, develop a structural Verilog®-HDL model for the design. Simulate the design in its' fault-free state and for each of the stuck-at-faults possible in the designs.

## Question 10

For the design identified in Question 7, develop a structural Spice model for the design. Simulate the design in its' fault-free state and for each of the stuck-at-faults possible in the designs.

## Question 11

Design a four-bit up/down counter using D-Type bistables that are positive edge triggered and have an active low reset signal. The reset state sets the counter in count state $0000_2$. For each of the combinational logic blocks feeding the D-inputs, develop a fault matrix (stuck-at-faults considered) for the block of logic and identify whether the faults can be detected. What is the fault coverage for the complete design?

## Question 12

For the design identified in Question 11, develop a structural VHDL model for the design. Simulate the design in its' fault-free state and for each of the stuck-at-faults possible in the designs.

## Question 13

For the design identified in Question 11, develop a structural Verilog®-HDL model for the design. Simulate the design in its' fault-free state and for each of the stuck-at-faults possible in the designs.

## Question 14

For the design identified in Question 11, develop a structural Spice model for the design. Simulate the design in its' fault-free state and for each of the stuck-at-faults possible in the designs.

*The following questions will require access to the PC based tester arrangement identified in Appendix E.*

## Question 15

With reference to the tester design in Appendix E, develop such a tester suitable for developing digital logic test programs. Repeat the above simulation exercises with the tester and compare the experiences.

## Question 16

Identify any limitations of the identified tester architecture in Appendix E and develop an improved design.

# Chapter 4
# Memory Test

*Semiconductor memories are to be found in many electronic/microelectronic applications from the everyday Personal Computer (PC) through to the latest generation Personal Digital Assistant (PDA), and embedded systems ranging from automotive electronics through to everyday household products. The memory is required to store data and program code that can be accessed and/or modified in a suitable manner.*

## 4.1 Introduction

### 4.1.1. Memory Overview

The basic rationale for memory is for the storage of, access to, and potential modification of, data and program code for use within a processor based electronic circuit/system. There are many processor (Microprocessor (µP), Microcontroller (µC), Digital Signal Processor (DSP)) based systems in existence and one of the key system building blocks is the memory. Memory is required to store data and program code that can be accessed and/or modified in a suitable manner. With the increases seen in the development and use of embedded systems, the role of this type of circuit component is increasingly important.

The key drivers for memory are essentially the need for increased capacity (capacity – amount of data that can be stored within a single memory circuit) and increased operating speed (reduced times to write data to and read data from the memory), but at a lower purchase cost for the user. The move towards larger capacity and faster memories is most noticeable in everyday life when specifying and purchasing a PC – the amount of Random Access Memory (RAM) is a selling point! In general, memory can be considered for use for one of the following three data/program storage purposes:

- **Permanent Storage** – Values are stored in memory that can be read only within the application and if the stored values are to be changed (if possible), would require the memory to be taken out of the application and replaced. A typical application would be the program code within a microprocessor based embedded system, where it is not anticipated that code would be required to be changed.

- **Semi-Permanent** – Values are stored and normally only read (as in the permanent storage application). However, it may be necessary that the stored values would have to be modified as and when required whilst the memory remains in the circuit. A typical application would be a microprocessor based system where the program code would need to be periodically and easily upgraded.

- **Temporary** – Values are stored for temporary use that requires fast access (*e.g.* program code within a computer system that is required to be used at the current (run) time, but can be removed once the use has finished with the program) and/or modification (*e.g.* the temporary storage of data used within a program).

The types of memory [1, 2] available fall into one of two types: **Read Only Memory** (ROM) and **Random Access Memory** (RAM). RAM is also sometimes referred to as **Read-Write Memory** or RWM. The basic types of memory are shown in Fig. 4.1.

**Memory bandwidth** is an increasingly important aspect to memory design and choice for use. The increase in processor performance, and ever demanding applications such as multimedia and 3D graphics, means that high bandwidth memory is essential to sustain electronic system performance. Memory bandwidth is the amount of information that can be transferred to and from memory per unit time. In current electronic systems, the problem arises in the connections between individual ICs on a PCB. The timing limitations due to the interconnect delays between a processor IC and a memory IC restricts the speed of information transfer. For faster memory access, the reduction in the interconnect length reduces the delay due to this interconnect. This is achieved by adopting a System on a Chip (SoC) [3] or System in a Package (SiP) approach to the circuit implementation. Such approaches allow for memory to be integrated within the same package as the controlling processor.

RAM can be used for temporary and/or semi-permanent storage purposes (a key point of the RAM is that when the power is removed, the memory contents are lost). SRAM (Static RAM) and DRAM (Dynamic RAM) are the common types used. SRAM operates faster than DRAM, but DRAM is physically smaller and cheaper. In computer systems, SRAM is typically used for on-chip fast cache memory in processors whilst DRAM for off-chip (discrete) temporary storage. ROM can be used for permanent and semi-permanent storage purposes (the memory contents are retained even when the power is removed).

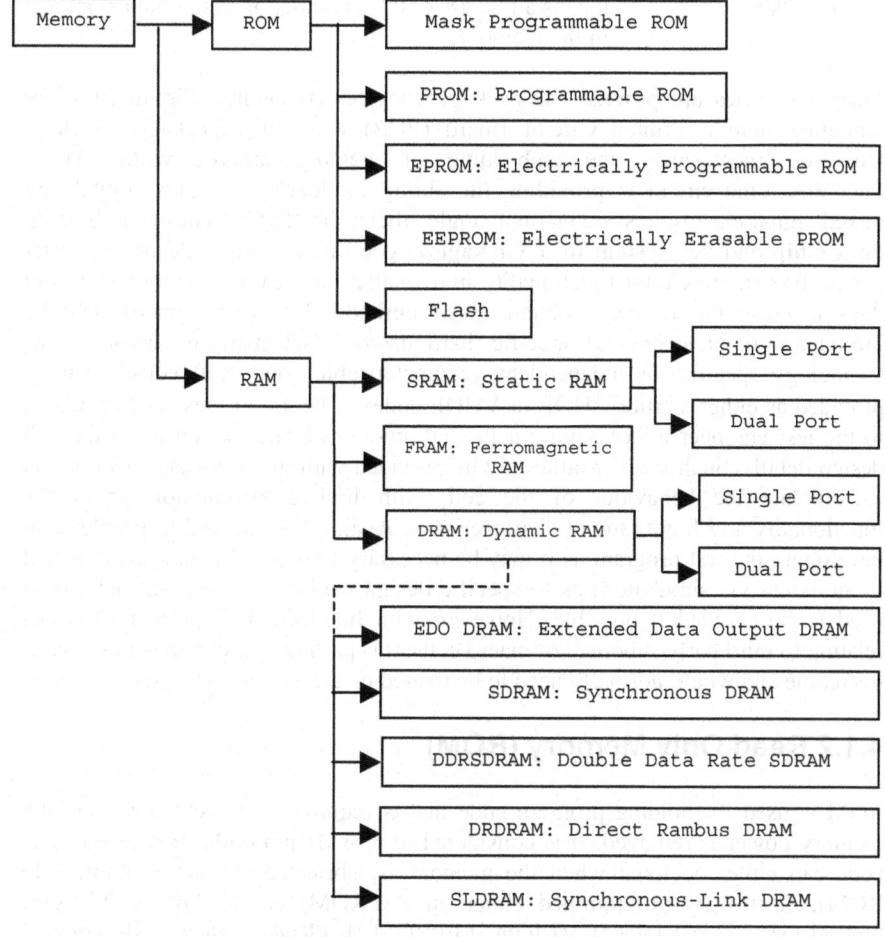

**Fig. 4.1.** Basic memory types

The basic terminology used for the memory is:

- **Memory Cell**    The part of the memory that stores a single bit of data.
- **Memory Word**  A group of bits, typically ranging from 8 to 64.
- **Byte**               A group of 8 bits.
- **Capacity**         The number of bits that a memory can store.

The capacity of the memory will be the number of memory addresses and the number of bits stored in each memory address. This may be stated in different ways and can lead to confusion. For example:

- 8k x 8-bit            means 8k (8 x 1024) memory addresses storing 8 bits of data in each memory address.

- 8kb                          means an 8k  (8 x 1024) number of bits of data storage
                               in the memory.

Many memories are provided as packaged devices (Integrated Circuit (IC)) for mounting onto a Printed Circuit Board (PCB) with other packaged devices. However, increasingly the embedding of memory **macros** within larger microelectronic circuit is providing the ability to develop complex digital and mixed-signal electronic systems on a single silicon die. The advent of the **System on a Chip** and the **System in a Package** is good news from a design and user perspective (more circuit functionality in a smaller physical circuit footprint), but does increase the testing problem. Such **embedded** memory macros can be provided as fixed, technology specific "hard macros" (schematic and layout), or as technology specific, or independent, parameterisable code "soft cells" (usually provided as either Verilog®-HDL or VHDL code).  The use of these cells can lead to the test engineer either having a limited amount of information as to the cell design details. Such a cell would only be provided with the necessary information to simulate the behaviour of the cell, with limited information as to the functionality and Input-Output characteristics. Again, this can lead to problems in developing the test program as it may be necessary to make assumptions (verified to an extent via simulation) as to specific design details – a situation that would need to be avoided if possible. However, with Intellectual Property (IP) issues relating to third party suppliers of macro cells, this particular problem is not easy to overcome and would normally need to be treated on a case-by-case basis.

## 4.1.2 Read Only Memory (ROM)

ROM is used for holding program code that is required to be retained when the memory power is removed. It is considered to provide **non-volatile storage**. The code can either be fixed when the memory is fabricated (Mask Programmable ROM), electrically programmed either once (PROM) or multi-times (EPROM (erased using UltraViolet (UV) light, EEPROM ($E^2$PROM - electrically erased), Flash (electrically erased)).

## 4.1.3 Random Access Memory (RAM)

RAM is used for holding data and program code that require fast access and the ability to modify the contents during normal operation. RAM differs from read-only memory (ROM) in that it can be both read from and written to in the normal application (although considering $E^2$PROM and Flash memory, this distinction is becoming somewhat limited). Flash memory can also be referred to as **Non-Volatile RAM** (NVRAM). **RAM** is considered to provide a **volatile storage** since unlike the ROM, the contents of RAM are lost when the power is turned off. There are two main types of RAM: Static RAM (SRAM) and Dynamic RAM (DRAM). Each type of memory may have Single Port or Dual Port architecture. In the Single Port RAM, data is written to, and read from the same port on the device. In dual port RAM, there are two ports that allow for data to be written to and read from the memory simultaneously. Care has to be taken, through suitable circuit timing, so that no attempt is made to write to, and read from, a cell at the same time. In addition to SRAM and DRAM, the Ferromagnetic RAM (FRAM) has recently emerged and allows for non-volatile storage.

# 4.2 SRAM Structure

The SRAM consists of a number of connections:

- Address lines ($A_0$ ... $A_{(m-1)}$):
  These define the memory location to be selected for reading or writing.
- Input/Output data lines ($D_0$ ... $D_{(n-1)}$):
  These define the data to write to the memory or read from memory.
- Write Enable ($\overline{WE}$):
  This selects between the memory read and write operations.
- Output Enable ($\overline{OE}$):
  This is used to enable the output buffer for reading data from the memory (usually active low).
- Chip Select ($\overline{CS}$):
  This is used to select the memory (usually active low).
- Power Supply.

When the SRAM is a discrete packaged device, the above connections are available directly at the package pins. When the SRAM is embedded within a larger circuit, the above connections (except for the power supply connections) would be available provided that adequate controllability and observability is built into the overall circuit design. The SRAM stores data in a memory cell consisting of a number of transistors. A complete memory design will consist of an array of individual memory cells and additional logic. Each memory cell will hold a logic level (0 or 1) as long as power is supplied to the circuit. The read operation for the SRAM is non-destructive: that is, the reading of the value stored in the memory cell does not cause the value to be lost. When the power is removed, then the contents of the memory cell are however lost.

In CMOS technology, the cell consists of both nMOS and pMOS transistors. Considering the SRAM as a discrete IC, see Fig. 4.2, it will have the following connections (device pins):

- Address bus (input): location of the data that is to be read or written.

- Data bus (bi-directional (input and output)): Data to be written to the memory or read from the memory.
- Control signals: chip select, read, write (input): Memory control signals.
- Power supply.

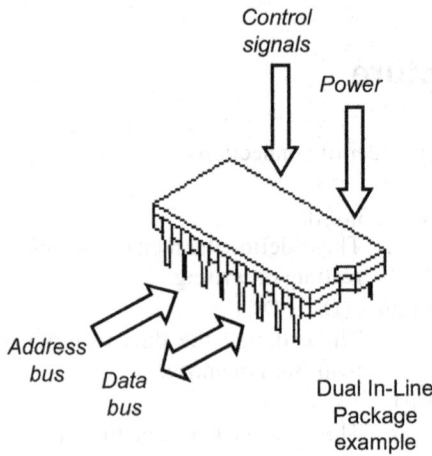

**Fig. 4.2.** Discrete SRAM IC

In this view, access to the pins of the device is possible and there are a number of key parts that require testing:

- The SRAM circuit in the core of the silicon die within the package
- The I/O cells in the periphery of the silicon die within the package
- The packaging – from the silicon die I/O through to the package pins

Within the core of the IC, the main part of the RAM circuit is contained. The general structure of the SRAM circuit is shown in Fig. 4.3. In this view, the memory is addressed by m-address lines (address bus - $A_0$ to $A_{(m-1)}$) providing $2^m$ addresses. The address bus is split to form the inputs to a row decoder and column decoder simultaneously. The row decoder will create the word lines that will access $2^x$ individual rows, where each row will contain a row of $2^{(m-x)}$ memory cells selected by the bit lines. A memory cell will be located at the intersection of a word line and bit line. The column decoder will select the memory cells to connect to the sense amplifiers. For a one-bit output, only one cell will be addressed. For a multi-bit output (*e.g.* 8), the array will be designed for all the required output bits to be selected (the memory array will contain a number of cells equal to the (number of word lines) x (number of bit lines) x (number of bits stored in each address). The sense amplifiers will amplify the small voltage change output from the memory array into the logic level voltages required by the external system. The output from the sense amplifiers is connected to tri-state buffers that will only pass the logic output to the device data I/O in read mode. In write mode, the data is applied to the input buffers and written to the memory array.

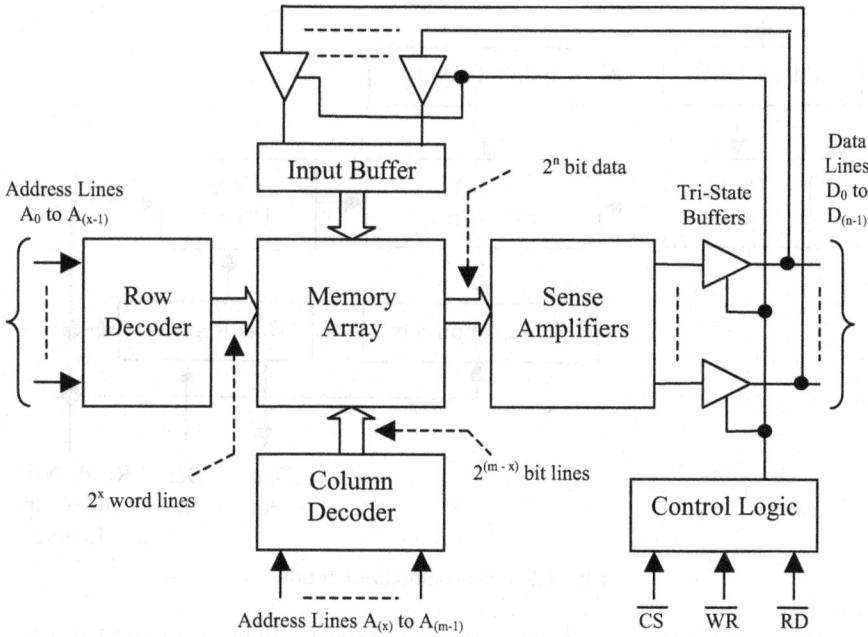

**Fig. 4.3.** SRAM structure

A CMOS implementation of a memory cell is shown in Fig. 4.4. This is an example of a six-transistor (6T) cell, in that it is made up of six transistors (two transistors in each inverter – one nMOS and pMOS transistor). Here, the two nMOS transistors act as switches. When the switches are closed under the control of the Word Line, the two inverters are connected to the Bit Line and $\overline{\text{Bit Line}}$. A value can either be written to or read from the cell. When the switches are open, then the inverters store and hold a value.

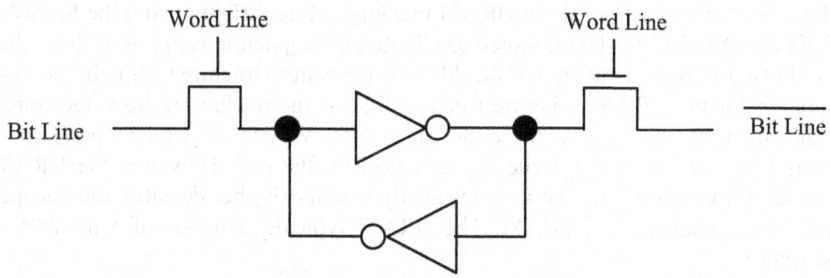

**Fig. 4.4.** 6T memory cell

The diagram in Fig. 4.3 provides detail of the SRAM architecture. A different way in which to view the SRAM is the functional model, as shown in Fig. 4.5.

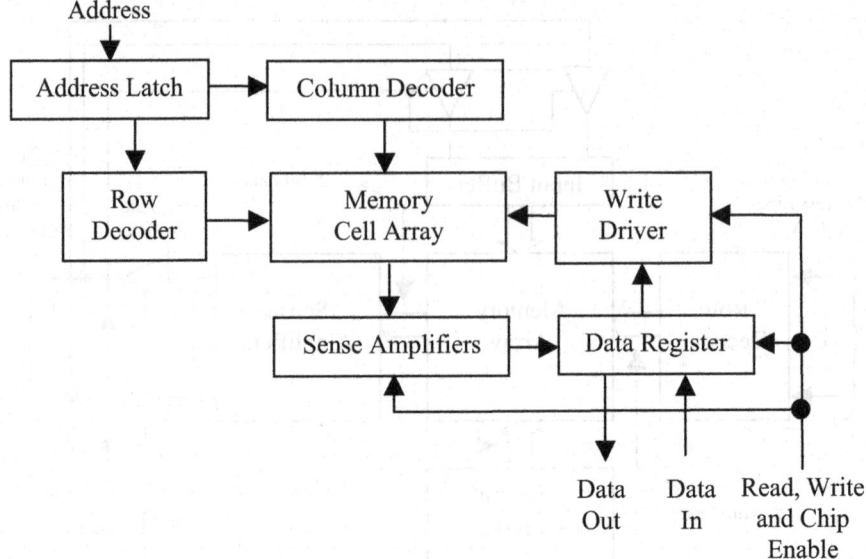

**Fig. 4.5.** SRAM functional model

It is also essential to ensure that the timing requirements of the memory (the access time) is adhered to and not violated, Each memory will have specific timing requirements for both read and write actions. The particular memory device datasheet will identify the timing requirements.

## 4.3 DRAM Structure

The DRAM was introduced in 1970 with a capacity of 1Kb (kilo-bit) - which is considered now as very small. Since then, the capacity has increased to 256Mb (Mega-bit). The structure of the DRAM memory cell is different from the SRAM - within the DRAM, the data is stored as charge on a capacitor as opposed to a value stored using transistors as in the SRAM. This capacitor structure results in the read operation for the DRAM being destructive: that is, the reading of the value stored in the memory cell causes the value to be lost. Whenever a read operation is undertaken, it is necessary write the data back to the cell. However, the DRAM provides a cost benefit in that it is physically smaller (higher density) and cheaper than the equivalent sized SRAM. The DRAM typically consists of a number of connections:

- Address lines ($A_0$ ... $A_{(m-1)}$):
  These define the memory location to be selected for reading or writing

- Input/Output data lines ($D_0$ ... $D_{(n-1)}$):

  These define the data to write to the memory or read from memory

- Output Enable ($\overline{OE}$):

  This is used to enable the output buffer for reading data from the memory

- Write Enable ($\overline{WE}$):

  This is used to enable the writing of data to the memory

- Row Address Strobe ($\overline{RAS}$):

  This signal clocks the row address

- Column Address Strobe ($\overline{CAS}$):

  This signal clocks the column address

- Power Supply

The structure of the DRAM is shown in Fig. 4.6. This is similar to the SRAM except for the inclusion of additional timing and control circuitry. Here, the data is stored in the DRAM Cell Array that is addressed through the Column and Row Address Decoders. The Cell Array contains the DRAM memory cells required to store the necessary number of data bits. The inputs to the memory are the control and address signals. The control signals tell the DRAM what operation is executed and the address signals tell the DRAM the cell location to perform the operation on. The outputs from the memory are the data lines to transfer data to and from the DRAM. The address and control signals are applied to the Command Decoder that generates the necessary Row and Column Decoder signals.

As with the SRAM, it is essential to ensure that the timing requirements of the memory (the access time) is adhered to and not violated. Each memory will have specific timing requirements for both read and write actions. The particular memory device datasheet will identify the timing requirements.

Each memory cell will consist of a capacitor and switch (transistor), see Fig. 4.7. This is a one-transistor (1T) cell design, as the cell contains one transistor is addition to the storage capacitor. Here, when the transistor "switch" is closed, the voltage on the Bit Line is stored on the capacitor (write operation) or the Bit Line voltage becomes the voltage already stored on the capacitor (read operation). The structure of the DRAM memory cell means that:

- Whenever the switch is closed, the charge on the capacitor can leak away into the circuitry connected to the switch (when a read operation is undertaken, so reducing the capacitor voltage over time)

- The charge can leak away into the surrounding circuitry in the physically realisation of the circuit even when the switch is open

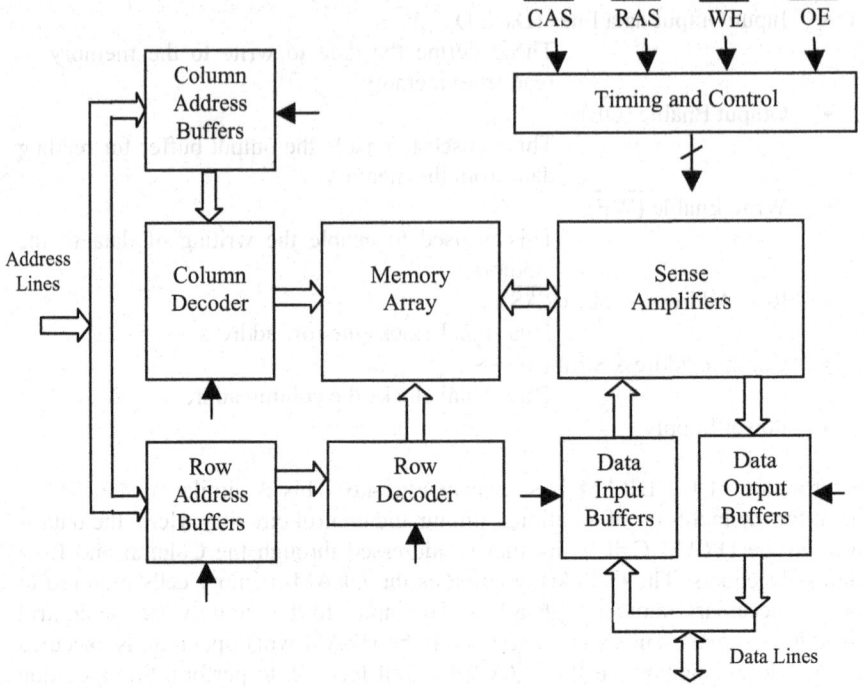

**Fig. 4.6**. DRAM structure

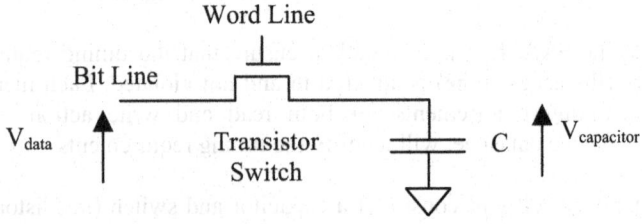

**Fig. 4.7**. DRAM memory cell

# 4.4 ROM Structure

The Ready Only Memory (ROM) is a **non-volatile memory** (NVM) which stores data on a permanent or semi-permanent basis, where the memory contents are retained even when the power has been removed from the circuit. The Mask Programmable ROM has the contents committed during the device fabrication process and cannot be changed. The Programmable ROM (PROM) is committed electrically once after the device has been fabricated and once committed the contents cannot be changed. The Erasable PROM (EPROM) (or Ultraviolet

EPROM (UVEPROM)) is committed electrically once after the device has been fabricated, but the contents can then be erased by exposure to ultraviolet light. The Electrically Erasable PROM (EEPROM or $E^2$PROM) is committed electrically once after the device has been fabricated, but the contents can then be erased electrically. Flash memory, based on EEPROM technology, can be electrically erased and reprogrammed, but have higher circuit densities than EEPROM.

The UVEPROM would need to be taken out of the application circuit for erasure using a special UV light source unit. The EEPROM and Flash memories can be erased when still in the application, allowing for greater ease and flexibility for device erasure and reprogramming.

The memory array within the ROM may be one of two forms:

- NOR-based array

- NAND based array

The basic structure for the ROM is shown in Fig. 4.8. In this, the m-bit address is decoded to produce $2^m$ horizontal address lines. Only one address line is selected at a time. The n-bit data lines run vertically through the memory array. At the cross-over of the address and data lines, the presence or absence of a transistor will set the data line to logic 0 (presence of a transistor) or 1 (absence of a transistor).

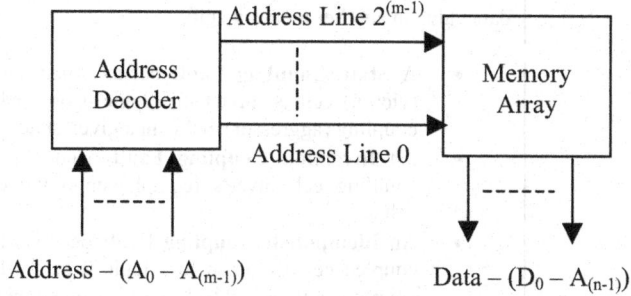

**Fig. 4.8.** Basic ROM structure

# 4.5 Fault Modelling in Memory

The following table, Table 4.1, identifies examples of the faults [3, 4] that may occur within the memory and their causes. Single or multiple faults may exist. Memories may have faults that are particular to the circuit structure and the dense packing of the circuitry within the circuit die required to provide for high-capacity memories within a small physical circuit. A number of the possible memory faults would not be seen in digital logic circuits.

**Table 4.1.** Example memory faults

| Fault | Description |
|-------|-------------|
| Addressing fault | A fault in the row or column decoder will cause errors in the address selection. There may be one of the following scenarios:<br><br>• For a certain memory address, no cell will be accessed<br>• For a certain address, multiple cells will be accessed<br>• A certain cell will never be addressed<br>• A certain cell will be accessed by multiple addresses |
| Bit-pattern fault | Occurs in programmable ROMs where programming faults may occur. A 0 may be programmed rather than a 1, or vice-versa |
| Bridging fault | An unwanted resistive path connects lines in the memory. Usually considered a low-resistance value |
| Cell coupling fault | The data in a memory cell is affected by either a transition or data value in another memory cell:<br><br>• A **State Coupling Fault** occurs when a coupled (victim) cell is forced to a logic 0 or 1 when the coupling (aggressor) cell is in a given state<br>• An **Inversion Coupling Fault** occurs when the coupling cell inverts (complements) the coupled cell<br>• An **Idempotent Coupling Fault** occurs when the coupled cell is forced to a logic 0 or 1 when the coupling cell has a 0 to 1 or 1 to 0 transition |
| Data retention fault | The memory is not able to hold data for a specified period of time |
| Neighbourhood pattern sensitive fault | Data in a memory cell is affected by either a transition or data value in another memory cell. The occurrence of the fault will be dependent on the value of the data (the pattern) |
| Open fault | A line is open–circuit due to missing interconnect |
| Parametric fault | The electrical parameters of the circuit components are outside their expected range of values (between the minimum and maximum range of fault-free operation) |

| Random corruption of data due to external effects | These are random effects that may cause the contents of a RAM cell to change state. This may be due to incident radiation or transients/noise on the memory power supply |
|---|---|
| Recovery fault | The **Sense Amplifier Recovery Fault** occurs in the RAM when the sense amplifier saturates after a long run of logic 0 or 1 read/write operations<br><br>The **Write Recovery Fault** occurs when a write operation is followed by a read/write operation at a different location but the operation results in a read/write operation at the original location |
| State transition fault | An error may occur when the data in a memory location in a RAM changes state (0 to 1, or 1 to 0) |
| Stuck-at-fault | A line within the memory (address, data, control or bit line) is stuck at either logic 0 or 1<br><br>A memory cell may be are stuck at either logic 0 or 1 |
| Timing Fault | Data access time may be slower than expected |
| Transistor stuck-on/ stuck-off fault | Considering the transistor as a switch, this switch may be either stuck-on or stuck-off |

# 4.6 RAM Test Algorithms

## 4.6.1 Introduction

In essence, the testing of the RAM [4-15] is based on writing a value to the cells located at an address in the RAM, reading back the contents of the cells and comparing the read-back value to the written value. If these values are the same, then the RAM passes the test for that particular memory address. If the values differ, then the RAM fails the test for that particular memory address. The order in which the memory addresses are accessed and the values written to the memory cell will be defined by a particular RAM test algorithm.

There are a number of algorithms that are used to test the RAM. It is possible, in theory, to develop and apply any algorithm that may be considered. However, each algorithm considered will have its own pros and cons, and specific algorithms have been developed and analysed in detail [25]. In the main, the fault coverage for each algorithm and the test time required (the number of read and write cycles) will differ between each algorithm and so the algorithm to use will need to be carefully considered.

## 4.6.2 Notation

It is common to present the algorithms in a form that presents the read/write operations in a mathematical type of notation. Table 4.2 summarises the notation commonly used.

**Table 4.2.** Memory test notation

| Notation | Action |
|----------|--------|
| r | Memory action: a read operation |
| w | Memory action: a write operation |
| r0 | Memory action: read a 0 from the selected memory location |
| r1 | Memory action: read a 1 from the selected memory location |
| w0 | Memory action: write a 0 to the selected memory location |
| w1 | Memory action: write a 1 to the selected memory location |
| $\Uparrow$ | Increasing memory address |
| $\Downarrow$ | Decreasing memory address |
| $\Updownarrow$ | Either increasing or decreasing address |
| $\uparrow$ | Write a 1 to a cell containing a 0 |
| $\downarrow$ | Write a 0 to a cell containing a 1 |
| $\rightarrow$ | Write a 0 to a cell containing a 0 |
| $\rightarrow$ | Write a 1 to a cell containing a 1 |

An example of its use is shown below. Here, the MATS algorithm (see next subsection) is shown:

$$\Uparrow (w0) \; \Uparrow (r0, w1) \; \Downarrow (r1, w0) \; \Uparrow (r0)$$

This is interpreted as:

- First, write 0s to the memory in ascending order of address (starting at the lowest address) - $\Uparrow$ (w0)

- Second, read the memory (expect to read a 0) and then write a 1 in ascending order of address (starting at the lowest address) - $\Uparrow$ (r0, w1)

- Third, read the memory (expect to read a 1) and write a 0 in descending order of address (starting at the highest address) - $\Downarrow$ (r1, w0)

- Fourth, read the memory (expect to read a 0) in ascending order of address (starting at the lowest address) - $\Uparrow$ (r0)

## 4.6.3 RAM Test Algorithms

A number of RAM test algorithms have been developed. These include:

- Memory Scan (MSCAN)
- Checker Patterns
- Galloping Patterns (GALPAT)
- Galloping Diagonal (GALDIA)
- Galloping Column (GALCOL)
- Galloping Row (GALROW)
- Walking Pattern (WALKPAT)
- March Algorithm

### Memory Scan (MSCAN)

This is the simplest pattern in which a 0 is written to a cell and read back for verification. Then a 1 is written to a cell and read back for verification. This is repeated for all cells in the memory. This is also referred to as the "Zero-One Algorithm".

For an n-address memory, the algorithm is:

```
For I = 0 to (n-1)
      Write 0 to address I
      Read data from cell I and check for correctness
      Write 1 to address I
      Read data from cell I and check for correctness
Next I
```

### Checker Patterns

With this test, the memory is filled with a pattern of alternating 0s and 1s (*e.g.* 01010101 at an address  followed by 10101010 at the next address). This is then read back and checked for correctness. In an addition to this basic test, the memory is then filled with the complementary pattern (e.g. 10101010 followed by 01010101), the contents read back and checked for correctness.

For an n-address memory, the algorithm is:

```
For I = 0 to (n - 1) step 2
      Write checker pattern to address I
      Write the complement of checker pattern to address (I+1)
Next I

For I = 0 to (n-1)
      Read data from cell I and check for correctness
Next I
```

```
For I = 0 to (n - 1) step 2
        Write the complement of checker pattern to address I
        Write the checker pattern to address (I+1)
Next I

For I = 0 to (n-1)
        Read data from cell I and check for correctness
Next I
```

## Galloping Patterns (GALPAT)

A number of variants of the GALPAT pattern exist for use. In one example of this test, the memory contents are initially set to 0. Then, starting at the lowest address, the contents of the address are read (expected 0) and then a 1 is written to the address. The contents of the remaining memory addresses are then read back and checked for correctness, and the address incremented. The read-write-read operations are repeated for all addresses up to the highest address. Then, starting at the highest address, the contents of the address are read (expected 1), then a 0 is written to the address. The whole memory is then read back and checked for correctness and the address decremented.

For an n-address memory, the algorithm is:

```
For I = 0 to (n-1)
        Write 0 to address I
Next I

For I = 0 to (n-1)
        Read data from cell I and check for correctness
        Write 1 to cell I
        Read data from cells (0 to (I-1)) and ((I+1) to (n-1)) and
                check for correctness
Next I

For I = (n-1) down to 0
        Read data from cell I and check for correctness
        Write 0 to cell I
        Read data from cells ((n-1) down to (I+1)) and ((I-1) down to
        0) and check for correctness
Next I
```

There are a number of variants on the GALPAT algorithm:

**Galloping Diagonal (GALDIA):**
A modification of the GALPAT except now a value is written to a diagonal

**Galloping Column (GALCOL):**
As with the GALPAT, except a value is written to a whole column

**Galloping Row (GALROW):**
As with the GALPAT, except a value is written to a whole row

**Walking Pattern (WALKPAT)**:
Another variant on the GALPAT

## March Algorithm

This is the most popular memory test algorithm. In this test, operations are performed on one cell before proceeding. There are a number of variations which have been developed. These include:

- MATS              (Modified Algorithmic Test Sequence)
- MATS+
- MATS++
- Marching-1/0
- March A
- March B
- March C
- March C-
- Extended March C-
- March G
- March X
- March Y

The MATS is the basic March Algorithm. In this test, the contents of each address are initially set to 0. Then, starting at the lowest address, the contents of each address is read and checked for correctness (expecting a 0) before writing a 1. This is repeated for all addresses. Then starting at the highest address, the contents of each address are read and checked for correctness (expecting a 1) before writing a 0. Then, starting from the lowest address, read the contents of each address and check for correctness (expecting a 0). For an n-address memory, the algorithm is:

```
For I = 0 to (n-1)
        Write 0 to address I
Next I

For I = 0 to (n-1)
        Read data from cell I and check for correctness
        Write 1 to address I
Next I

For I = (n-1) down to 0
        Read data from cell I and check for correctness
        Write 0 to address I
Next I

For I = 0 to (n-1)
        Read data from cell I and check for correctness
Next I
```

This particular (MATS) algorithm is referred to as a **6N test** where N is the number of addresses in the memory. It requires 6 read/write operations for each address to complete the test. In general, the application of different algorithms will result in different number of read/write operations and fault detection.

# 4.7 Memory Access for Test

Memories will be tested by one of a number of methods:

- For **discrete packaged devices**, then an external tester will have direct access to the memory via the package pins. The memory itself may also include BIST.

- For **embedded memories**, the additional problems for test are associated with the access to the memory (the controllability and observability issues). The memory can be tested using one or a combination of the following methods, as summarised in Table 4.3.

**Table 4.3.** Access for embedded memory test

| Functional test | The memory is written to and read from using the normal functions within the design |
|---|---|
| Direct access | The RAM is connected to the package pins using a multiplexer arrangement (see Fig. 4.9) |
| Scan chain access | An internal scan chain is added to surround the memory inputs and outputs. A number of possible arrangements for the scan chain configuration can be implemented |
| Memory Built-In Self-Test (MBIST) | Self-test circuitry is added to the circuitry so that the device can generate test signals and analyse the results itself. This is discussed in the following section |
| On-chip processor | If the design contains a processor ($\mu$P, $\mu$C or DSP), then the processor can be used to act as the tester (in the same manner as an external tester arrangement) |

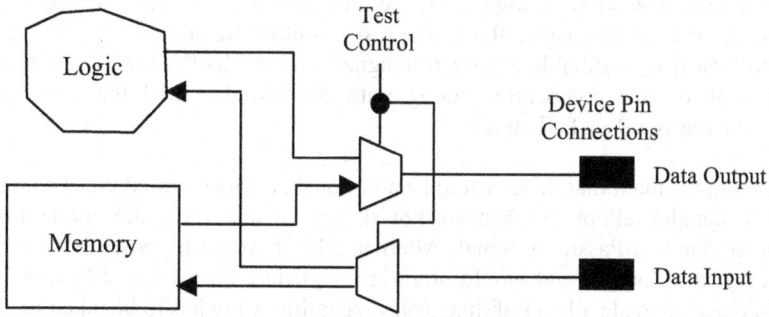

**Fig. 4.9.** Memory test via direct access (simplified view)

## 4.8 Memory BIST

The purpose of Memory BIST (MBIST) [16-22] is to provide the ability for on-chip signal generation and signal analysis. The rationale for the inclusion of MBIST is due to the significant increase in the inclusion of embedded memory macros within SoC designs and the difficulties in obtaining access for test purposes. These memories will vary in size and configuration, but will all require high quality and cost effective test procedures to be developed.

The basic architecture of MBIST for the RAM is shown in Fig. 4.10. There are two modes of operation:

- **Mode 1**: Normal Operating Mode:

  The circuit acts in its end-user operational mode (mission mode). The Normal Data Inputs (NDIs) come from other logic within the design and the Normal Data Outputs (NDOs) are applied to further logic within the design.

- **Mode 2**: Self-Test Mode:

  The Normal Data Inputs are disconnected from the memory (via the multiplexer) and Test Data Inputs (TDIs) generated on-chip are applied using the MSBIST logic.

The Normal Data Inputs (Address, Data and Control) come from logic within the design. These signals pass through a multiplexer that select the normal input signals (in normal operating mode) or test signals from the MBIST logic (in self-test mode). The MBIST logic generates the necessary address, data and control signals, and in this view, the multiplexer control signal. The MBIST logic is controlled from a suitable test control signal. The results from the RAM (data out) are applied to a comparator, along with the test data and the output of the comparison is a Pass/Fail signal.

It should be noted that there would be additional signals required (such as reset and clock signals), although these are not shown. In this view, the output from the comparator is a Pass/Fail signal. Alternatively, it would be possible to include a test signature output that would produce a signature (sequence of 0s and 1s) that would also provide a level of diagnostic capability. Care has to be taken in order to ensure that the BIST circuitry is of a suitable size (it would be unhelpful if the BIST circuitry for any design was of comparable physical size, or larger than the circuit to test), and that the inclusion of the multiplexers in the path of the normal signals (and the resulting signal delays) does not affect the normal speed of operation. MBIST circuits can either be designed "by-hand" or by the use of BIST compilers. In the case of the compiler, RTL level HDL descriptions would typically be generated and inserted into the overall design description. The aim however would be to automate as much as possible the BIST generation and insertion tasks in order to reduce development time and minimise the potential for errors.

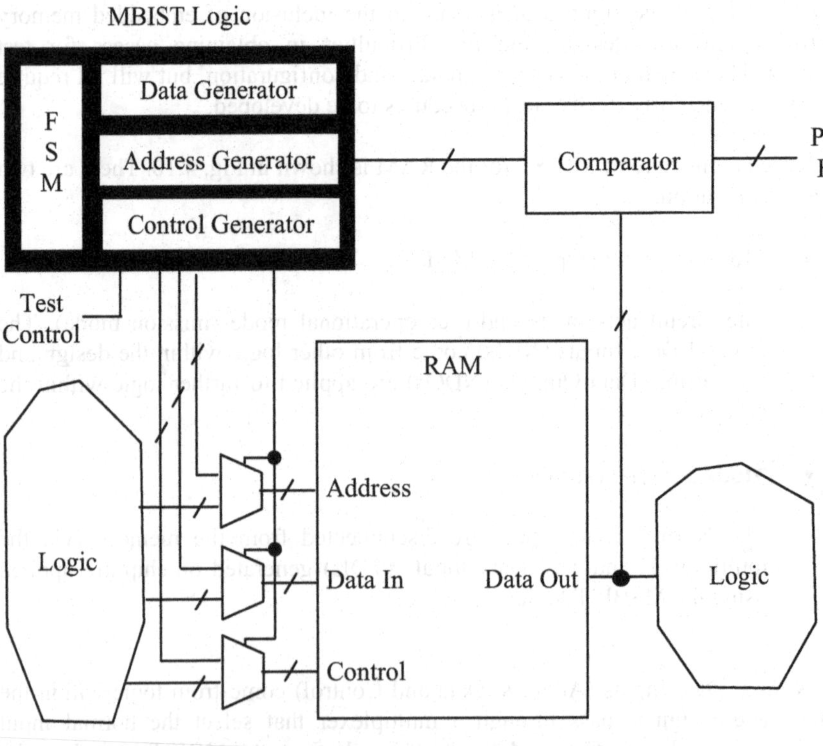

**Fig. 4.10.** Generic RAMBIST structure

# 4.9 ROM Test

Functional ROM testing [23] is based on cycling through the addresses of the ROM and checking the data contents for correctness. For large ROMs, then the amount of data that has to be retrieved and checked will become large and potentially problematic to deal with. In order to reduce the amount of data to deal with, the data is applied to a Single-Input Signature Register (SISR) for single data bits, or a Multiple-Input Signature Register (MISR) for multiple data bits, see Fig. 4.11. Once all the addresses in the ROM have been cycled through, the result (signature) from the SISR or MISR is compared to that of a known good ROM. If they are the same, the ROM passes the test. If they differ, the ROM fails the test. The address generator, control signal generator and SISR/MISR functions can be undertaken either using an external ATE or using on-chip BIST.

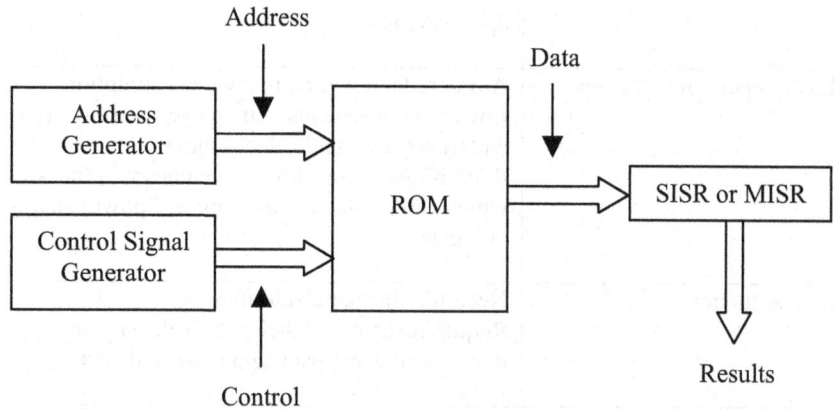

**Fig. 4.11.** ROM test using the SISR/MISR

# 4.10 Future Directions

Memory is required as a support circuit for processor based systems. Both volatile and non-volatile memories are required. The future trends [24] for memories impacts on the ability to effectively test these types of circuit. Table 4.4 provides a summary of the technology drivers for memories.

**Table 4.4.** Future memory test Issues

| Technology driver | Description |
|---|---|
| Faster operating speeds | Shorter read/write times leading to faster applications (*e.g.* computer graphics, |

| | communications systems) New memory architectures to improve memory bandwidth The need for memory bandwidth to catch-up with processor bandwidth |
|---|---|
| Higher levels of integration (higher density) | Larger memory capacity within a physically small package. Required for portable electronics (e.g. communications portable PCs, PDAs) More complex software that requires a large amount of temporary storage. This includes increasingly complex user interfaces Increases in the number of users of an electronic system, for example in file servers, telecommunications and networking applications |
| Lower operating voltages | Aid to reducing circuit power consumption Lower requirements on power supply (e.g. batteries for portable electronics) Requirement for lower geometry processes which cannot tolerate the "higher" power supply voltages |
| Cost reduction | Need for cheaper electronics Requirement for higher circuit density in order to reduce die and packaging size and cost |
| Memory redundancy and built-in self-repair | Enhancement of memory yield by identifying faults and allowing for particular memories to be repaired by using redundant circuitry within the memory Hard or Soft repair. Hard repair replaces faulty bits with redundant bits using fuses, antifuses or laser programming. Soft repair uses address mapping to bypass faulty address locations |
| Multi-site testing | The testing of multiple memories devices in parallel. By undertaking multi-site testing, the test time is decreased. However, there are requirements on the ATE to enable the devices to be connected in parallel and for the ATE to enable the handling of multiple devices |

# 4.11 Summary

This chapter has discussed requirements for the testing of memories. Random Access Memory (RAM) is used for temporary or semi-permanent storage of data and program code. Read Only Memory (ROM) is used for semi-permanent or permanent storage of data and program code. There are a number of memory circuit designs in use today and require specific tests in order to verify that they are fully functional. With the increase in memory capacity and speed, the requirements on the test equipment specification is ever increasing. Additionally, with the move towards embedded memories within SoC designs, access to the memories for test purposes is becoming increasingly difficult. Memory Built-In Self-Test (MBIST) is used to enable signal generation and signal analysis to be performed on-chip and to reduce the external tester requirements.

# 4.12 References

[1]     Prince B., "High Performance Memories, New architecture DRAMs and SRAMs evolution and function", Wiley, England, 1996, ISBN 0-471-95646-5

[2]     Bellaouar A. and Elmasry M., "Low-Power Digital VLSI Design, Circuits and Systems", Kluwer Academic Publishers, The Netherlands, 1995, ISBN 0-7923-9587-5

[3]     Rajsuman, R., "System-on-a-Chip Design and Test", Artech House Publishers, USA, 2000, ISBN 1-58053-107-5

[4]     Bushnell M. and Agrawal V., "Essentials of Electronic Testing for Digital, Memory and Mixed-Signal VLSI Circuits", Kluwer Academic Publishers, The Netherlands, ISBN 0-7923-7991-8

[5]     Piotr R. Sidorowicz and Janusz A. Brzozowski, "A Framework for Testing Special-Purpose Memories, IEEE Transactions on Computer-Aided Design of Integrated Circuits and Systems, Vol. 21, No. 12, December 2003, pp1459-1468

[6]     Cheng K., Tsai M. and Wu C., "Neighbourhood Pattern Sensitive Fault Testing and Diagnostics for Random-Access Memories", IEEE Transactions on Computer-Aided Design of Integrated Circuits and Systems, Vol. 21, No. 11, November 2002, pp1328-1336

[7]     van de Goor A. and Offernan A., "Towards a Uniform Notation for Memory Tests", Proceedings of the European Design and Test Conference, 1996, pp420-427

[8]     van de Goor, A. J., "Testing Semiconductor Memories", Wiley, New York, 1991, ISBN 0-4719-2587-x

[9]     Sharma A. K., "Semiconductor Memories: Technology, Testing, and Reliability", Wiley-IEEE Press, 2002, ISBN 0-7803-1000-4

[10]    Wilkins B.R. "Testing Digital Circuits An Introduction", Van Nostrand Reinhold (UK), UK, 1986, ISBN 0-442-31748-4

[11]    Al-Ars Z. and van de Goor A., "Test Generation and Optimization for DRAM Cell Defects Using Electrical Simulation", IEEE Transactions on Computer-Aided Design of Integrated Circuits and Systems, Vol. 22, No. 10, October 2003, pp1371-1384

[12]    Barth J. et al., "Embedded DRAM design and architecture for the IBM 0.11μm ASIC offering", IBM Journal of Research and Development, Vol. 46, No. 6, November 2002, pp675-689

[13]    Vollrath J., "Testing and Characterization of SDRAMs", IEEE Design and Test of Computers, January-February 2003, pp42-50

[14]    van de Goor, "Using March Tests to Test SRAMs", IEEE Design & Test of Computers, Vol. 10, Issue 1, March 1993, pp8-14

[15]    van de Goor A., "An Industrial Evaluation of DRAM Tests", IEEE Design and Test of Computers, September-October 2004, pp430-440

[16]    Der-Cheng Huang and Wen-Ben Jone, "A Parallel Built-In Self-Diagnostic Method for Embedded Memory Arrays", IEEE Trans. on Computer-Aided Design of Integrated Circuits and Systems, Vol. 21, No. 4, April 2002, pp449-465

[17]    D. C. Huang and W. B. Jone, "A Parallel Transparent BIST Method for Embedded Memory Arrays by Tolerating Redundant Operations", IEEE Transactions on Computer-Aided Design of Integrated Circuits and Systems, Vol. 21, No. 5, May 2002, pp617-628

[18]    Tehranipour M. and Navabi Z., "Zero-Overhead BIST for Internal SRAM Testing", Proceedings of the 12th International Conference on Microelectronics, 2000, pp109-112

[19]   Park S., et al., "Designing Built-In Self-Test Circuits for Embedded Memories Test", Proc. of the Second IEEE Asia Pacific Conference on ASICs, 2000, pp315-318

[20]   Powell T. et al., "BIST for Deep Submicron ASIC Memories with High Performance Application", Proc. of the International Test Conference, 2003, pp386-392

[21]   Zorian Y. and Shoukourian S., "Embedded-Memory Test and Repair: Infrastructure IP for SoC Yield", IEEE Design and Test of Computers, May-June 2003, pp58-67

[22]   Rajsuman R., "Rambist builder: a methodology for automatic built-in self-test design of embedded RAMs", Records of the 1996 IEEE International Workshop on Memory Technology, Design and Testing, 1996, pp50-56

[23]   Barth R., "Selective Optimization of Test for Embedded Flash Memory", Proceedings of the International Test Conference, 2002, pp1222

[24]   International Technology Roadmap for Semiconductors (ITRS), 2003 Edition, Test and Test Equipment

[25]   Riedel M. and Rajski J., "Fault coverage analysis of RAM test algorithms", Proceedings of the 13th IEEE VLSI Test Symposium, 1995, pp227-234

# Exercises

### Question 1

Consider an 8k x 8-bit **single port** SRAM with active low Chip Select, Read and Write signals. For the SRAM, in VHDL:

- Model the SRAM cell and verify the operation through simulation.
- Apply each of the following RAM test algorithms to the fault-free design:
    - Memory Scan
    - Checker Patterns
    - Galloping Patterns
    - March Algroithm (MATS)
- Within the model of the memory cell, model stuck-at-faults (both SA0 and SA1 faults) for the memory primary I/O
- Apply the RAM test algorithms to the faulty design and identify whether the particular algorithm will detect the fault or not

Within the above simulation exercises, store relevant simulation and analysis results in output text files for documentation purposes and further analysis.

## Question 2

Repeat Question 1 using Verilog®-HDL.

## Question 3

Repeat Question 1 except now consider a dual port SRAM.

## Question 4

Repeat Question 3 using Verilog®-HDL.

## Question 5

Consider an 8k x 8-bit ROM with a user set contents and with active low Chip Select and Read signals. For the ROM in VHDL:

- Model the ROM and verify the operation through simulation
- Within the model of the memory cell, model stuck-at-faults (both SA0 and SA1 faults) for the memory primary I/O
- Simulate to the faulty design and identify whether the particular faults are detected as would be predicted

Within the above simulation exercises, store relevant simulation and analysis results in output text files for documentation purposes and further analysis. The content of the memory address (*i.e.* the stored data value) should be (the value of the memory address + $5_{10}$).

## Question 6

Repeat Question 6 using Verilog®-HDL.

## Question 7

For the SRAM model in Question 1, consider how it may be tested as follows:

- Functional test
- Direct Access
- Scan Chain Access
- Memory Built-In Self-Test (MBIST)
- On-chip processor

Develop and simulate a VHDL testbench in order to model the circuitry and signals required to implement the above five memory test access methods. Clearly identify any assumptions made.

## Question 8

For commercially available discrete SRAM and DRAM ICs, identify the memory capacity and the package types that they are provided in.

*The following questions will require access to the PC based tester arrangement identified in Appendix E.*

## Question 9

Identify a suitable 8k x 8-bit **single port** SRAM and using the PC based tester, repeat the experiments identified in question 1.

## Question 10

For the SRAM in Question 9, develop a BIST structure as shown in Fig. 4.10 in order to implement a MATS algorithm. Implement the BIST using the PC based tester. Identify any specific implementation issues arising.

## Question 11

Identify a suitable 8k x 8-bit ROM and using the PC based tester, repeat the experiments identified in Question 5.

## Question 12

For the ROM in Question 11, develop a BIST structure as shown in Fig. 4.11 in order to test the complete memory. Implement the BIST using the PC based tester. Identify any specific implementation issues arising.

# Chapter 5

# Analogue Test

*Analogue ICs perform an important role in electronic circuits and systems in their ability to process analogue signals in the continuous time domain, in particular performing actions such as signal amplification and filtering. Despite the size of the analogue circuits being small in comparison to the digital electronic circuits and systems found today, the testing for analogue specifications is challenging and can be time consuming, and hence of high cost.*

## 5.1 Introduction

In many applications, digital electronic circuits and systems [1-4] are being demanded by the end user and are driving many of the advances in integrated circuit design, fabrication and test. However, the world is analogue in nature. Electronic circuit functions can therefore be undertaken in either the analogue or digital domains:

- *The **analogue domain** is a representation of a signal that varies continuously over a range of values.*

- *The **digital domain** represents a signal that varies in discrete levels over a range of values.*

Given that electronic circuit functions can be undertaken in either the analogue or digital domains, integrated circuits will be classified to be one of three types of circuit:

- **Analogue**: this type of circuit manipulates continuously varying signals (voltages and currents) over a range of values.

- **Digital:** this type of circuit manipulates signals which are in the form of discrete (usually **binary** – a logical 0 or 1 value) values that change at discrete points in time.

- **Mixed-Signal**: this type of circuit combines the functionality of analogue and digital circuits, usually for interfacing analogue signals to a digital processor system.

In the test environment, analogue signals will be required to be suitably generated, captured and analysed. Analysis of signal characteristics resulting from a device under test (DUT) in both the **time domain** and the **frequency domain** will contain information relevant to determining whether an analogue circuit passes or fails a particular test. In the main, **functional testing** of the analogue circuit is undertaken in the test environment. This differs from digital device test in that, during production testing, structural testing of digital logic is well understood and universally implemented. In the analogue domain, structural test is not as established as in the digital domain and the application specific nature of the analogue circuit behaviour means that the development of structural tests is not as straightforward to consider and automate as it is with digital.

Testing will be undertaken by applying specific analogue test signal waveform types by applying either voltage or current to the DUT:

- DC (constant) value
- Sine wave (single tone (*i.e.* single frequency value signal)) with variable frequency, amplitude, offset and phase, see Fig. 5.1 for a single frequency sine wave plot
- Sine wave (multi-tone (*i.e.* two or more sine wave signals)) with variable frequency, amplitude, offset and phase
- Ramp
- Triangle
- Sawtooth
- Step
- Arbitrary waveform (a user defined signal shape which is used to represent a complex signal not represented in any of the above)

Signals may be generated by using either a **DC source** or **signal generator** (for the common signal waveforms (DC, sine, triangular, sawtooth, step)), or by using an **Arbitrary Waveform Generator** (AWG). In the signal generator, a signal of a predefined type can be selected with specific amplitude, offset, phase and frequency. For example, the sine wave shown in Fig. 5.1 shows the key parameters of this type of signal which is defined by the following equation:

$$v(t) = V_{OFFSET} + V_{AMPLITUDE}.\sin(2.\prod.f.t)$$

where v(t) is the instantaneous value of the signal at time t, but with no phase shift.

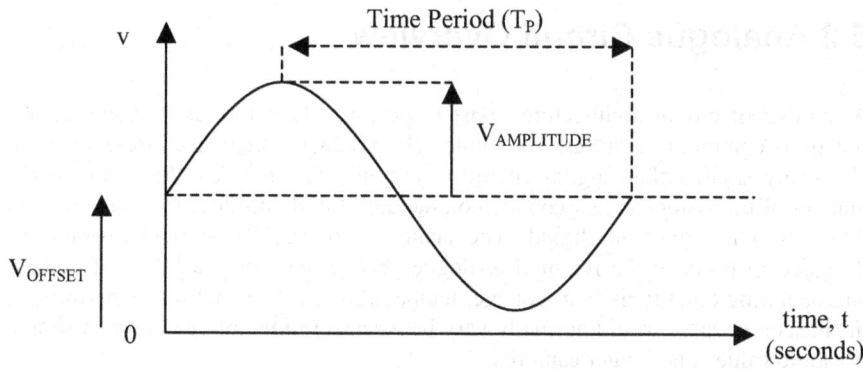

V<sub>OFFSET</sub> : DC offset voltage    T<sub>P</sub> : Sine wave time period
V<sub>AMPLITUDE</sub> : Sine wave amplitude   f : Signal frequency = (1/ T<sub>P</sub>)

**Fig. 5.1.** Sine wave signal parameters

A typical AWG arrangement can be used to develop the same signals as the signal generator, but with the additional benefit of generating waveforms of arbitrary value. For example, the same electronics would be used to create a single tone (single frequency) sine wave as well as multi-tone (multiple sine wave) signals. Figure 5.2 shows the basic architecture of an AWG [5].

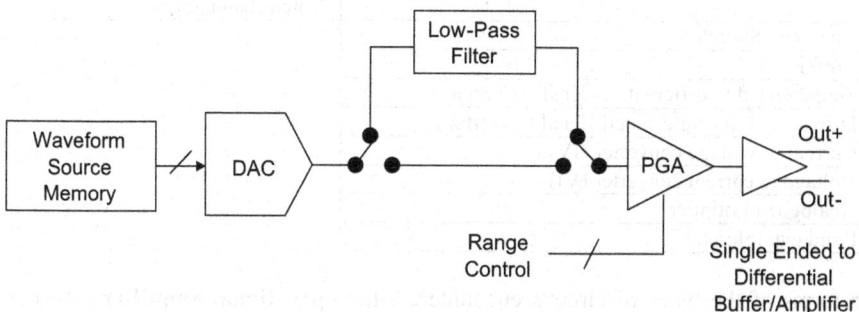

**Fig. 5.2.** Basic AWG arrangement

In this arrangement, the **waveform source memory** (*i.e.* RAM) would be initially loaded with a set pattern (*i.e.* the memory contents) of bits that digitally represents the required analogue signal. In order to create the output signal, the memory is continuously read at a set rate and the digital pattern applied to the Digital to Analogue Converter (DAC). The output of the DAC will be optionally low-pass filtered to remove high frequency components in the DAC output signal (*e.g.* voltage) and passed through a programmable gain amplifier (PGA). This is a digitally controlled analogue amplifier. The last stage is a single ended to differential signal buffer/amplifier, although this stage may not necessarily be included. The size of the memory (number of addresses), the DAC resolution (number of bits) and the DAC update rate would need to be selected in order to produce signals of the right quality.

## 5.2 Analogue Circuit Overview

A number of circuit architectures exist to perform the range of analogue circuit functions commonly required, see Table 5.1. In general, analogue circuits will be physically smaller than digital circuits commonly encountered (in terms of the number of transistors and silicon area on the die), but do produce problems for test that would not exist in digital. The analogue circuit, for example, would be designed to produce the required analogue performance over a range of process and operating conditions (*e.g.* voltage, temperature) and when a test is performed, the measured value would normally vary between set limits of operation. A simple 0/1 logic value is no longer captured.

**Table 5.1.** Analogue circuit types

| Type | Sub-type |
|---|---|
| Amplifier (single ended or differential) | Audio |
| | Power |
| | Operational amplifier |
| Filter (continuous time [6], switched-capacitor [7] and switched-current [8, 9]) | Low-pass |
| | High-pass |
| | Band-pass |
| | Notch (band reject) |
| Analogue Switch | |
| Buffer | |
| Single ended to differential signal converter | |
| Differential to single ended signal converter | |
| Current to voltage converter (IV) | |
| Voltage to current converter (VI) | |
| Analogue multiplier | |
| Band gap reference | |

In terms of the types of circuits encountered, the **operational amplifier** [10, 11] should be noted in further detail since this differential amplifier is used along with external components to create a range of analogue signal processing functions, both linear and non-linear. The operational amplifier is a differential amplifier, see Fig. 5.3, with a number of key characteristics, see Table 5.2. The purpose is to amplify a difference voltage (differential mode signal) at the two input terminals and reject common signals (common mode signal). The op-amp parameters are defined as **DC performance, dynamic performance, noise/distortion performance, input characteristics, output characteristics** and **power supply performance**. For example, the amplifier itself has an open loop gain ($A_O$) which is ideally infinite, but in reality is high but finite (*e.g.* 100,000) and which decreases with increasing signal frequency. The output voltage ($V_{OUT}$) is given by:

$$V_{OUT} = A_O.(V_{POS} - V_{NEG})$$

The input node impedance is ideally infinite (in reality finite but high) and the output impedance is ideally zero (in reality greater than zero). In linear applications, by using negative feedback and passive components (resistors and capacitors), the gain of the overall circuit can be accurately controlled and made to be selective to signal frequencies (*i.e.* produce linear operations such as signal amplification and filtering). Positive feedback however will produce non-linear functions.

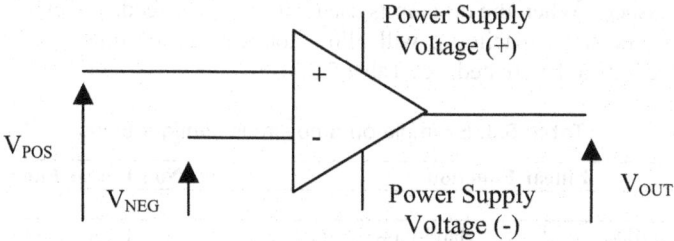

**Fig. 5.3.** Operational amplifier connections

The operational amplifier (op-amp) developed from the need to implement high gain differential (voltage) amplifiers for use within analogue computers. The analogue computer itself was a computational machine that performed a range of mathematical calculations – *i.e.* addition, subtraction, *etc.*

**Table 5.2.** Operational amplifier parameters

| Parameter | Notation | Units |
|---|---|---|
| **DC performance** | | |
| Input offset voltage | $V_{OS}$ | V |
| Input bias current | $I_B$ | A |
| Open loop voltage gain | $A_O$ | --- |
| **Input characteristics** | | |
| Input resistance | $R_I$ | $\Omega$ |
| Input capacitance | $C_I$ | F |
| Input common-mode voltage raange | $V_{CM}$ | V |
| Common-mode rejection ratio | CMRR | dB |
| **Output characteristics** | | |
| Positive voltage swing | $V_{SW+}$ | V |
| Negative voltage swing | $V_{SW-}$ | V |
| Output resistance | $R_O$ | $\Omega$ |
| Short-circuit current | $I_{SC}$ | A |
| **Dynamic performance** | | |
| -3dB bandwidth | $f_{-3dB}$ | Hz |
| Slew rate | SR | V/s |
| Settling time | $t_S$ | s |
| **Noise/distortion performance** | | |
| Total harmonic distortion | THD | dB |
| Input voltage noise | $v_{noise}$ | V/$\sqrt{Hz}$ |

| Input current noise | $i_{noise}$ | $A/\sqrt{Hz}$ |
|---|---|---|
| **Power supply performance** | | |
| Operating range | $V_{SUPPLY}$ | V |
| Quiescent current | $I_{DDQ}$ | A |
| Power supply rejection ratio | PSRR | dB |

The values identified in Table 5.2 will be guaranteed over a range of values and provided as minimum (min), typical (typ) and maximum (max) values within the device datasheet. When the op-amp is used with external feedback elements (in the main resistors and capacitors), will allow for a range of linear and non-linear circuit functions to be created, see Table 5.3.

**Table 5.3.** Example op-amp circuit configurations

| Linear Function | | Non-Linear Function |
|---|---|---|
| Amplifier | Band-Pass Filter | Comparator |
| Attenuator | Notch (Band Reject) Filter | Precision Half-Wave Rectifier |
| Adder (summer) | Integrator | Precision Full-Wave Rectifier |
| Subtractor | Differentiator | Oscillator |
| Low-Pass Filter | Unity Gain Buffer | Peak Detector |
| High-Pass Filter | Voltage Reference | Sample and Hold |

As with all ICs, the device will have defined **absolute maximum ratings** that if exceeded, may cause permanent damage to the device. The device data sheet (*e.g.* [12]) for the particular device will identify these ratings. For an operational amplifier, these will be:

- Power supply voltage
- Power dissipation
- Common-mode input voltage (referenced to the power supply voltage)
- Differential input voltage
- Storage temperature range
- Operating temperature range
- Device soldering: maximum temperature and duration

The signal frequencies that would be encountered by analogue circuits will range from DC into the GHz range. The general trend is for higher signal frequencies to be encountered. With the trend also for mobile communications, the development of RF (radio frequency) circuits that require high-speed analogue tests, is pushing the limits of the tester technology. In the broad picture, the **electromagnetic spectrum** encompasses all possible wavelengths of electromagnetic radiation. Electromagnetic energy at a particular wavelength $\lambda$ (in vacuum) has an associated frequency $v$ and photon energy $E$. At the lower frequencies, the electromagnetic spectrum forms the **radio spectrum**, see Table 5.4.

**Table 5.4.** Radio spectrum

| Frequency | | Band |
|---|---|---|
| **From** | **To** | |
| 3 | 300 | Extremely Low Frequency (ELF) |
| 300 | 3 kHz | Voice Frequency (VF) |
| 3 kHz | 30 kHz | Very Low Frequency (VLF) |
| 30 kHz | 300 kHz | Low Frequency (LF) |
| 300 kHz | 3 MHz | Medium Frequency (MF) |
| 3 MHz | 30 MHz | High Frequency (HF) |
| 30 MHz | 300 MHz | Very High Frequency (VHF) |
| 300 MHz | 3 GHz | Ultra High Frequency (UHF) |
| 3 GHz | 30 GHz | Super High Frequency (SHF) |
| 30 GHz | 300 GHz | Extremely High Frequency (EHF) |

At the higher frequencies, infrared light, visible light, ultraviolet light, X-rays and gamma rays are defined. Table 5.5 identifies the visible light spectrum and associated wavelength range.

**Table 5.5.** Visible light spectrum

| Colour | Wavelength range |
|---|---|
| Red | 630 to 700 nm |
| Orange | 590 to 630 nm |
| Yellow | 530 to 590 nm |
| Green | 480 to 530 nm |
| Blue | 440 to 480 nm |
| Violet | 400 to 440 nm |

The SI multiples used are identified in Table 5.6.

**Table 5.6.** SI multiples

| Prefix | Symbol | Value |
|---|---|---|
| 1 yocto | y | $1 \times 10^{-24}$ |
| 1 zepto | z | $1 \times 10^{-21}$ |
| 1 atto | a | $1 \times 10^{-18}$ |
| 1 femto | f | $1 \times 10^{-15}$ |
| 1 pico | p | $1 \times 10^{-12}$ |
| 1 nano | n | $1 \times 10^{-9}$ |
| 1 micro | μ | $1 \times 10^{-6}$ |
| 1 milli | m | $1 \times 10^{-3}$ |
| --- | --- | $1 \times 10^{0}$ |
| 1 kilo | k | $1 \times 10^{3}$ |
| 1 mega | M | $1 \times 10^{6}$ |
| 1 giga | G | $1 \times 10^{9}$ |
| 1 tetra | T | $1 \times 10^{12}$ |
| 1 pera | P | $1 \times 10^{15}$ |
| 1 exa | E | $1 \times 10^{18}$ |
| 1 zetta | Z | $1 \times 10^{21}$ |
| 1 yotta | Y | $1 \times 10^{24}$ |

Analogue ICs will currently operate in the frequency range from DC to the low GHz frequencies, with the higher frequencies for devices such as high-performance digital processor designs, video applications, RF transmitter and receiver circuits, and high-speed serial communications on and off-chip. With the integration of RF and optical components and higher digital clock frequencies, the move is for high-frequency, multi-technology devices. The **operating temperature range** is also a factor that would need to be taken into account when looking at the device operation. The circuit operation would vary with temperature and given device specifications will be quoted based on the electrical and thermal conditions. Depending on whether the IC is for commercial, industrial or military use, the device operating temperature range will vary:

- Commercial     $0^{O}C$ to $+70^{O}C$
- Industrial       $-40^{O}C$ to $+85^{O}C$
- Military         $-55^{O}C$ to $+125^{O}C$

## 5.3 Measuring Analogue Parameters

A range of test equipment will be required to measure analogue parameters of an IC. The equipment requirements will be dependent on the required tests to be undertaken, speed of operation and accuracy requirements. For non-production based test activities, high-speed operation may be less important than flexibility and high-accuracy in the equipment use, and ability for undertaking a broad range of signal analysis operations. In production test, the need is for tests to be undertaken that are suitably comprehensive for the particular product and product application area, but at the lowest cost possible. This means using the most cost-effective equipment possible and minimising the test time per device. The role of the test equipment used is essential to achieving this. In general, a (software) test program is run on a tester to implement the required test procedure. The semiconductor testers used require both hardware and software parts in order to set-up and control the tester and test program execution. During the production test stage, **Automatic Test Equipment** (ATE) is used to reduce the test times by automating as much of the test process as practical. The equipment used may be categorised as:

**Dedicated test equipment**:

Specially designed to measure specific parameters for a device. This will be dedicated to a particular device or small set of devices.

**General purpose testers**:

Used to test a range of devices, where the devices may have vastly different operational parameters. With this type of tester, it is temporarily customised to a particular IC to test.

In analogue test, the types of values to be measured will be:

- Voltage
- Current
- Frequency
- Resistance
- Capacitance
- Inductance
- Circuit input resistance/impedance
- Circuit output resistance/impedance

Both **time** domain and **frequency** domain information will be determined through the test process. These can be determined through the use of analogue (continuous time) test equipment or digital based test equipment. Digital test equipment will be based on **digital signal processing** (DSP) [13] operations typically using a processor based system and interfacing through Analogue to Digital (A/D) and Digital to Analogue (D/A) data converters. A suitably developed tester platform will allow for fast operation of complex signal processing operations. For example, support for analogue signal sampling and undertaking a **Fast Fourier Transform** (FFT) would be integral to analogue and mixed-signal testers. DSP techniques in analogue and mixed-signal test will be based on the generation, sampling and analysis of analogue signals. The basic arrangement required is shown in Fig. 5.3.

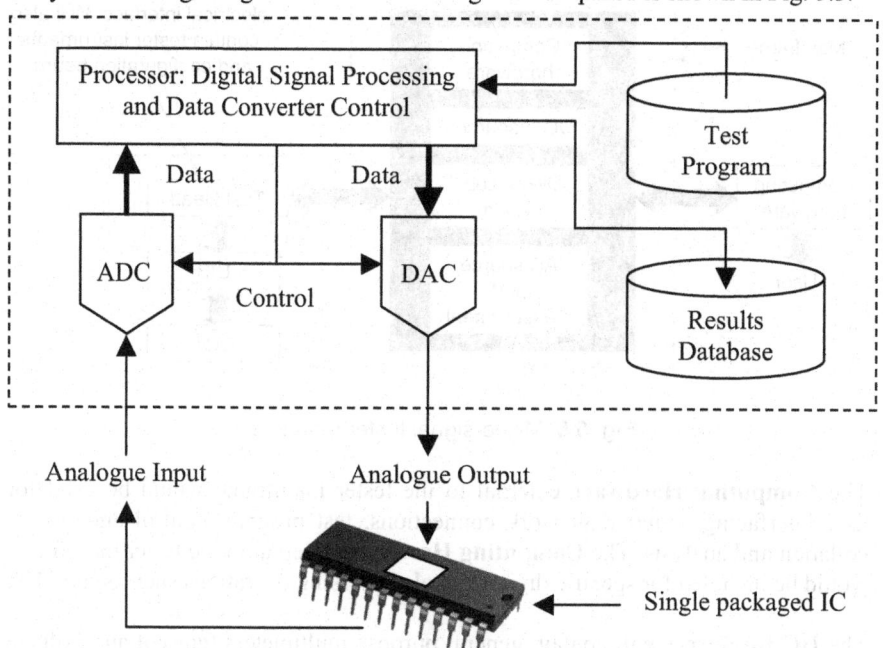

**Fig. 5.4.** Basic DSP based tester arrangement

Here, the tester uses ADC [14] and DAC devices with suitable speed (sample and update rates), resolution (number of bits) and range (range of output current/voltage from a minimum to a maximum value). These will be controlled via the processor running the test program. Analogue signals from the device under test (DUT) will be sampled, converted into digital and stored in memory. There will be suitable results analysis routines built into the test program and the raw and/or processed results will be stored in the results database. The choice of processor will be made based on the supported functions and speed of operation, in addition to cost(!). In general, DSP ICs will provide special hardware macros for performing high-speed mathematical operations such as addition/subtraction and multiplication. These special hardware macros allow for such operations to be undertaken in a short time and do not require the main processor program to perform time consuming operations via software. For a **mixed-signal ATE** [5, 15] system, the tester hardware may be categorised, see Fig. 5.5, in the following sub-systems:

- Computing hardware
- DC resources
- Digital sub-system
- AC source and measurement

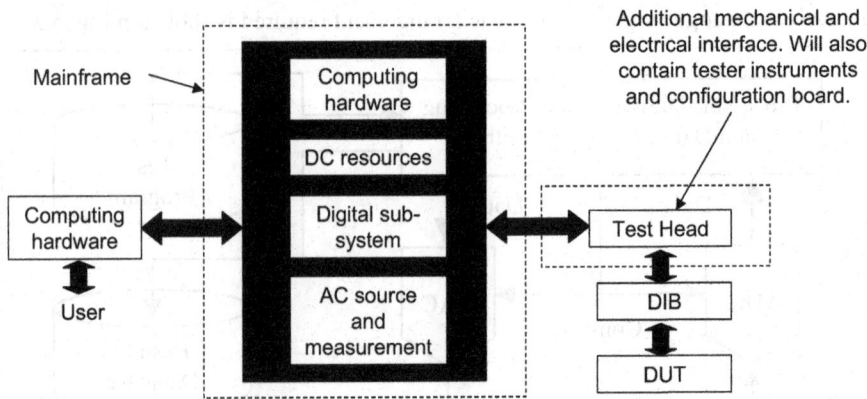

**Fig. 5.5.** Mixed-signal tester resources

The **Computing Hardware** external to the tester mainframe would be used for user interfacing, external network connections, test program control and results collation and analysis. The **Computing Hardware** internal to the tester mainframe would be included for specific digital signal processing operations such as the FFT.

The **DC Resources** will contain general purpose multimeters (current and voltage measurement), general purpose voltage/current sources, precision voltage references (for use where the general purpose sources would not provide the necessary accuracy), power supplies, calibration sources (for periodic calibration of the tester), relay matices (for low-impedance switching of signals) and relay control lines (for control of the relay operation).

The **Digital Sub-System** would be for application and monitoring of digital patterns. This resource would also contain memory for digital signal storage. **Vector memory** would be used for the storage of digital vectors – both the vector (output) and digital input (result) signals would be stored. **Source memory** would be used for the storage of digital signal samples – the digital signals representing DAC output waveforms. **Capture memory** would be used to capture digitised analogue waveforms (the opposite operation to source memory).

The **AC Source and Measurement** resource is used to generate and capture AC signals. The **Continuous Waveform Source** would be used to generate single tone AC signals with user set signal amplitude and frequency. The **RMS Voltmeter** is used to measure the RMS (Root Mean Squared) of single tone signals. The **Arbitrary Waveform Generator** (AWG) will be used to generate signals of arbitrary shape. **Waveform digitisers** will capture analogue signals (using an ADC) and store the digitised waveform in the capture memory.

# 5.4 Coherent Sampling

Analysis of the circuit under test output (voltage or current) can be undertaken in the analogue or digital domains. In the analogue domain, measurement equipment such as RMS voltmeters, etc., would be used and the measurement is undertaken solely using analogue test and measurement equipment. Increasingly, the test equipment analogue signal generation and capture/analysis is undertaken using DSP techniques and this requires suitable sampling of the analogue signal. The principle of operation of signal sampling is shown in Fig. 5.6. An input signal $(v(t))$ is sampled at a sampling frequency of $F_s$ (Hz) and sampled values $(v(n))$ are available at the sampler output. The time between the samples is fixed value of $T_s$, where $T_s = (1/ F_s)$ (seconds).

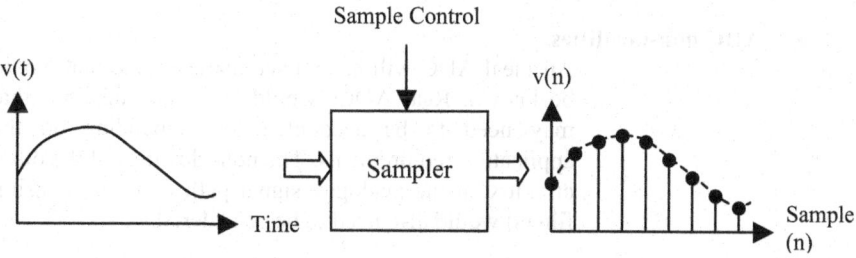

**Fig. 5.6.** Sampling of a continuous time signal

With the sampling of an analogue signal (a circuit under test output) for analysis purposes, this will be achieved through the use of an analogue to digital converter (ADC). In this arrangement, it will be necessary to consider the following:

- **Signal sampling frequency**:

  The analogue signals will be sampled at a certain sampling frequency ($F_s$), and the choice of this frequency will be dependent on the input signal frequency (or range of frequencies). In order to sample the signal correctly, the sampling frequency must be at least twice the maximum signal frequency ($F_{max}$) – this maximum signal frequency is the **Nyquist Frequency** – the ADC must sample at least twice the Nyquist Frequency if **aliasing** effects are to be avoided. If the signal frequency exceeds this value, aliasing will occur. In many cases, an analogue low-pass filter is used at the ADC input to cut-off high frequency signals (referred to as an **anti-aliasing filter**).

- **ADC resolution**:

  The ADC converts an analogue signal (continuous time) into a digital (discrete time and discrete levels) representation. This is a **many to one mapping** and results in a **quantisation error**. The resolution of the ADC will be the number of bits (n). For example, with an 8-bit ADC (n=8), there will be 256 ($2^n$) possible output combinations. With a 16-bit ADC (n=16), there will be 65,536 possible output combinations. The converter will be capable of converting a voltage (or current) between a lower and upper limit (full-scale voltage ($V_{FS}$) or full-scale current ($I_{FS}$)), and the change required in the input signal to produce a 1-bit change in the output code will be dependent on the ADC resolution and the full-scale range.

- **ADC non-idealities**:

  An ideal ADC will have a set characteristic that would be known. Real ADCs would have non-idealities that may need to be accounted for, depending on the application requirements. The non-idealities of the other circuitry in the analogue signal path (*e.g.* anti-aliasing filters) would also need to be considered.

In many converters, the sampling is undertaken at twice the **Nyquist Frequency** (at the **Nyquist Rate** which is twice the Nyquist Frequency). It is however also possible to sample at greater than the Nyquist Rate (**oversampling**), or less than the Nyquist Rate (**undersampling**), for specific requirements. **DSP techniques** will allow for both time domain and frequency domain signal analysis to be undertaken. In the frequency domain, the FFT (see Chap. 6) is used to determine the frequency components and their magnitudes from samples of an analogue signal. An important consideration when sampling a signal is the number of samples taken for a complete cycle of the signal. With reference to a sine wave of frequency $F_{in}$ (Hz), this will be sampled at a number of points within a complete cycle of the signal. The sampling can be either **coherent** or **non-coherent**, and the result in the frequency domain will be different. In **coherent sampling**, see Fig. 5.7, N samples are used to represent exactly one cycle of the signal (sine wave), and will repeat without disjunction from one cycle to the next. Here N=4 and the sampling repeats at the same point on the sine wave.

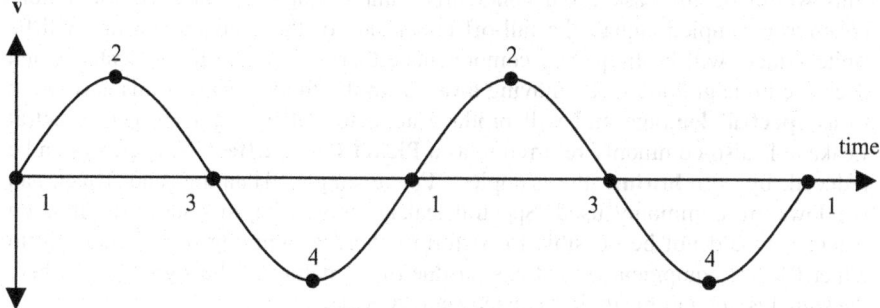

**Fig. 5.7.** Coherent sampling

If N samples are taken for one cycle of the signal frequency ($F_{in}$), then the sampling frequency ($F_s$) [16], is given by:

$$F_{in} = \frac{F_s}{N}$$

When samples are taken on a single cycle of $F_{in}$, then $F_{in}$ is the **primitive (or Fourier) frequency, $F_f$** – that is, in the frequency domain, signals will be identified as being multiples of this frequency. This is also the **spectral resolution**. The period ($T_{in}$) of the input signal ($F_{in}$) is referred to as the **Unit Test Period** (UTP) [16]. Samples of the input signal can however be taken over either single or multiple periods of the input signal. Where **multiple periods** of the input signal are taken, then for coherent sampling:

$$\frac{F_{in}}{F_s} = \frac{M}{N}$$

where **M** is the number of complete cycles of the input signal, **N** is the number of samples over the UTP, and the **UTP** is now the total sampling interval time. The values of M and N will normally however be restricted to certain ratios. For an unknown signal, then M and N should be relatively prime. The **primitive frequency** is given as ($F_s/N$) (or ($F_{in}/M$)). Therefore, the situation will exist in that:

- A sample set can be derived from a single complete cycle of the input.

- A sample set can be derived from multiple complete cycles of the input.

The decision would need to be made as to the number of samples to take and the number of cycles of the input signal to sample over. In general, if a signal is sampled with N samples over a sampling interval T (seconds), the spectral resolution will be 1/(N.T). In **non-coherent sampling**, this smooth transition in sampling from one cycle of the sine wave to the next does not occur. The effect of non-coherent sampling can be seen when an FFT is performed on the sampled signal. The expected plot would be a single ray at the fundamental frequency ($F_f$). This would be the case for a coherently sampled signal. However, for a non-coherently sampled signal, the fall-off either side of the fundamental ray will be finite -- there will be frequency components either side of the fundamental which decrease in magnitude when moving away from the fundamental -- this is referred to as **spectral leakage** and will produce an output that is not correct. Spectral Leakage is also commonly referred to as a **Picket Fence Effect**. This effect can be reduced by **windowing** the samples. For example, Hanning and Blackman windows are commonly used. Spectral leakage would be an issue in a situation where it would not be possible to perform coherent sampling of a signal due to either the test equipment limitations, or due to the nature of the signal (*e.g.* where the signal is not a pure single frequency sine wave).

## 5.5 Functional vs Structural Test

Functional testing of analogue ICs [17] is still the preferred approach in order to identify and guarantee the wide range of required design specifications. For example, to identify the op-amp specifications detailed in Table 5.2. This differs from the digital domain in that the development of structural tests for analogue is not as straightforward to consider and automate as it is with digital, (as discussed in Chap. 3). A number of studies have been undertaken to investigate structural testing [18-20] outside the purely digital domain. However, the uptake of structural testing methods is still limited.

## 5.6 Fault Modelling in Analogue

In order to develop effective structural test programs for analogue circuits, there is the need for fault simulation and the use of suitable fault models that need to reflect the process defect that is represented by the electrical fault. Defects occurring both in the circuit components and the interconnect would need to be modelled. Typical analogue fault models [20-25] are shown and discussed in Fig. 10.4 (Chap. 10). Here, the models are based around resistive open/short faults on the nodes of specific circuit components, and rely on representative values (minimum/maximum values of fault resistance) to be used. The fault must be detectable at these limits. However, up-to-date and accurate process defect information can be difficult to acquire, and models may need to be developed based on public-domain (published) material unless access to specific foundry information can be obtained. This however may not be possible due to commercial requirements and certain company specific information may not be accessible outside the company. Additionally, the rapid progress in fabrication processes moving along the technology roadmap introduces additional problems given the introduction of new defect mechanisms that may not have previously existed.

## 5.7 Future Directions

The future for analogue test has many parallels with digital circuit and system test. A summary is provided in Table 5.7.

**Table 5.7.** Future analogue test issues

| Technology driver | Description |
|---|---|
| Design styles | Increasing complexity of designs<br>Increased integration – SoC and SiP test issues<br>The use of multiple power supply voltage rails on a single IC<br>Use of designs with multiple threshold voltage ($V_T$) MOS transistors<br>Move to mixed-technologies incorporating logic, memory, mixed-signal and RF circuit elements<br>Integrated optical and MEMs devices<br>Inclusion of more DfT and BIST into the device |
| Faster operating speeds | Higher signal frequencies<br>Driven by communications applications<br>Reduced device geometries (gate length and gate oxide thickness)<br>Reduced separation of interconnect tracks (track pitch) – capacitive and inductive coupling problems and signal cross-talk – signal integrity issues<br>Moving onto the next technology node to improve performance of components and reduce interconnect delays<br>Move into the nanotechnology domain |
| Higher levels of integration (higher | Moving onto the next technology node to improve performance of components and reduce interconnect |

| density) | delays<br>Newer failure mechanisms - limitations of existing fault models and need for models more representative of process defects<br>New test issues emerging for SoC and SiP devices |
|---|---|
| Lower operating voltages | A need to improve device reliability by reducing electric field strength in dielectrics<br>Aim to lower device power consumption<br>Portable, battery operated circuits<br>Multi-voltage power supply device operation and interfacing |
| Test equipment and costs | Need for reduced test time and test costs<br>Need for reduced test pattern generation time and maintenance<br>More formal approaches to test program generation – test re-use<br>Problem of increasingly long test pattern generation and application times<br>Increasing requirements for test data storage in ATE<br>Increasingly difficult to perform at-speed testing of devices using current ATE<br>Need for reduced external ATE requirements and costs – simpler and less expensive ATE<br>Multi-site test<br>Inclusion of more DfT and BIST into the device |

## 5.8 Summary

This chapter has discussed a number of issues relating to the testing of analogue integrated circuits. With the ever increasing demands on analogue circuit design specifications, the need to test effectively for these specifications is increasingly difficult and time consuming, and hence costly. Analysis of signal characteristics resulting from a device under test (DUT) in both the time domain and the frequency domain will contain information relevant to determining whether an analogue circuit passes or fails a particular test. In the main, functional testing of the analogue circuit is undertaken in the test environment. This differs from digital device test in that, during production testing, structural testing of digital logic is well understood and universally implemented. In the analogue domain, structural test is not as established as in the digital domain and the application specific nature of the analogue circuit behaviour means that the development of structural tests is not as straightforward to consider and automate as it is with digital.

# 5.9 References

[1]     Smith M., "Application Specific Integrated Circuits", Addison-Wesley, 1999, ISBN 0-201-50022-1

[2]     Bellaouar A. and Elmasry M., "Low-Power Digital VLSI Design Circuits and Systems", Kluwer Academic Publishers, The Netherlands, 1995, ISBN 0-7923-9587-5

[3]     Kang S. and Leblebici Y., "CMOS Digital Integrated Circuits Analysis and Design", McGraw-Hill International Editions, Singapore, 1996, ISBN 0-07-114423-4

[4]     International Technology Roadmap for Semiconductors (ITRS), 2003 Edition, "Design"

[5]     Burns M. and Roberts G.W., "An Introduction to Mixed-Signal IC Test and Measurement", Oxford University Press, New York, 2001, ISBN 0-19-514016-8

[6]     Schaumann R. and Valkenburg M., "Design of Analog Filters", Oxford University Press, USA, 2001, ISBN 0-19-511877-4

[7]     Cheung V. and Luong H., "Design of Low-voltage CMOS Switched-opamp Switched-capacitor Systems", Kluwer Academic Publishers, 2003, ISBN 1-4020-7466-2

[8]     Toumazou C., Lidgey F. and Haigh D., "Analogue IC design: the current-mode approach", IEE Circuits and Systems Series 2, IEE, UK, 1993, ISBN 0-86341-297-1

[9]     Toumazou C., Hughes J. and Battersby N., "Switched-Currents an analogue technique for digital technology", IEE Circuits and Systems Series 5, IEE, UK, 1993, ISBN 0-86341-294-7

[10]    Laker K. and Sansen W., "Design of Analog Integrated Circuits and Systems", McGraw-Hill International Editions, USA, 1994, ISBN 0-07-113458-1

[11]    Franco S., "Design with Operational Amplifiers and Analog Integrated Circuits", McGraw-Hill International Editions, Singapore, 1988, ISBN 0-07-100435-1

[12]    ADA4851 Low Cost, High Speed Rail-to-Rail Output Op Amp, dataheet, Analog Devices Inc., USA

[13]    Ifeachour E. and Jervis B., "Digital Signal Processing, A Practical Approach", Prentice Hall, UK, 2002, ISBN 0-201-59619-9

[14]    Jespers P., "Integrated Converters, D to A and A to D Architectures, Analysis and Simulation", Oxford University Press, USA, 2001, ISBN 0-19-856446-5

[15]    Hurst S., "VLSI Testing digital and mixed analogue/digital techniques", IEE, 1998, ISBN 0-85296-901-5

[16]    Roberts G. and Lu A., "Analog Signal Generation for Built-In Self-Test of Mixed-Signal Integrated Circuits", Kluwer Academic Publishers, USA, 1995, ISBN 0-7923-9564-6

[17]    Van Spaandonk J. and Kevenaar T., "Selecting Measurements to Test the Functional Behavior of Analog Circuits", Journal of Electronic Testing: theory and Applications, No. 9, 1996, pp9-18

[18]    Sachdev M., "Defect Oriented Testing for CMOS Analog and Digital Circuits", Kluwer Academic Publishers, 1998, ISBN ISBN 0-7923-8083-5, 1998

[19]    Spinks S. et al., "Generation and Verification of Tests for Analogue Circuits Subject to Process Parameter Deviations", Proceedings of the IEEE International Symposium on Defect and Fault Tolerance in VLSI Systems, 1997, pp100-108

[20]    Walker H. and Stephen W., "VLASIC: A Catastrophic Fault Yield Simulator for Integrated Circuits", IEEE Transactions on Computer-Aided Design, Vol.CAD-5, No. 4, October 1986, pp541-556

[21]    Ferguson F. and Shen J., "A CMOS Fault Extractor for Inductive Fault Analysis", IEEE Transactions on Computer-Aided Design, Vol. 7, No. 11, November 1988, pp1181-1194

[22]    Hawkins C. Soden J., Righter A. and Ferguson F., "Defect Classes – An Overdue Paradigm for CMOS IC Testing", Proceedings of International Test Conference, USA, 1994, pp 413-425

[23]    Hawkins C. and Segura J., "Failure Modes in Nanometer Technologies", Tutorial D2, Design and Automation in Europe Conference (DATE), 2003

[24]    Gaitonde D. and Walker D., "Hierarchical Mapping of Spot Defects to Catastrophic Faults – Design and Applications", IEEE Transactions on Semiconductor Manufacturing, Vol. 8, No. 2, May 1995, pp167-177

[25]    Chess B., Roth C. and Larrabee T., "On Evaluating Competing Bridge Fault Models for CMOS ICs", Proceedings of the 12th IEEE VLSI Test Symposium, 1994, pp446-451

# Exercises

*The following questions will require access to an analogue circuit simulator. In the following questions, the HSPICE® circuit simulator is considered, although with suitable modifications, a suitably available simulator could be used.*

## Question 1

For the LM741 operational amplifier, identify the device specifications in terms of:

- DC performance
- Dynamic performance
- Noise/distortion performance
- Input characteristics
- Output characteristics
- Power supply performance

Identify and develop a Spice simulation model for this device and simulate the operation of the device, with the simulation arranged to determine the above specifications. For this simulation exercise, identify and justify any assumptions made. For the simulation model developed, identify and discuss any specifications that cannot be determined here. Can they be determined through simulation and if so, how?

Identify and discuss the following absolute maximum ratings for the device above:

- Power supply voltage
- Power dissipation
- Common-mode input voltage (referenced to the power supply voltage)
- Differential input voltage
- Storage temperature
- Operating temperature range
- Device soldering maximum temperature and duration

## Question 2

Repeat Question 1 for the ADA4851 operational amplifier.

## Question 3

Using the LM741 op-amp model, develop Spice simulation models and verify the operation for the op-amp based circuits identified in Table 5.3.

## Question 4

Using the ADA4851 op-amp model, develop Spice simulation models and verify the operation for the op-amp based circuits identified in Table 5.3.

*The following questions will require access to the PC based tester arrangement identified in Appendix E.*

## Question 5

Using the LM741 op-amp, build a tester arrangement that will allow for the verification of the operation for the op-amp based circuits identified in Table 5.3. Compare the response to the simulation results in Question 3.

## Question 6

Using the ADA4851 op-amp, build a tester arrangement that will allow for the verification of the operation for the op-amp based circuits identified in Table 5.3. Compare the response to the simulation results in Question 4.

# Chapter 6
# Mixed-Signal Test

*The need to interface between the analogue and digital domains exists for a number of reasons. A key reason is that the real-world is analogue in nature whilst the digital domain provides a convenient means in which to simplify the analogue, continuous time, domain into a manageable form for the design and analysis of complex circuit and systems. This is achieved by modeling the signals as discrete-time, discrete level events. The interface between the analogue and digital, the mixed-signal interface, provides many challenges to test and the requirement to develop efficient, cost-effective and comprehensive test procedures.*

## 6.1 Introduction

The mixed-signal interface provides the electronic circuit means in which to transfer electrical signals between the analogue and digital domains. Such circuits will contain both analogue and digital circuitry, and so require test procedures in which both the analogue and digital parts are effectively tested. An example of a large, "discrete" mixed-signal system uses a PC acting to perform Digital Signal Processing (DSP) operating on an audio music source, and outputs the result to an audio amplifier and speaker system. Such as system is shown in Fig. 6.1

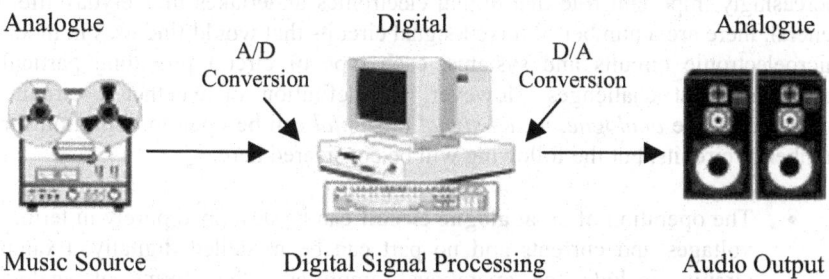

**Fig. 6.1.** Mixed-signal system example

143

Here, an analogue signal (music stored on tape or vinyl record) is played and sampled at regular time intervals by the PC. The PC itself is digital and so the analogue signal is required to be converted from analogue to digital (using an Analogue to Digital Converter (referred to as an ADC or A/D Converter)). A software program running on the PC manipulates the digitised signal (using DSP techniques) before the results are sent out to an audio amplifier and ultimately the audio speakers. In this case, the audio amplifier is analogue and so the digital signal sent out from the PC is converted back to analogue (using a Digital to Analogue Converter (referred to as a DAC or D/A Converter)). Many of these circuit functions however can be placed on a single IC. For example, microcontoller and DSP ICs can provided with ADC and DACs within a single packaged device. The basic arrangement for such a system is shown in Fig. 6.2

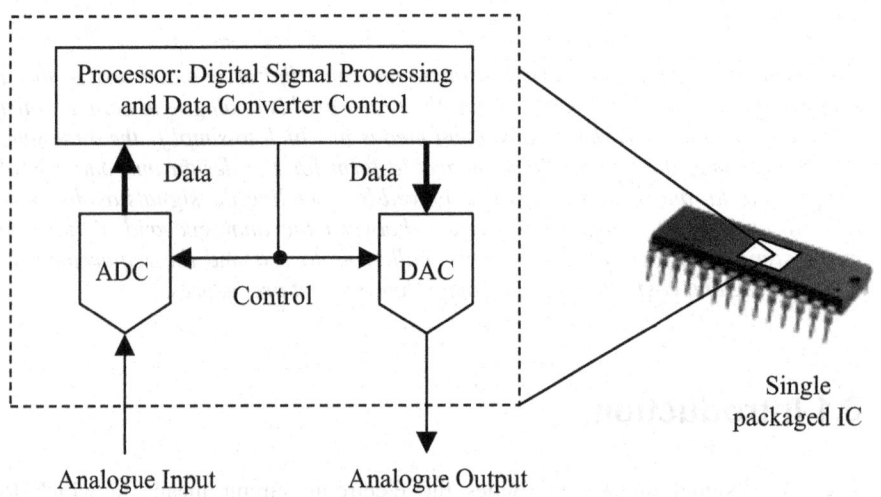

**Fig. 6.2.** Digital processor with on-board ADC and DAC

The ADC and DAC are two examples of mixed-signal circuits that would be found in many electronic and microelectronic circuits and systems. Their use is becoming ever more important given the revolution in digital systems design and the increasingly important role that digital electronics undertakes in everyday life. In general, there are a number of mixed-signal circuits that would find uses in modern microelectronic circuits and systems, each type of circuit providing particular design and test challenges. However, the definition of whether a circuit is considered to be *analogue*, *mixed-signal* or *digital* can be open to definition for a number of circuits, but the following will be considered here:

- The operation of an **analogue circuit** can be described purely in terms of voltages and currents and no part can be modelled digitally. Example circuits include the operational amplifier, other forms of analogue amplifier and analogue filters

- The operation of a **mixed-signal circuit** can be described in terms of a mixture of voltages and currents and digital logic simultaneously

- The operation of a **digital circuit** can be described purely in terms of digital logic levels and time delays for logic levels changes. Examples include microcontrollers, microprocessors and Digital Signal Processors

Examples of the types of mixed-signal circuits that are commonly used include the:

- Analogue to Digital Converter (ADC) or A/D Converter [1]
- Digital to Analogue Converter (DAC) or D/A Converter [2]
- Phase-Locked Loop (PLL) [3, 4]
- Digitally Programmable Analogue Amplifier
- Comparator
- Analogue Switch

These circuits would be found as discrete (packaged) ICs, or as macro cells within larger System on a Chip (SoC) [5] designs. The ADC and DAC will be considered for further discussion in this chapter.

# 6.2 Mixed-Signal Circuit Test Overview

The need to develop efficient test procedures for mixed-signal designs combines the challenges of digital and analogue circuit test. Increasing design performance requires:

- **Structured approaches** to test procedure development: the need to introduce formal methods in the development of the test procedure, including the re-use of existing test methods, tester hardware and test program software. Efficient test programs (software code to run on the ATE): including test program **software re-use**

- The inclusion of **Design for Testablity** (DfT) techniques and circuitry within the Device Under Test (DUT)

- Consideration for the inclusion of **structural test** to supplement specific functional test, as opposed to a functional only test

- Increasingly sophisticated test equipment (hardware): including tester hardware re-use and the use of more flexible **ATE** architectures

The nature of mixed-signal circuit operation and the need to test for a range of functional parameters has meant that the test costs for mixed-signal parts of an IC can be a significant part of the overall IC product costs. It has been quoted that this can be as high as 50% of the overall device cost. With such overheads resulting from the test requirements, the need to develop more cost effective test procedures is driving the development of more efficient implementations of existing, and the development of new, test methods. Figure 6.3 shows the basic arrangement required for testing a mixed-signal circuit. The *Stimulus Generator* is required to create by digital logic (logical 1s and 0s with correct timing), along with analogue signals (both current and voltage). This requires transmission to the **Device Under Test** (DUT), with a **Transmission Channel** that is required to present the generated signal at the DUT without significant modification of the signal. The resulting output from the DUT is transmitted back to a **Results Analyser** for data capture and analysis. Again, the transmission channel to the Results Analyser must not significantly modify the signal.

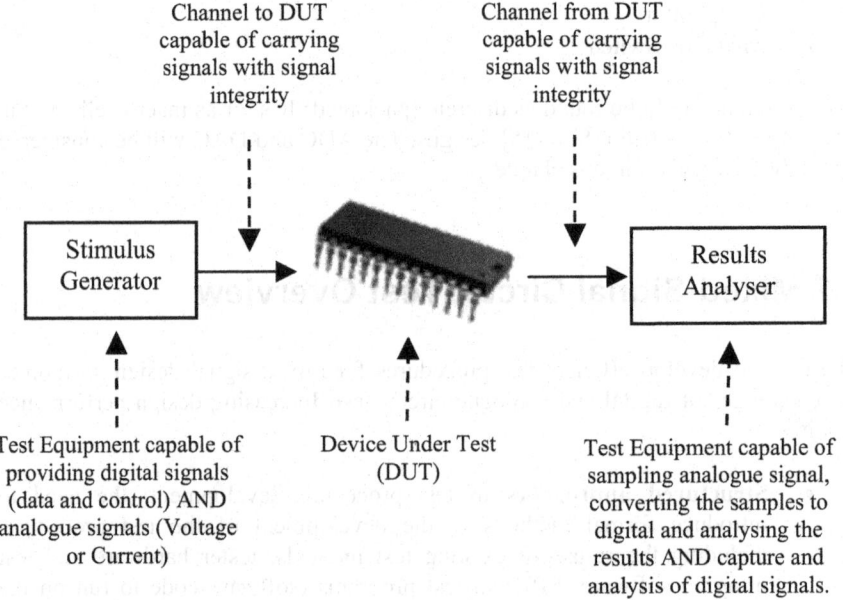

**Fig. 6.3.** Mixed-signal I/O

Considering the data converter, be it A/D or D/A, there are a number of end-user demands that are requiring the circuit designs to have an:

- Increase in resolution (number of bits). The typical resolution of available data converters is 8, 10, 12, 14, 16, 18, 20 and 24

- Increase in speed of operation, in particular driven by the communications applications

- Operation on lower power supply voltages and with lower power consumption, in particular driven by the portable electronics applications

Applications range from control systems through to audio, communications systems and test equipment (!) applications. A number of data converter architectures are available to support particular application requirements and choosing the right data converter architecture is required for a particular application. With the move towards the use of data converters with higher resolution (number of bits), see Table 6.1, the increase in the number of codes that would need to be tested for, is a cause for concern. Where each code would need to be tested (either applied as in the D/A or sampled as in the A/D), the ability to test for all codes in the higher resolution converters is not economic, and a means to address this problem is an on-going activity. For example, the model-based approach [6-8] has been identified as a means to reduce the number of codes to be tested. However, functional rather than structural testing of analogue and mixed-signal ICs is dominant.

**Table 6.1.** Data converter codes

| Number of bits (n) | Number of codes ($2^n$) |
|:---:|:---:|
| 8 | 256 |
| 10 | 1024 |
| 12 | 4096 |
| 14 | 16,384 |
| 16 | 65,536 |
| 18 | 262,144 |
| 20 | 1,048,576 |
| 22 | 4,194,304 |
| 24 | 16,777,216 |

# 6.3 Fault Modelling in Mixed-Signal Circuits

In the development of structural test programs, there will be the need to develop vectors to detect considered and specific faults. In digital [9-12], structural test based on logical and defect oriented fault models is commonplace and well accepted. In production test, structural test is initially undertaken and where specific parts of the circuit cannot be tested with the structural test, a functional test is then undertaken on that specific part. Functional testing of analogue and mixed-signal ICs [13, 14] on the other hand is still the preferred approach in order to identify and guarantee the wide range of required design specifications. In order to develop effective structural test programs for analogue and mixed-signal circuits, there is the need for suitable fault simulation and the use of suitable fault models that need to reflect the process defect that is represented by the electrical fault. Commonly considered fault models are based around resistive open/short faults on the nodes of specific circuit components, and rely on representative values (minimum/maximum values of fault resistance) to be used. The fault must be detectable at these limits. Due to concerns with the relevance of these models and the potential removal of the ability to test for specific analogue/mixed-signal circuit parameters with structural test, the uptake in analogue and mixed-signal test has been limited.

# 6.4 DAC Architectures

## 6.4.1 Introduction

The basic operation of the DAC is to convert a digital word (a binary code) to an analogue output. This analogue output may be either a current or a voltage. The value of the output signal is a factor based on the value of the digital input code and a static analogue reference signal (either current or voltage). For an n-bit DAC with a reference and output voltage, see Fig. 6.4, the n-bit digital input (where n is the size of the digital input data given by the number of bits), produces an output signal ($V_{OUT}$).

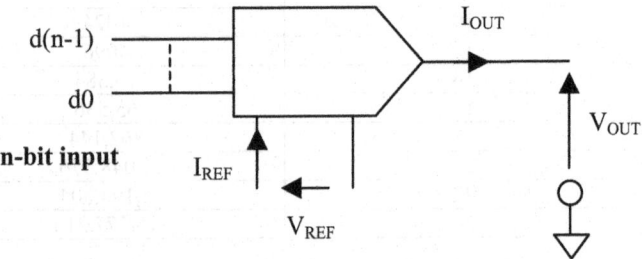

**Fig. 6.4.** DAC set-up

Within a typical DAC, there are also control signals to operate, see Fig. 6.5. These are **Chip Select** (to enable the DAC so that the device is set-up to allow digital data to be written to it) and **Write** (to enable the digital input code to be stored (latched) in the DAC in order to update the analogue output) signal.

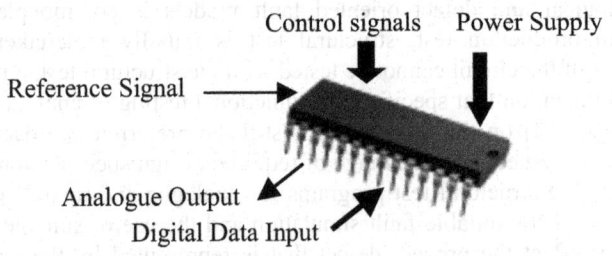

**Fig. 6.5.** DAC signals

The DAC output will vary from a predefined minimum value to a predefined maximum value that is a function of the reference signal. For the DAC, the **Full-Scale Voltage** ($V_{FS}$) **or Full-Scale Current** ($I_{FS}$) defines the limit of operation. The output of the DAC can be written mathematically as:

$$V_O = V_{FS}.(b_1.2^{-1} + b_2.2^{-2} + b_3.2^{-3} + ...... + b_n.2^{-n}) + V_{OS}$$

for a voltage output DAC with a referene voltage ($V_{REF}$).

$$I_O = I_{FS}.(b_1.2^{-1} + b_2.2^{-2} + b_3.2^{-3} + ...... + b_n.2^{-n}) + I_{OS}$$

for a current output DAC with a reference current ($I_{REF}$).

where $b_1$ is the binary value (1 or 0) of the **Most Significant Bit** (MSB) and $b_n$ is the binary value of the **Least Significant Bit** (LSB). A change in the LSB creates the smallest single change in the output signal. It is common for the digital input code to be an unsigned binary count, starting at $0_{10}$ and incrementing in unit steps. However, with the use of suitable digital signal encoding, any digital code (*e.g.* signed binary) count could be implemented. A change in the MSB creates the largest single change in the output signal. The above equations include an **offset voltage** ($V_{OS}$) or **offset current** ($I_{OS}$) if the output signal is not zero for an input code of $0_{10}$. The output of the DAC can be either **unipolar** (either a positive or negative value only) or **bipolar** (both positive and negative values can be generated). The **resolution of the converter** is given above as the number of digital bits at the converter input. It can also be quoted to identify the minimum output voltage change, which occurs when there is a change of 1 Least Significant Bit (LSB) in the input code. This voltage change is given by:

$$V_{LSB} = 2^{-n} . V_{FS}$$

For example, with a unipolar 8-bit DAC with a full-scale voltage of 5V, $V_{LSB}$ = 19.53mV. For a 16-bit DAC with a full-scale voltage of 5V, $V_{LSB}$ = 76.29μV. For a 20-bit DAC with a full-scale voltage of 5V, $V_{LSB}$ = 4.77μV. The reduced step size requires careful consideration in test as noise introduced into the circuit or tester transmission channel will add to this, and if comparable with the step size change, will cause incorrect digital values to be captured. Noise issues with the higher resolution converters would not necessarily have been a problem with the lower resolution converters. Given that a change in the input code creates a change in the output signal, this can be graphically represented as in the transfer characteristic shown in Fig. 6.6. Here, a 3-bit DAC creates an output voltage. In this example, a $000_2$ input code creates a 0V output. This is the **ideal transfer curve**, and a range of tests can be undertaken to identify the deviation of an actual DAC from the ideal device. The output voltage will vary from 0V to 1LSB less than the full-scale voltage. This is typical of the types of DACs considered here, although specific DAC designs (*e.g.* the segmented resistor DAC [15]) can be designed to produce an output equal to the full-scale voltage.

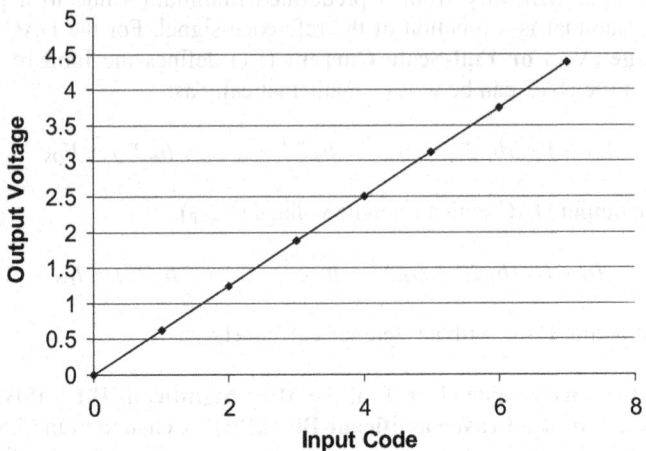

**Fig. 6.6.** Ideal 3-bit DAC input-output (transfer curve) characteristic

For the above DAC, the discrete points are joined with a straight line. Operating with a full-scale voltage of +5V, the idealised output voltage for this device is given in Table 6.2.

**Table 6.2.** Idealised 3-bit DAC

| Input code | | | Output voltage (V) |
|---|---|---|---|
| b2 | b1 | b0 | |
| 0 | 0 | 0 | 0 |
| 0 | 0 | 1 | 0.625 |
| 0 | 1 | 0 | 1.25 |
| 0 | 1 | 1 | 1.875 |
| 1 | 0 | 0 | 2.5 |
| 1 | 0 | 1 | 3.125 |
| 1 | 1 | 0 | 3.75 |
| 1 | 1 | 1 | 4.375 |

There are a number of architectures that have been developed for the DAC, with each architecture providing its own unique operating characteristic and limitations. The particular DAC will fall into one of two categories:

- Nyquist Rate DAC
- Oversampling DAC

With the **Nyquist Rate** DAC, the input signal bandwidth is equal to the Nyquist frequency. The Nyquist frequency is ½ the DAC update frequency (the Nyquist Rate). With the **Oversampling** DAC, the converter update frequency is much greater than the Nyquist Rate. A number of DAC architectures (Nyquist Rate and Oversampling) are available to use. These include the:

- Binary-Weighted Resistor DAC
- Binary-Weighted Capacitor DAC
- Binary weighted Current DAC
- R-2R Ladder
- Resistor String
- Segmented Resistor String
- Sigma-Delta ($\Sigma\Delta$) … or also referred to as Delta-Sigma ($\Delta\Sigma$)
- Hybrid DAC

A DAC will be for use in one of a number of applications, and the importance of specific DAC parameters will be dependent on the particular application. For example **audio**, **video** and **signal processing** applications.

## 6.4.2 Binary-Weighted Resistor DAC

With this implementation, see Fig. 6.7, resistors with binary weighted values are required to be switched between the circuit common voltage (here 0V) and the reference voltage ($V_{REF}$). This produces a variable input (current) to a current to voltage converter (IV) converter – the op-amp with feedback resistor. The output voltage from the operational amplifier is proportional to the current flowing through the resistor R.

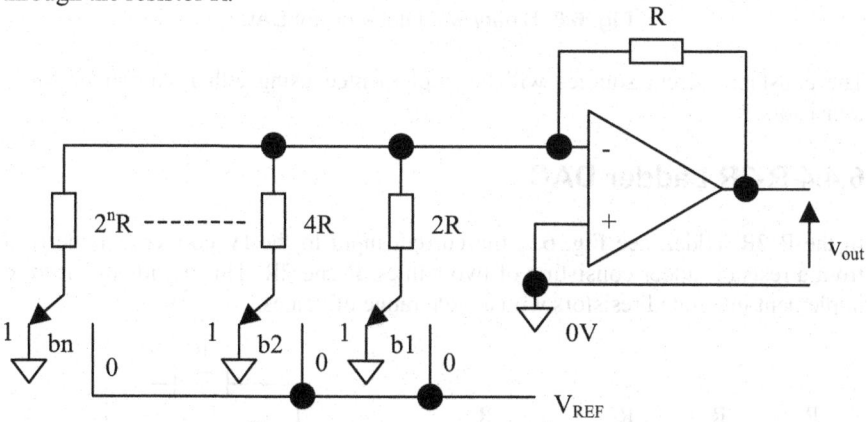

**Fig. 6.7.** Weighted resistor DAC

This design requires resistors with accurately controlled values. Additionally, as the number of bits increases, the resistor values can become large, resulting in physically large designs. Additionally, the switches would be implemented using transistors. Ideally, the ON resistance of the switch should be 0Ω, although in practice, the ON resistance is finite, and adds to errors within the design. This type of DAC is only suited to low-resolution converters due to variations in the fabricated resistor values. A variant on this converter is the **binary-weighted capacitor DAC**. This uses **switched capacitor** techniques rather than resistors to implement the circuit function.

### 6.4.3 Binary-Weighted Current DAC

With this implementation, see Fig. 6.8, this is basically the same architecture as the binary weighted-resistor DAC, except now the resistors have been replaced with constant current sources. This produces a variable input (current) to a current to voltage converter (IV) converter. The output voltage from the operational amplifier is proportional to the current flowing through the resistor R.

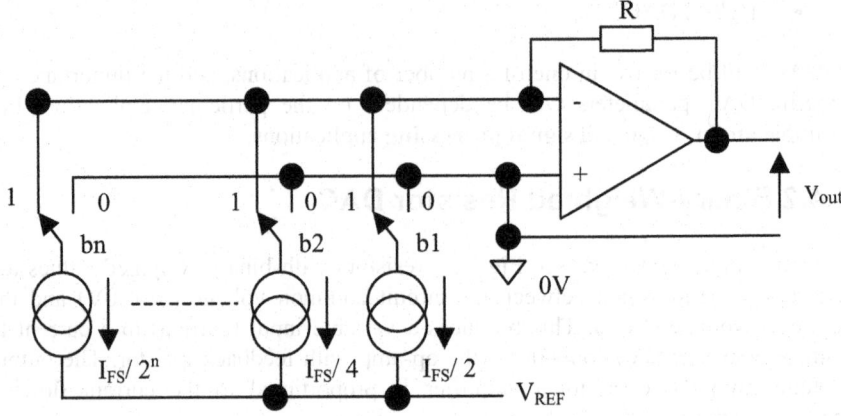

**Fig. 6.8.** Binary weighted-current DAC

The constant current sources will be implemented using either bipolar or MOS transistors.

### 6.4.4 R-2R Ladder DAC

In the R-2R ladder, see Fig. 6.9, the current input to the IV converter is derived from a resistor ladder consisting of two values, R and 2R. This avoids the need to implement integrated resistors with a wide range of values.

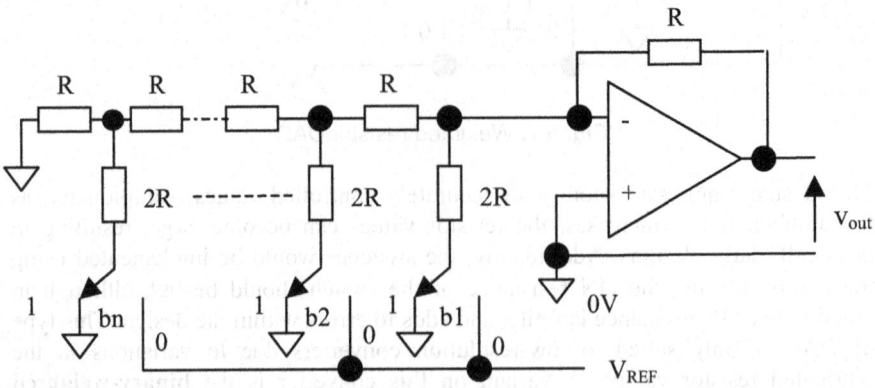

**Fig. 6.9.** R-2R DAC

## 6.4.5 Resistor String DAC

A simple DAC that in inherently monotonic is the resistor string DAC, see Fig. 6.10. In this circuit, a reference voltage ($V_{REF}$) is connected across a resistor string of equal value resistance. For an n-bit DAC, this requires $2^n$ resistors and the taps between the resistors are connected to switches controlled by the digital logic input values (and their complement). At the output node, an output voltage ($V_{OUT}$) is generated, the value set by the positions of the switches. The output would need to be suitably buffered (with for example, a unity gain op-amp buffer as shown here) in order to prevent electrical loading of the converter output.

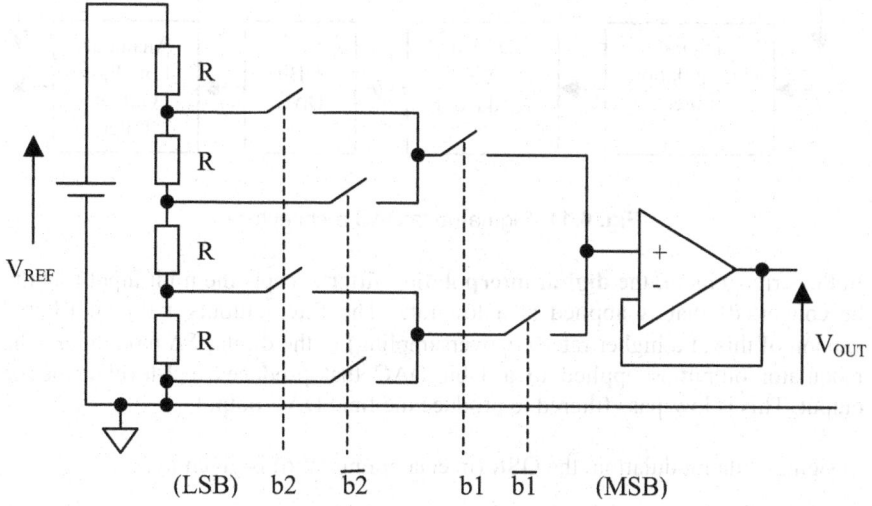

Fig. 6.10. Resistor string DAC

## 6.4.6 Segmented Resistor String DAC

With the segmented resistor DAC [15, 16], a resistor string DAC (the resistor string part of Fig. 6.10) is used and a second, sub-dividing resistor string is connected (via switches) across any one resistor, depending on the input code. The main resistor string is a main-DAC, and the sub-dividing resistor string is the sub-DAC. This allows for the ability for increased resolution than would be possible with a single resistor string. Care must be taken however in the choice of resistor values in order to minimise loading effects of the sub-resistor string on the main resistor string. Buffering using a unity gain buffer of the first string can be included.

## 6.4.7 Sigma-Delta ($\Sigma\Delta$) DAC

This type of DAC uses oversampling and noise shaping methods in order to achieve high resolution. The noise shaping reduces noise components in the low-frequency spectrum of the reconstructed signal. Both single-bit and multi-bit delta-sigma modulation can be used. The architecture of a single-bit modulator is shown in Fig. 6.11.

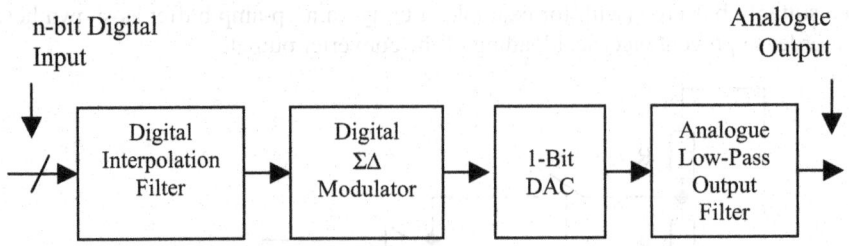

**Fig. 6.11.** Sigma-delta DAC architecture

In this arrangement, the **digital interpolation filter** accepts the n-bit input (data to be converted) that is applied at a low-rate. The filter outputs a digital filtered version of this at a higher rate (*i.e.* oversampling) to the digital $\Sigma\Delta$ modulator. The modulator output is applied to a 1-bit DAC that produces a 2-level analogue output. This is low-pass filtered to produce the final DAC output.

In sigma-delta modulation, the OSR (oversampling ratio) is given by:

$$OSR = \frac{F_{os}}{2F_{bw}}$$

where $F_{os}$ is the oversampling frequency and is the and $F_{bw}$ is the signal bandwidth.

The single-bit DAC can be modified to become a multi-bit DAC by using a digital multi-bit $\Sigma\Delta$ modulator and an n-bit DAC.

## 6.4.8 Hybrid DAC

This uses a combination of D/A conversion techniques within a single converter in order to attain specified performance.

# 6.5 DAC Test

## 6.5.1 Introduction

Real data converters will deviate from the ideal performance, see Fig. 6.6, and these deviations must be identified. Any device operating outside the measured performance limits will fail the test. The testing of the DAC must consider a number of DAC characteristics:

- Data-input, signal-output relationship
- Correct operation of the circuit control signals – controlling the enabling of the device and writing of new digital data in order to update the DAC output
- Correct operation of the Input-Output cells (analogue parameters)
- Power supply current values (both static and dynamic currents)

The testing of the DAC [13, 14] is predominantly undertaken using functional tests. By taking measurements of the resulting analogue output (voltage or current), and considering the data-input, signal-output relationship, DAC testing is considered to fit into the following three categories:

- Static (DC) Tests
- Transfer Curve Tests
- Dynamic Tests

The significance of particular tests would be dependent on the device applications: some parameters may be more important than others within a particular application.

## 6.5.2 Static (DC) Tests

In this test, the DC performance of the DAC is determined, see table 6.3.

**Table 6.3.** Static (DC) tests

| | Test | | Description |
|---|---|---|---|
| 1 | Code specific parameters | | Measure of the output signal value at specific input codes |
| 2 | Full-scale range ($V_{FSR}$) | | The full-scale range is the difference between the maximum and minimum output signal values |
| 3 | DC gain error | | A measure of the deviation of the slope of the straight-line approximation of the actual converter output from the ideal converter straight line output |
| 4 | Offset error | | Offset of the actual converter output. The offset error may be taken at the lowest input code, the mid-code, or the highest input code point |

| 5 | LSB step size | Measure of the average step size between input codes (quoted in volts per bit) |
| 6 | DC PSS | DC Power Supply Sensitivity. This is a measure of the sensitivity of the DAC circuitry to variations in the power supply voltage |

Fig. 6.12 shows a graph of a 3-bit (voltage output) DAC showing gain and offset errors. In this figure, the values measured are denoted with the points, and a straight-line approximation passes through these points. The offset error is taken at the first code and the converter has a gain error as denoted by the difference in the slopes of the straight-line approximation and the ideal converter straight-line.

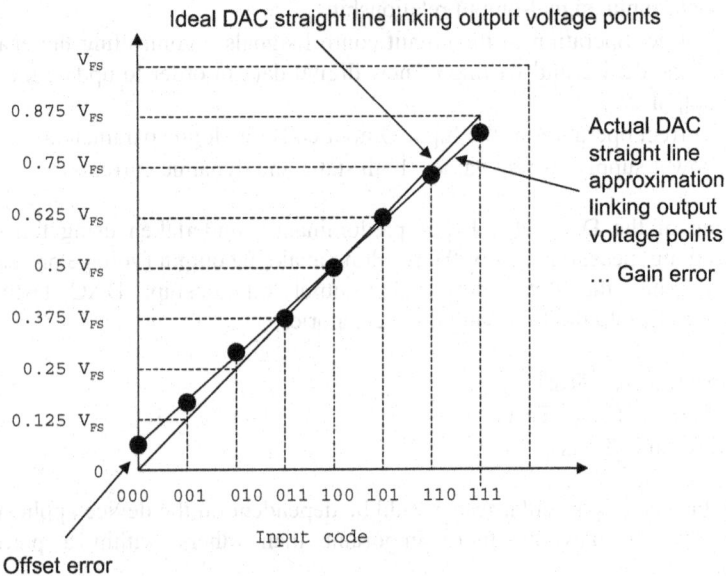

**Fig. 6.12.** 3-Bit DAC with gain and offset errors

In general, the straight-lines on this plot will be:

- The ideal converter relationship
- End-points – a straight-line approximation is drawn between the first and last data points for the actual converter
- Best-Fit – a fitting procedure to find the straight-line approximation that best fits all the data points for the actual converter is generated

## 6.5.3 Transfer Curve Tests

Transfer curve testing looks at the transfer curve plot at each of the data points and derives tests based on the deviation of the converter by comparison to either the ideal DAC or the straight-line approximation of the actual converter data points (end-points or best-fit), see Table 6.4.

**Table 6.4.** Transfer curve tests

| | Test | Description |
|---|---|---|
| 1 | Absolute error | Difference between the ideal DAC output curve and actual DAC output curve. This is identified for each input code |
| 2 | Monotonicity | When the input code increments by 1-bit, there should be an increment in the output signal. This situation occurs when the DAC is monotonic. In a non-monotonic DAC, an increase in the input code may result in a decrease in the output signal |
| 3 | Integral Non-Linearity (INL) | The INL is the deviation of the actual converter data point from the point on the straight-line approximation. The ideal converter, end-points or best-fit straight-line approximation can be used. Where the ideal converter is used, then this value will be the same as the absolute error. This is normally quoted in Least Significant Bits (LSBs) |
| 4 | Differential Non-Linearity (DNL) | Where a binary input code change of 1 bit occurs, the output should change by 1 LSB. The DNL is the difference between each output step size of the converter and an ideal step size of 1 LSB. For a given input code, the output step size is taken between the current input code and the previous code. This is normally quoted in Least Significant Bits (LSBs) |

Fig. 6.13 shows the transfer curve for an example 3-bit DAC. In this figure, the ideal straight line (*left*) can be compared to the measurements from an actual (non-ideal) DAC (*right*). For each input code, there would normally be a deviation of the output voltage (or current) of the actual DAC from the ideal.

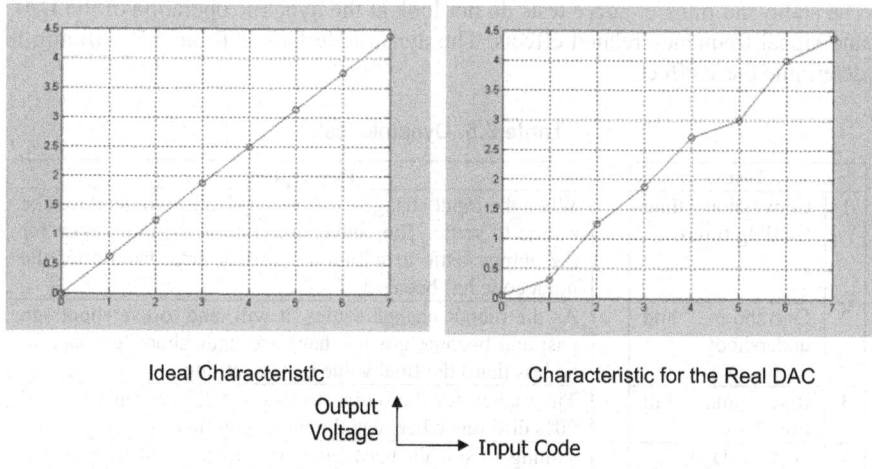

Ideal Characteristic                    Characteristic for the Real DAC

Output
Voltage ⌐→ Input Code

**Fig. 6.13.** Ideal and non-ideal DAC transfer curve

For the 3-bit DAC transfer curve in Fig. 6.13, the INL and DNL plots are shown in Fig. 6.14. With the move towards the use of data converters with higher resolution (number of bits), see Table 6.1, the increase in the number of codes that would need to be tested for, is a cause for concern. Where each code would need to be tested (either applied as in the D/A or sampled as in the A/D), the ability to test for all codes in the higher resolution converters is not economic and a means to address this problem is an on-going activity [6-8]. It is however possible to test the DAC operation for **all possible codes,** or to test for a selected sub-set of all codes – **selected codes testing.** Selected codes testing can reduce the test effort, but the performance of the DAC for the untested codes would also need to be considered.

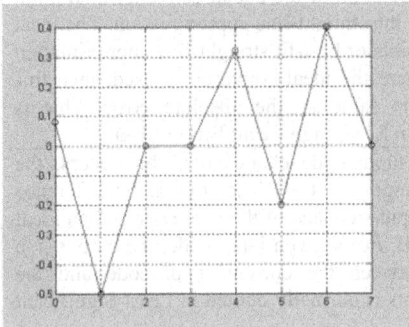 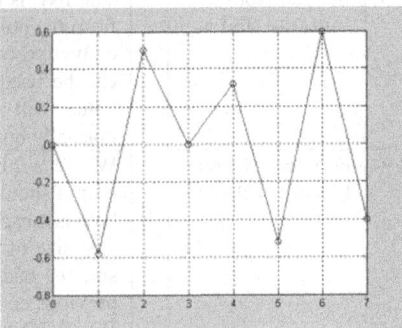

**Fig. 6.14.** 3-Bit DAC INL (*left*) and DNL (*right*) plots

## 6.5.4 Dynamic Tests

The static and transfer curve tests do not look at the dynamic operation of the DAC and signal frequency related effects. The dynamic tests, see Table 6.5, will aim to determine these effects.

**Table 6.5.** Dynamic tests

| | Test | Description |
|---|---|---|
| 1 | Conversion time (settling time) | When the input changes, the output change will tend to take a time to settle. The conversion time is the time taken for the output settle to within a specified error band after the input code has been set |
| 2 | Overshoot and undershoot | As the output change settles, it will tend to overshoot (go past and become greater than) and undershoot (go back to be less than) the final value before settling |
| 3 | Rise and fall times | Time taken for the output to rise or fall between 10% and 90% difference between the initial and final values |
| 4 | DAC-to-DAC skew | Timing mismatch between DACs to be used in matched groups |
| 5 | Glitch Energy | Specification common to high frequency DACs |
| 6 | Clock and data feedthrough | A measure of the cross-talk of the digital signals to the analogue output signal |

## 6.5.5 FFT, SNR, SFDR and THD

Where samples of an analogue signal are taken, then DSP techniques can perform analysis of the signal in both the time and frequency domains. The **Fast Fourier Transform** (FFT) [17] is undertaken on sampled signals in order to identify the frequency components of a complex signal in terms of the signal magnitude/RMS value and phase at different frequencies. The FFT is an efficient implementation of the **Discrete Fourier Transform** (DFT). This is a discrete form of the Fourier Transform using samples of a signal. However, the Fourier Transform can be implemented as a **Discrete Time Fourier Transform** (DTFT) or as a **Continuous Time Fourier Transform** (CTFD). By analysis of the FFT plot, then it is possible to identify a number of converter characteristics:

- Signal to Noise Ratio (**SNR**): This is the ratio of the signal power (in the fundamental frequency) and the noise power over the frequency band of interest. For an ideal converter, the SNR is given by:

$$SNR_{dB} = 6.02N + 1.76$$

  where $SNR_{dB}$ is the SNR quoted in decibels (dB) and N is the resolution of the converter (number of bits).

- Spurious Free Dynamic Range (**SFDR**): This is the difference (in dB) between the fundamental frequency and the largest ray (of all other frequencies identified on the FFT plot). It is the usable dynamic range of the converter before noise effects become noticeable

- Total Harmonic Distortion (**THD**): This is the ratio of the sum of the power in the signal harmonics to the power in the fundamental signal (in dB)

- Signal to Noise-Plus-Total-Harmonic Distortion (**S/(N + THD)**): This is the plot of the actual SNR curve vs input signal magnitude

- Signal to Noise And Distortion (**SINAD**): A combination of SNR and THD

- Effective Number of Bits (**ENOB**): This is a measure of how close the actual converter is to the theoretical model

# 6.6 ADC Architectures

## 6.6.1 Introduction

The basic operation of the ADC is to convert an analogue input signal (voltage or current) to a digital code. The action is to produce an output code that represents a range of input signal values that can be correctly converted. As the input signal increases from a minimum value to a maximum value, the output code increments at discrete points in value of the input signal. This is a **many-to-one mapping** and a result of the conversion is to produce a **quantisation** error. The value of the output code, usually unsigned binary although any digital code may be considered (e.g. 2s complement signed binary), is a factor based on the value of the input signal and a static analogue reference signal (either current or voltage). For an n-bit ADC with a reference and input voltage, see Fig. 6.15, the input voltage is converted to an n-bit output code.

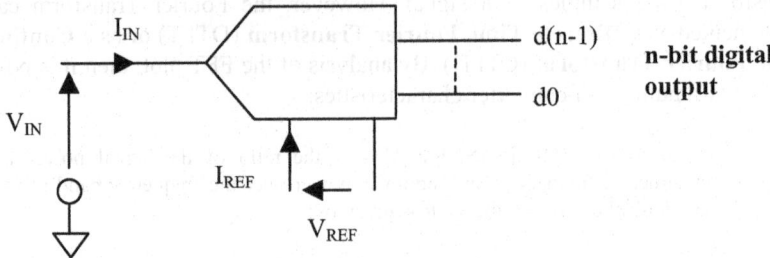

**Fig. 6.15.** ADC set-up

Within a typical ADC, there are also control signals to operate, see Fig. 6.16. These are **Chip Select** (to enable the ADC so that the device is enabled to allow a conversion to take place), **Clock** (for internal control of the sequential logic within the ADC circuit), **Read** (to read the digital output when a conversion has been completed) and **Ready** (to identify that a conversion has been completed and the results are ready to be read).

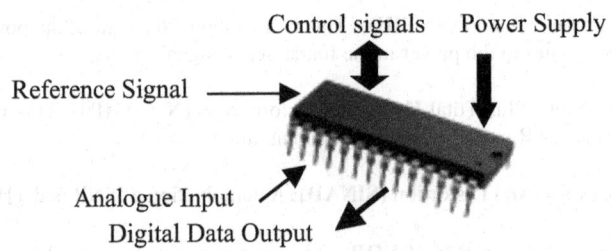

**Fig. 6.16.** ADC signals

The ADC input signal range for conversion will vary from a predefined minimum value to a predefined maximum value. Figure 6.17 shows a representation of the input-output (transfer curve) characteristic for a 3-bit ideal ADC. In this arrangement, the input signal conversion range is divided into $2^n$ ($2^3 = 8$) equal segments and the code transition point (between one code and the next) occurs in the middle of the segment. A straight-line can be drawn to join the transition point corners. The output code is an unsigned binary code from $000_2$ to $111_2$. There are a number of architectures that have been developed for the ADC, and as with the

DAC, these will be either **Nyquist Rate** or **Oversampling** converters. The ADC architectures include:

- Successive Approximation
- Integrating
- Flash
- Sigma-Delta ($\Sigma\Delta$)

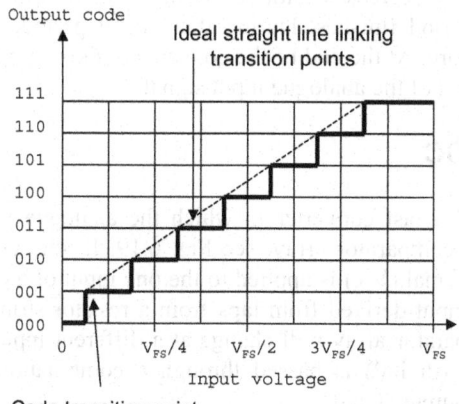

**Fig. 6.17.** Ideal 3-bit ADC input-output (transfer curve) characteristic (1)

An ADC will be for use in one of a number of applications, and the importance of specific ADC parameters will be dependent on the particular application.

## 6.6.2 Successive Approximation ADC

This is a common architecture, see Fig. 6.18, which uses an n-bit DAC in a feedback loop. The circuit compares the output of the DAC (representing the digitised input signal) with the original input, and the conversion continues whilst the DAC output is less than the input signal (the sampled input voltage within the sample and hold circuit). The Successive Approximation Register (SAR) increments/decrements the data output in a binary search pattern.

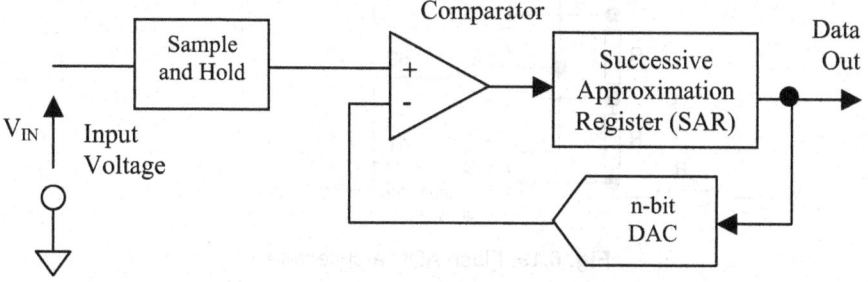

**Fig. 6.18.** Successive approximation ADC

### 6.6.3 Integrating (Single and Dual Slope) ADC

An ADC architecture that separates the conversion cycle into two separate integrating periods. In the first period, the input voltage (unknown value) is connected to the input of an analogue integrator and the signal is integrated over a set time period. During this time, an n-bit counter is incremented. During the second period, the integrator input is connected to a known reference voltage and the integrator output decreases until it reaches a zero point, at which time the conversion time period finishes. During the second period, the n-bit counter is incremented as before. At the end of the conversion, the value of the counter is the digital representation of the analogue input signal.

### 6.6.4 Flash ADC

The Flash ADC is a fast converter in which the analogue input is applied to a resistor string and comparator array, see Fig. 6.19. In this view, a 3-bit ADC is shown. The input signal ($V_{IN}$) is applied to the one input of a comparator, with the other comparator input derived from taps from a resistor string. Each comparator output in the comparator array will change at a different input voltage level. The comparator output (digital) is passed through a combinational logic block that creates the binary output signal.

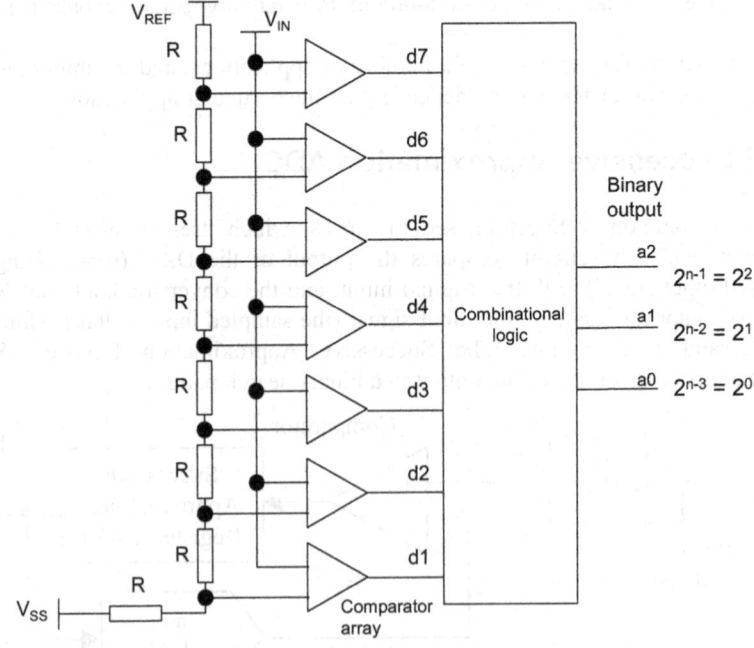

**Fig. 6.19.** Flash ADC architecture

## 6.6.5 Sigma-Delta (ΣΔ) ADC

This sigma-delta (ΣΔ) converter uses oversampling and noise shaping techniques in order to provide for A/D conversion.

# 6.7 ADC Test

## 6.7.1 Introduction

Real data converters will deviate from the ideal performance, see Fig. 6.17, and these deviations must be identified. Any device operating outside the measured performance limits will fail the test. The testing of the ADC must consider:

- Signal-input, data-output relationship
- Correct operation of the circuit control signals – controlling the enabling of the device and writing of digital control signals in order for the ADC to perform a conversion
- Correct operation of the Input-Output cells (analogue parameters)
- Power supply current values (both static and dynamic currents)

The testing of the ADC is predominantly undertaken using functional tests. By taking measurements of the resulting output (digital code), and considering the signal-input data-output relationship, ADC testing is considered to fit into the following three categories:

- Static (DC) and Transfer Curve Tests (considered together)
- Dynamic Tests

The significance of particular tests would be dependent on the device applications - some parameters may be more important than others.

An important point to note will be the **code transition points**, or **code edges**. The deviation of the actual converter code transition point will need to be compared to the ideal converter code transition point. This can be achieved by using a **step search** (incrementing/decrementing the input signal in small steps until the output code dithers equally between two codes), by using a **binary search** (faster than a step search using the algorithm as used in the successive approximation ADC, until the output code dithers equally between two codes), or by using a **servo loop test** [1]. When undertaking this test, then the need to perform a large number of conversions in order to attain a statistically significant number of samples per code edge [13]. This becomes very time consuming and hence costly. For production test, this would be an issue that would need to be addressed. The **code transition**

**point** identified in Fig. 6.17 shows the ideal placement in the centre of each segment dividing the input signal range. An alternative, see Fig. 6.20, places the code transition point on the edge of the segment. Care would need to be taken with the particular ADC in order to identify the code transition point of the particular device. Alternatively, the **code centre** point (mid-point between code edges) may be of interest rather than the code edge.

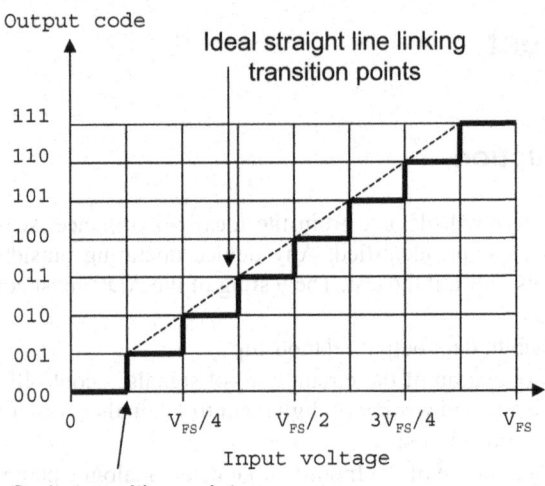

**Fig. 6.20.** Ideal 3-bit ADC input-output (transfer curve) characteristic (2)

## 6.7.2 Static (DC) and Transfer Curve Tests

These tests refer to the static conditions and the input-output relationship as identified in the transfer curve. The tests are identified in Table 6.6.

**Table 6.6.** Static (DC) and transfer curve tests

| | Test | Description |
|---|---|---|
| 1 | DC gain error | A measure of the deviation of the slope of the straight-line approximation of the actual converter output from the ideal converter straight-line output. The best-fit straight-line approximation would be used for the actual converter straight-line approximation, see Fig. 6.21 |
| 2 | Offset | The deviation of the first code transition point from the expected. The ideal converter or best-fit straight-line approximation would be used. This is normally quoted in Least Significant Bits (LSBs) |
| 3 | Integral Non-Linearity (INL) | The INL is a measure of the deviation of the actual converter code transition point from the straight-line approximation for each code. The best-fit straight-line approximation would be used. This is normally quoted in Least Significant Bits (LSBs) |

| 4 | Differential Non-Linearity (DNL) | The DNL is the difference between the width (range of input signal) between converter output code changes and an ideal step size of 1 LSB. For a given input code, the output step size is taken between the current input code and the previous code. This is normally quoted in Least Significant Bits (LSBs) |
|---|---|---|
| 5 | Monotonicity | The output code should increase with an increase in input signal: this is a monotonic ADC. A non-monotonic ADC has an output code that decreases (at particular codes) as the input signal increases, see Fig. 6.22 |
| 6 | Missing codes | The converter output (digital) should generate $2^n$ codes where n is the resolution of the converter. Problems may occur within the converter where certain codes are not generated |

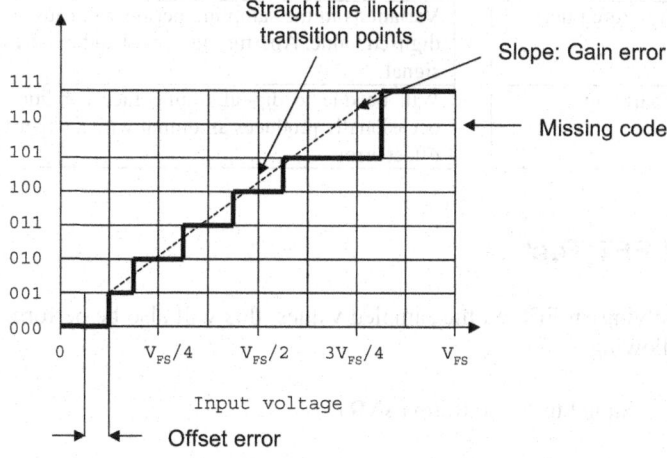

**Fig. 6.21.** Gain error, offset error and missing codes

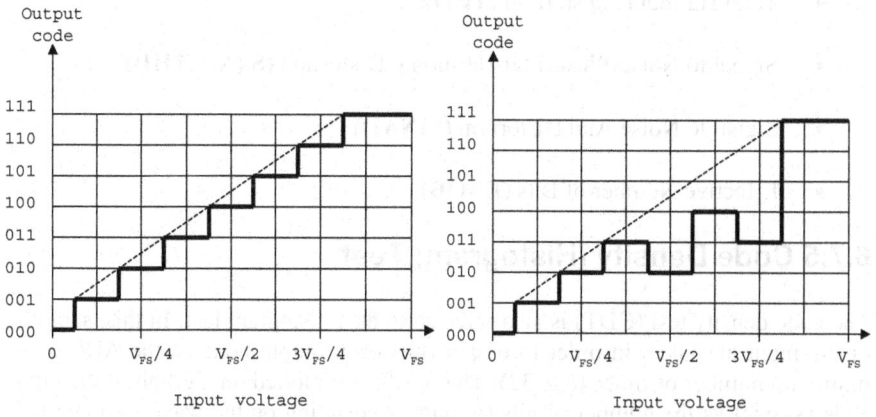

**Fig. 6.22.** Monotonic and non-monotonic ADC

## 6.7.3 Dynamic Tests

The static and transfer curve tests do not look at the dynamic operation of the ADC and the effects of signal changes and frequency related effects. The dynamic tests, see Table 6.7, will determine these effects.

**Table 6.7.** Dynamic tests

| | Test | Description |
|---|---|---|
| 1 | Conversion time | There must be a guaranteed maximum conversion time (time from start of conversion to conversion completed) |
| 2 | Recovery time | Some ADCs require a minimum time after a conversion has been completed before the next conversion may start |
| 3 | Sampling frequency | Testing of ADC at maximum sampling frequency and ensuring that no errors occur |
| 4 | Aperture jitter | Variations in the sampling period will cause an error in the digitsed value. Aperture jitter will add noise to the digitised signal |
| 5 | Sparkling | Will be due to digital timing race conditions. The ADC occasionally produces an output with a larger than expected offset error |

## 6.7.4 FFT Test

By applying an FFT on the sampled values, this will also be performed to identify the following:

- Signal to Noise Ratio (**SNR**)

- Spurious Free Dynamic Range (**SFDR**)

- Total Harmonic Distortion (**THD**)

- Signal to Noise-Plus-Total-Harmonic Distortion (**S/(N + THD**)

- Signal to Noise And Distortion (**SINAD**)

- Effective Number of Bits (**ENOB**)

## 6.7.5 Code Density (Histogram) Test

The code density test (CDT) is also referred to as a histogram test. In this, samples of the input are taken in order to obtain (hit) each output code of the ADC a set minimum number of times (*e.g.* 32). The results are plotted on a graph of the input code (x-axis) vs the number of hits (y-axis). Depending on the waveform applied, the graph would be expected to be a characteristic shape. The two waveform types used would be:

- Sine Wave

- Ramp

Fig. 6.23 shows the expected graph for each of the above waveforms.

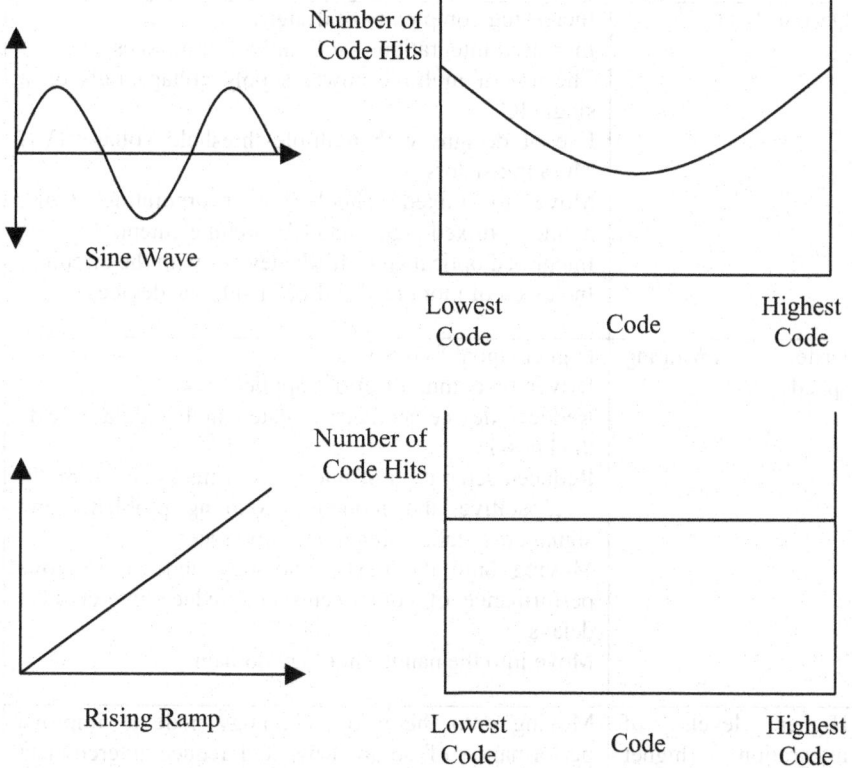

**Fig. 6.23.** Histogram test, sine wave (*top*) and ramp (*bottom*)

For the **sine wave**, the plot is a curve with a higher number of hits at the lower and higher codes denoting the lower rate of change of the sine wave at the upper and lower limits. For the **ramp**, this is a flat line denoting the constant rate of change of the ramp waveform. Problems with the converter can be obtained from the histogram by comparison of the expected (ideal) converter with the actual converter. It is particularly good for identifying DNL errors.

## 6.8 Future Directions

The future for mixed-signal test has many parallels with analogue circuit and digital circuit and system test. A summary is provided in Table 6.8.

**Table 6.8.** Future mixed-signal test issues

| Technology driver | Description |
|---|---|
| Design styles | Increasing complexity of designs<br>Increased integration – SoC and SiP test issues<br>The use of multiple power supply voltage rails on a single IC<br>Use of designs with multiple threshold voltage ($V_T$) MOS transistors<br>Move to mixed-technologies incorporating logic, memory, mixed-signal and RF circuit elements<br>Integrated optical and MEMs devices with the circuit<br>Inclusion of more DfT and BIST into the device |
| Faster operating speeds | Higher signal frequencies<br>Driven by communications applications<br>Reduced device geometries (gate length and gate oxide thickness)<br>Reduced separation of interconnect tracks (track pitch) – capacitive and inductive coupling problems and signal cross-talk – signal integrity issues<br>Moving onto the next technology node to improve performance of components and reduce interconnect delays<br>Move into the nanotechnology domain |
| Higher levels of integration (higher density) | Moving onto the next technology node to improve performance of components and reduce interconnect delays<br>Newer failure mechanisms - limitations of existing fault models and need for models more representative of process defects<br>New test issues emerging for SoC and SiP devices<br>Move into the nanotechnology domain |
| Lower operating voltages | A need to improve device reliability by reducing electric field strength in dielectrics<br>Aim to lower device power consumption<br>Portable, battery operated circuits<br>Multi-voltage power supply device operation and interfacing |

| Test equipment and costs | Need for reduced test time and test costs |
|---|---|
| | Need for reduced test pattern generation time and maintenance |
| | More formal approaches to test program generation – test re-use |
| | Problem of increasingly long test pattern generation and application times |
| | Increasing requirements for test data storage in ATE |
| | Increasingly difficult to perform at-speed testing of devices using current ATE |
| | Need for reduced external ATE requirements and costs – simpler and less expensive ATE |
| | Multi-Site test |
| | Inclusion of more DfT and BIST into the device |

# 6.9 Summary

This chapter has discussed the types of mixed-signal circuits commonly encountered and their basic operation. These types of circuits combine the challenges of digital and analogue circuit testing. In the main, functional testing of mixed-signal circuits is undertaken and structural testing has had limited acceptance. This is the same as for analogue circuit test and differs from digital test where structural test is commonplace and well accepted. In this chapter, the digital to analogue converter and analogue to digital converter designs were introduced and the main test methods identified.

# 6.10 References

[1]   Jespers P., "Integrated Converters, D to A and A to D Architectures, Analysis and Simulation", Oxford University Press, USA, 2001, ISBN 0-19-856446-5

[2]   Haskard M.R. and May I.C., "Analog VLSI Design nMOS and CMOS", Prentice Hall Silicon Systems Engineering Series, Australia, 1988, ISBN 0-7248-0027-1

[3]   Stoffels R., "Cost effective frequency measurement for production testing: new approaches on PLL testing", Proceedings of the International Test Conference, 1996, pp708-716

[4]     Burbidge M. et al., "Motivations towards BIST and DfT for embedded charge-pump phase-locked loop frequency synthesisers", IEE Proceedings on Circuits, Devices and Systems, Vol. 151, Issue 4, August 2004, pp337-348

[5]     Rajsuman, R., "System-on-a-Chip Design and Test", Artech House Publishers, USA, 2000, ISBN 1-58053-107-5

[6]     Wegener C. and Kennedy M., "Implementation of Model-Based Testing For Medium-To-High-Resolution Nyquist-Rate ADCs", Proceedings of the International Test Conference, 2002, pp851-860

[7]     Wegener C. and Kennedy M., "Model-Based Testing of High-Resolution ADCs", Proceedings of the IEEE International Symposium on Circuits and Systems, 2000, ppI-335-I-338

[8]     Carroll B., Wegener C. and Kennedy M., "Lemma-ADC: The Linear Error Mechanism Modelling Algorithm Applied to A/D-Converters", Proceedings of the Advanced A/D and D/A Conversion Techniques and their Applications Conference, 1999, pp145-148

[9]     Bushnell M. and Agrawal V., "Essentials of Electronic Testing for Digital, Memory & Mixed-Signal VLSI Circuits", Kluwer Academic Publishers, 2000, ISBN 0-7923-7991-8

[10]    Rajsuman, R., "System-on-a-Chip Design and Test", Artech House Publishers, USA, 2000, ISBN 1-58053-107-5

[11]    Hurst S., "VLSI Testing digital and mixed analogue/digital techniques", IEE, 1998, ISBN 0-85296-901-5

[12]    Needham W., "Designer's Guide to Testable ASIC Devices", Van Nostrand Reinhold, 1991, ISBN 0-442-00221-1

[13]    Burns M. and Roberts G.W., "An Introduction to Mixed-Signal IC Test and Measurement", Oxford University Press, New York, 2001, ISBN 0-19-514016-8

[14]    Van Spaandonk J. and Kevenaar T., "Selecting Measurements to Test the Functional Behavior of Analog Circuits", Journal of Electronic Testing: theory and Applications, No. 9, 1996, pp9-18

[15]    Tuthill M., "High Resolution digital-to-analogue converter", U.S. Patent 4 338 591, 6th July 1982

[16] Cummins T. et al., "An IEEE 1451 Standard Transducer Interface Chip with 12-b ADC, Two 12-b DAC's, 10-kB Flash EEPROM, and 8-b microcontroller", IEEE Journal of Solid-State Circuits, Vol. 33, No. 12, December 1998, pp2112-2120

[17] Ifeachour E. and Jervis B., "Digital Signal Processing, A Practical Approach", Prentice Hall, UK, 2002, ISBN 0-201-59619-9

[18] DAC 08 "8-Bit, High Speed, Multiplying D/A Converter (Universal Digital Logic Interface" datasheet, Analog Devices Inc., USA

[19] AD7575 "LC$^2$MOS 5μs 8-Bit ADC with Track/Hold" datasheet, Analog Devices Inc., USA

# Exercises

*The following questions will require access to MATLAB®.*

## Question 1

Using the data points of the DAC in Table H.2 (Appendix H), and with the aid of MATLAB®, derive the static and transfer curve characteristics of the DAC when using:

- The ideal converter
- End-points straight-line
- Best-fit curve

For the best-fit curve, identify the possible best-fit algorithms that can be used and compare the results for each of these algorithms.

## Question 2

Using the tester arrangement in Appendix E, develop a test for the DAC08 [18] 8-bit DAC. Perform a test and suitably save the results in order to undertake the tests as identified in Question 1.

## Question 3

Using MATLAB®, model an ideal 8-bit ADC ($V_{REF}$ = +5V). Simulate the design in order to derive the static and transfer curve characteristics of the ADC.

## Question 4

Using the tester arrangement in Appendix E, develop a test for the AD7575 [19] 8-bit ADC. Perform a test and suitably save the results in order to undertake the tests as identified in Question 3.

*The following questions will require access to an analogue circuit simulator. In the following questions, the HSPICE® circuit simulator is considered, although with suitable modifications, a suitably available simulator could be used.*

## Question 5

Model and simulate the full operation of a 4-bit Flash ADC (resistor string, CMOS analogue switches, comparators (using the Voltage Controlled Voltage Source (VCVS)) and the decoding logic (using static CMOS logic gates). Choose a suitable sampling frequency and size the MOS transistors accordingly. Use 10kΩ resistors throughout the design. Use the simulation results to run an FFT and compare with the theoretical results. If necessary, write a C-program to aid the analysis.

## Question 6

Repeat Question 5 for an 8-bit Flash ADC.

## Question 7

For the segmented resistor DAC design identified in [16], model and simulate the full operation of the DAC. Choose a suitable update frequency and size the MOS transistors accordingly. Use the simulation results to run an FFT and compare with the theoretical results. If necessary, write a C-program to aid the analysis.

## Question 8

Model and simulate the full operation of a 4-bit resistor string DAC (resistor string, CMOS analogue switches and op-amp (using the Voltage Controlled Voltage Source (VCVS)). Choose a suitable update frequency and size the MOS transistors accordingly. Use 10kΩ resistors throughout the design. Use the simulation results to run an FFT and compare with the theoretical results. If necessary, write a C-program to aid the analysis.

## Question 9

Model and simulate the full operation of a 4-Bit successive approximation ADC using the 4-bit DAC in Question 8. Choose a suitable sampling frequency and size the components accordingly. Use the simulation results to run an FFT and compare with the theoretical results. If necessary, write a C-program to aid the analysis.

## Question 10

Modify the simulation file in Question 8 to allow for a histogram test to be undertaken for both a sine wave and ramp input. If necessary, write a C-program to aid the analysis.

## Question 11

Using the 4-bit flash ADC design in Question 5, run a fault simulation for open-short faults in the components and interconnect. Identify the fault coverage figure and any faults that cannot be detected (if any).

## Question 12

Using the 4-bit resistor string DAC design in Question 8, run a fault simulation for open-short faults in the components and interconnect. Identify the fault coverage figure and any faults that cannot be detected (if any).

# Chapter 7

# Input-Output Test

*Within an Integrated Circuit, the main circuit functionality is performed within the core of the circuit die. However, there is the need to provide for signal I/O and power supply connections between the core of the die and the package. The cells within the periphery of the device provide for this operation.*

## 7.1 Introduction

A **circuit die** (layout floorplan) consists of two main parts – the **core** and the **periphery**. The basic structure of the IC is shown in Fig. 7.1. Here, the circuit is fabricated within the die and this is mounted in a suitable package, see Chap. 2.

The die will be square or rectangular in shape and will be fabricated on a wafer, where:

- The **core** contains the bulk of the circuitry

- The **periphery** will contain Input/Output cells and Power Supply cells. These cells are physically much larger than the core cells and will be required to bond the die to the package.

The majority of the testing undertaken on an IC is to ensure the correct operation of the core circuitry. However, since the signals pass through the periphery circuitry into the core of the circuit and out again through the periphery, see Fig. 7.2, core test routines will also test the functionality of the periphery cells. However, there is an additional need to perform parametric tests on the periphery to ensure that the cells operate according to specific voltage and current parameters -- to ensure that the periphery circuitry has been fabricated without defect.

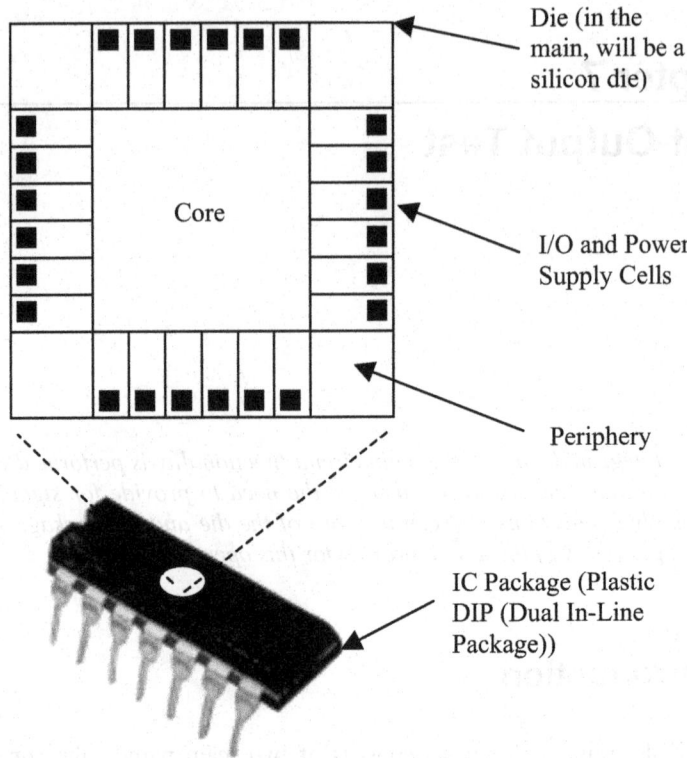

**Fig. 7.1.** Structure of the IC

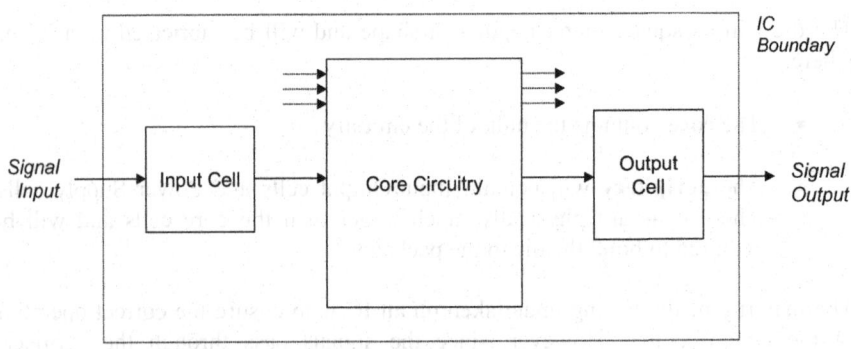

**Fig. 7.2.** Signal I/O path

The types of cells found in the periphery [1-4] will fall into one of the following types of cell:

- Power Supply (positive - $V_{DD}$ / $V_{CC}$):
    o $V_{DD}$ for CMOS, $V_{CC}$ for bipolar circuitry

    o   The power supply cell may provide a power connection for the core and periphery together, for the core only, or for the periphery only

    o   There may be different power supply cells for the analogue and digital circuitry on the IC

- Power Supply (negative - $V_{SS}$ / $V_{EE}$):
  - $V_{SS}$ for CMOS, $V_{EE}$ for bipolar circuitry
  - The power supply cell may provide a power connection for the core and periphery together, for the core only, or for the periphery only
  - There may be different power supply cells for the analogue and digital circuitry on the IC

- Digital Input
- Digital Output
- Digital Bidirectional (selected to be an input or output using an internal **enable** signal)

- Crystal Oscillator cell
- RC Oscillator cell
- Clock Input

- Analogue Input/Output cell (the same cell design may be used for both applications)

The digital cells will additionally have various forms:

- Basic cell (input, output or bidirectional)
- Cell with pull-up or pull-down device on-chip
- Cell with slew rate control
- Cell with Schmitt Trigger
- Cells (output and bidirectional) with different output current drive capabilities (in terms of mA) for different output circuit load conditions
- Cell with open collector (Bipolar) or open drain (CMOS) output

The cell design will be viewed by the user in a number of forms depending on the user requirements:

- **Circuit schematic**:

  This is the level that the cell is designed at. It consists of a circuit diagram (and circuit netlist) showing the basic circuit components (transistors, diodes and resistors), along with the component interconnection. This is an analogue circuit schematic and its operation will be simulated using a suitable analogue circuit simulator such as Spice

- **Circuit symbol**:
    This is the view that the cell is used within the overall (final) circuit design. The symbol is a simplified view of the circuit schematic and shows the circuit input-output connections and a suitable outline diagram representing the particular cell

- **Layout**:
    This is the level that the cell is designed at for the particular fabrication process. It consists of geometries identifying the different layers that combine in the process to make up the circuit components and interconnect. The circuit netlist of the layout will be the same as that of the circuit schematic

- **Abstract**:
    This is a simplified view of the layout, showing a simple cell physical outline (shape) and input-output connections. It is used as a simple representation of the layout within a layout editor tool – it removes the need to redraw the complex shapes that would be seen in the layout view

- **Simulation model**:
    This is a (textural) description of the cell operation in the logic (typically using VHDL or Verilog®-HDL) or analogue (typically using Spice) domains

The design engineer would use these views at various stages in the design process from schematic/netlist level design through to layout generation and data generation for circuit fabrication.

For the test engineer, these views would be used for activities ranging from DfT strategy development and test circuit insertion, along with general test development activities such as test program simulation to fault list generation (netlist and/or layout) and fault simulation purposes.

# 7.2 Electrical Overstress and Electrostatic Discharge

**Electrical Overstress** (EOS) [5-11] is a major concern relating to the failure of ICs and covers a broad category of electrical threats to semiconductor devices including **electrostatic discharge** (ESD, although ESD is normally considered separately, as below). EOS occurs when pins on an IC are subjected to electrical signals (voltage/current) that exceed the **Absolute Maximum Rating** of the device. EOS testing of a design uses applied electrical values up to and sometimes past the circuits' electrical specifications. Common parameters that might be tested include input voltage, input current and input signal frequency.

**Electrostatic discharge** (ESD) is a single, fast transfer of electrical charge. Static charges can accumulate on the surfaces of objects due to touching and rubbing, these charges can discharge through an IC due to contact, and the result can be damage to the IC itself. At the IC level, ESD damage can cause increased leakage currents at the I/O pins, increased stand-by (quiescent) current of the circuit, and in extreme cases, complete circuit failure. The effect of ESD due to device handling is commonly modelled by either the the human body model (HBM), the machine model (MM), or the charged device model (CDM). The human body model is intended to represent the ESD caused by human handling of ICs. The HBM test circuit is shown in Fig. 7.3. Here, the 100pF capacitor is charged with a high-voltage supply (in the range +/-4kV) and then discharged through a 1.5kΩ resistor onto the pin of the device under test (DUT). ESD-HBM is the most widely used method of qualifying the ESD performance of on-chip protection circuits.

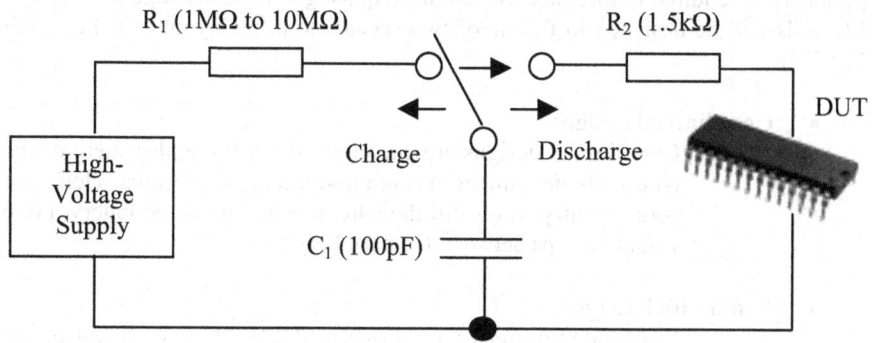

**Fig. 7.3.** Human body model

The potential for ESD damage can be reduced by careful design of the IC I/O circuitry, by careful handing of the IC, by suitable "grounding" (connecting to an earth connection) of equipment and personnel, by control of the relative humidity of the environment, and by the use of electrically conductive surfaces to conduct electrical charges to "ground".

# 7.3 Digital I/O Structures

## 7.3.1 Introduction

Early digital I/O circuits were designed and used according to a limited number of I/O standards. The early digital ICs in bipolar technology were based on TTL

(Transistor-Transistor Logic) operating at a +5V power supply level, and the I/O characteristics well known. When CMOS emerged for digital logic, still working at a +5V power supply level, the problem emerged in the interfacing of TTL and CMOS logic. Today, there are a number of I/O standards in existence and that, coupled with the move towards reduced device power supply operation (+3.3V, moving down towards, and beyond, +1.0V operation), the need to interface ICs with different power supply ratings and the range of digital I/O circuits available, has extended the test problem. With the move towards the smaller process geometries, this has resulted in the need to reduce the device power supply voltage. This does have a benefit in reducing power consumption of the IC, but produces the additional problem that different ICs may work at different power supply voltage levels (mixed-voltage systems) and also, the IC itself may have more than one power supply voltage requirement. It is common for many complex digital ICs to operate on dual power supply voltage, with the core of the IC operating at a lower voltage than the periphery (I/O). This can be due to the core circuitry requiring a lower operating voltage for device reliability reasons whilst the periphery is required to interface to a circuit requiring a higher voltage level range. I/O cells will be designed to fit one of two scenarios in the layout of the die, see Fig. 7.4:

- **Core limited** designs:

  Core limited designs are encountered when the placement of the core cells determines the minimum size of die area. Here, the core circuitry would fill the core area and in the periphery, there would be gaps between the periphery cells

- **Pad limited** designs:

  Pad limited designs are encountered when the placement of the pads determines the minimum size of die area. Here, the core circuitry would be smaller than the area used in the final design

Core Limited IC                                           Pad Limited IC

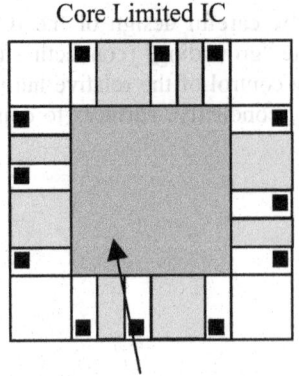    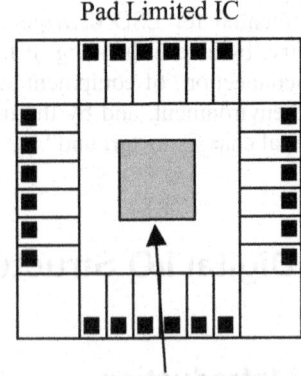

Circuitry within the core will
completely fill the available core area

Circuitry within the core will fill only
part of the available core area

**Fig. 7.4.** Core and pad limited designs

The basic **digital I/O** cells available for use are:

**Input**:

- Non-inverting input
- Inverting input
- Input (non-inverting or inverting) with pull-up
- Input (non-inverting or inverting) with pull-down
- Schmitt Trigger input

**Output**:

- Non-inverting output
- Inverting output
- Tristate output
- Output with open collector (in TTL)
- Output with open drain (in CMOS)

**Bidirectional**:

- Non-inverting bidirectional cell
- Inverting bidirectional cell
- Bidirectional (non-inverting or inverting) cell with pull-up
- Bidirectional (non-inverting or inverting) cell with pull-down

With the input cells, it is important to ensure that any unused inputs on the IC are not left unconnected (floating inputs). This may cause unwanted behaviour, particularly in CMOS designs. With reference to the static CMOS inverter (see Sect. 7.3.2), the input of the logic gate is capacitive (the gate inputs of the individual transistors). If left unconnected, a charge stored on the capacitive input will result in an input voltage at the transistor gate. If this voltage is somewhere between the power supply levels, then current may pass through the circuit. This will result in higher power consumption and unexpected operation of the IC may follow.

The digital cell will consist of the logic operation and protection circuitry, see Fig. 7.5. In this example, specific designs may vary, the bond pad is connected to the logic gate (inverter/ buffer) through a protection network. This protection network is in a **H-network** configuration in that four protection devices are connected through a resistor (R). The protection devices are connected to the power supplies ($V_{DD}$ and $V_{SS}$). An EOS/ESD situation is prevented from reaching the core of the device by the protection devices. Both positive (protection devices 1 and 2 operate) and negative (protection devices 3 and 4 operate) transients are accounted for. The resistor limits the amount of current that can flow through the protection devices.

The protection network will consist of a diode structure.

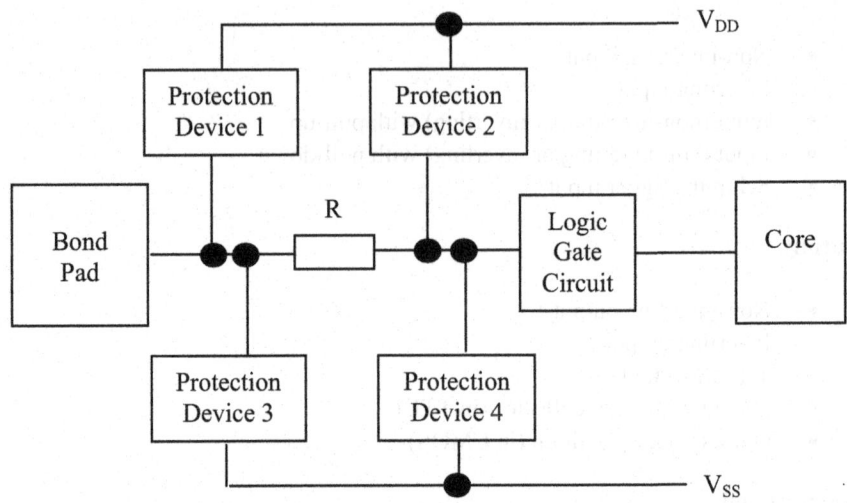

**Fig. 7.5.** Digital I/O cell structure

## 7.3.2 CMOS Inverter

When considering the digital inverter, the same operation of the inverter in the I/O cell is considered as for an inverter logic gate in the core. With static operation, the input-output relationship is considered by applying a DC voltage at the gate input and observing the output voltage and power supply current, see Fig. 7.6. There are five regions of operation (A to E) identifying the operation of each transistor from cut-off through the linear region to the saturation region.

In addition to the static characteristics, the CMOS inverter exhibits dynamic characteristics. Here, the output voltage change due to a dynamic change in the input voltage is noted. The dynamic characteristics will determine the maximum operating frequency of the logic gate and will be dependent on the output load conditions.

The dynamic characteristics for the inverter are shown in Fig. 7.7. Here, a step change at the input (with zero rise and fall times) is applied. The voltages vary from $V_{OL}$ (output voltage, low value) to $V_{OH}$ (output voltage, high value).

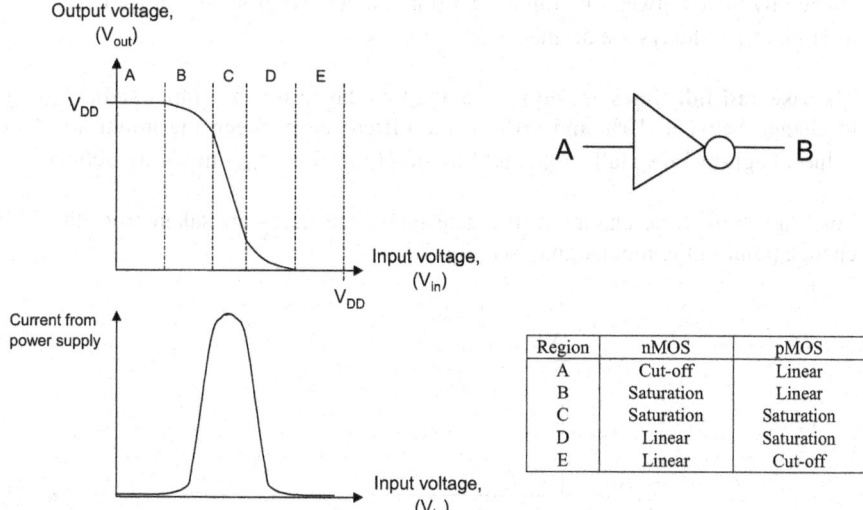

**Fig. 7.6.** Static CMOS inverter – static characteristics

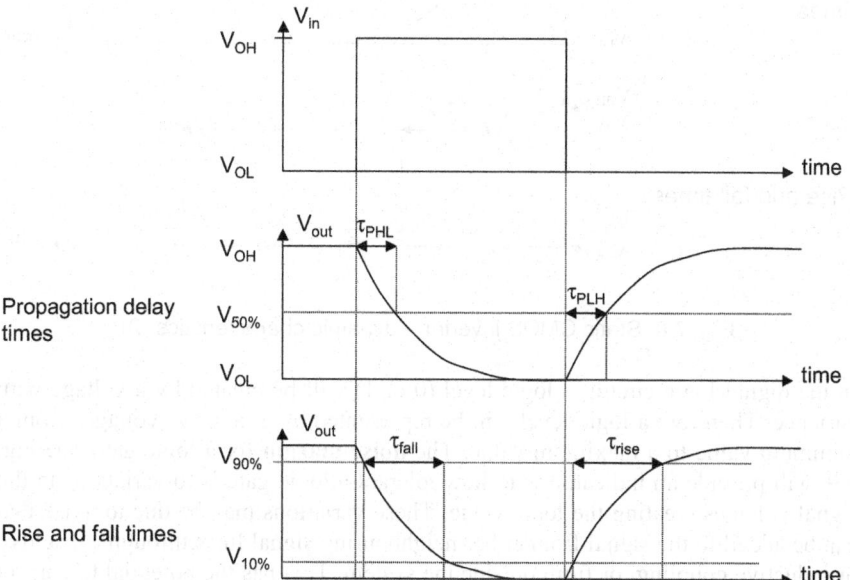

**Fig. 7.7.** Static CMOS inverter – dynamic characteristics (1)

The **propagation delay** is defined as the time taken for the output signal voltage to change by 50% between its initial and final values. High-to-Low ($\tau_{PHL}$) and Low-to-High ($\tau_{PLH}$) delays are defined.

The **rise and fall times** are defined as the time taken for the output signal voltage to change between 10% and 90% of the difference between the initial and final values. High-to-Low (fall - $\tau_{fall}$) and Low-to-High (rise - $\tau_{rise}$) times are defined.

For a non-zero time change at the gate input, the times are taken from the 50% change point in the input signal, see Fig. 7.8.

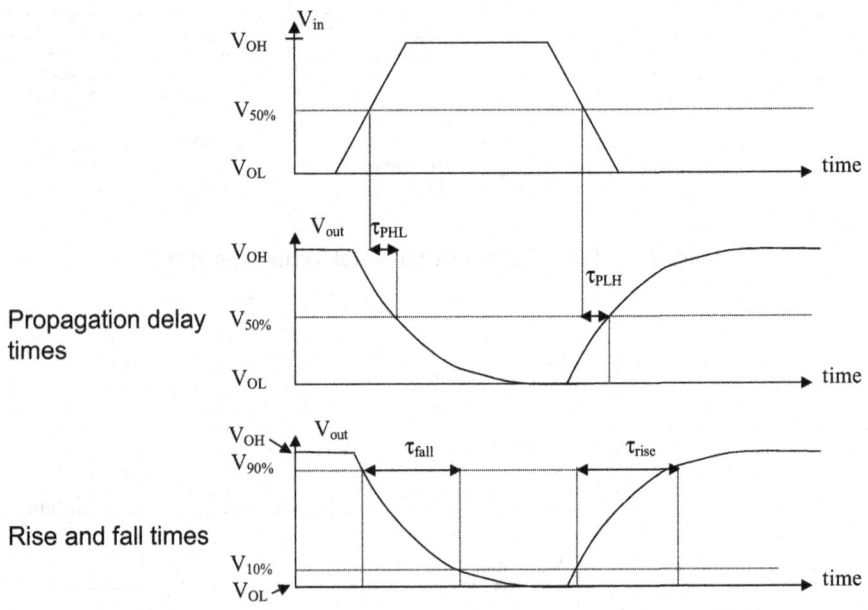

**Fig. 7.8.** Static CMOS inverter – dynamic characteristics (2)

In the digital logic circuit, a logic level (0 or 1) will be created by a voltage with variance. Therefore a logic level will be represented by a range of voltages from a minimum value to a maximum value. The **noise margin** for a logic gate, see Fig. 7.9, will provide an indicator as to how tolerant a logic gate is to variations in the signal voltages creating the logic value. These variations may be due to noise that can be added to the signal from either neighbouring signal lines through capacitive or inductive coupling, or from outside the system. This has the potential to corrupt the signal.

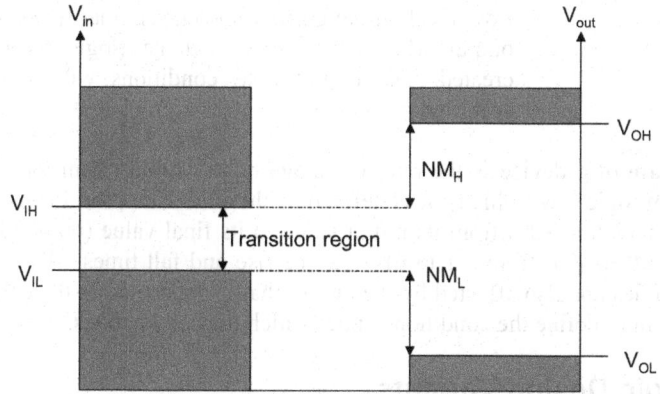

**Fig. 7.9.** Noise margin

Here, two voltages are identified ($V_{in}$ and $V_{out}$) which represent the input and output voltages. For each voltage, the following are defined:

- $\mathbf{V_{IL}}$   Maximum input voltage which can be interpreted as a logic 0
- $\mathbf{V_{IH}}$   Minimum input voltage which can be interpreted as a logic 1
- $\mathbf{V_{OL}}$   Maximum output voltage when the output is a logic 0
- $\mathbf{V_{OH}}$   Minimum output voltage when the output is a logic 1

Two values for noise margin are identified-

- $\mathbf{NM_L}$   (Noise margin for low levels)      $NM_L = V_{IL} - V_{OL}$
- $\mathbf{NM_H}$   (Noise margin for high levels)     $NM_H = V_{OH} - V_{IH}$

The noise margin and tolerance for digital logic becomes increasingly important for low-voltage systems (moving down to and below 1V $V_{DD}$) as the noise margin decreases and the potential for noise to corrupt values can increase. In addition to the voltages defined above, the logic gate will also have low-level and high-level input and output currents:

- $\mathbf{I_{IH}}$       High-level input current: the current that flows into an input when a high-level voltage (value to be specified) is applied.

- $\mathbf{I_{IL}}$       Low-level input current: the current that flows into an input when a low-level voltage (value to be specified) is applied.

- $\mathbf{I_{OH}}$       High-level output current: the current that flows out of an output when a high-level voltage (logic 1 output) is created. The output load conditions will need to be specified.

- $I_{OL}$       Low-level output current: the current that flows out of an output when a low-level voltage (logic 0 output) is created. The output load conditions will need to be specified.

The **slew rate** of a device is the rate of change of its output (from logic high to low, or from logic low to high). It is defined as the time taken for a cell output to change from 10% to 90% from its initial value to its final value (output changing from a logic 0 to 1 or from a 1 to 0) – *i.e.* the rise and fall times. The slew rate /(rise/fall) times are also affected by the circuit that is connected to the cell output. It is important to define the conditions under which the cell operates.

## 7.3.3 Logic Design Variants

With the increasing complexity of digital systems designs encountered, digital devices may encounter a range of I/O standards adopted by different devices. For fixed functionality design devices, this standard will be built into the device. However, programmable digital logic devices may allow for the designer to select a particular I/O standard for the device which is programmed into the device along with the circuit design. For example, the FPGA and CPLD devices provided by Xilinx Inc. [13] allow for a range of I/O standards to be adopted in addition to the basic TTL and CMOS variants. The basic logic families in use are:

- TTL
- ECL
- CMOS
- BiCMOS

## 7.3.4 TTL Family Variants

Transistor-Transistor Logic (TTL) [4] is a circuit family in bipolar process. The original TTL devices were developed for a +5V power supply operation with interfacing to other TTL devices, see Table 7.1. With the advent of CMOS, variants on the basic TTL device allows for compatibility with CMOS devices.

**Table 7.1.** Selected TTL standard variants

| TTL Family Variant | Description |
|---|---|
| 74 | Standard TTL |
| 74AS | Advanced Schottky |
| 74ALS | Advanced low-power Schottky |
| 74F | Fast |
| 74H | High-speed |
| 74L | Low-power |
| 74LS | Low-power Schottky |
| 74S | Schottky |
| LVTTL | Low-voltage |

The voltage parameters for selected TTL variants of a 74x74 (dual D-Type bistable IC [14]) are provided in Table 7.2

**Table 7.2.** Selected TTL voltage parameters (+5V power supply operation)

| Parameter | Variant | | | |
|---|---|---|---|---|
| | 74LS | 74AS | 74ALS | 74F |
| $V_{OH}$ (min) | 2.7 | 3.0 | 3.0 | 2.5 |
| $V_{OL}$ (max) | 0.4 | 0.5 | 0.4 | 0.35 |
| $V_{IH}$ (min) | 2.0 | 2.0 | 2.0 | 2.0 |
| $V_{IL}$ (max) | 0.8 | 0.8 | 0.8 | 0.8 |

## 7.3.5 CMOS Family Variants

The original CMOS [4] devices were designed to operate on the same power supply voltage as TTL (+5V). Now, CMOS designs have been designed to operate at lower power supply voltages. Table 7.3 identifies CMOS family variants.

**Table 7.3.** Selected CMOS family variants

| CMOS Family Variant | Description |
|---|---|
| 4000 | True CMOS (non-TTL levels) |
| 74C | CMOS with pin compatibility to TTL with same number |
| 74HC | Same as 74C but with improved switching speed |
| 74HCT | As with 74HC but can be connected directly to TTL |
| 74AC | Advanced CMOS |
| 74ACT | As with 74AC but can be connected directly to TTL |
| 74AHC | Advanced high-speed CMOS |
| 74AHCT | As with 74AHC but can be connected directly to TTL |
| 74FCT | Fast - CMOS - TTL inputs |
| LVCMOS | Low-voltage CMOS |

The voltage parameters for selected CMOS [15] variants are provided in table 7.4.

**Table 7.4.** Selected CMOS voltage parameters (+5V power supply operation)

| Parameter | Variant | | | | | |
|---|---|---|---|---|---|---|
| | 74HC | 74HCT | 74AC | 74ACT | 74AHC | 74AHCT |
| $V_{OH}$ (min) | 4.44 | 2.4 | 4.44 | 2.4 | 4.44 | 2.4 |
| $V_{OL}$ (max) | 0.5 | 0.4 | 0.5 | 0.4 | 0.5 | 0.4 |
| $V_{IH}$ (min) | 3.5 | 2.0 | 3.5 | 2.0 | 3.5 | 2.0 |
| $V_{IL}$ (max) | 1.5 | 0.8 | 1.5 | 0.8 | 1.5 | 0.8 |

With the low-voltage (LVCMOS) family variants, examples of the available devices are shown in Table 7.5, provide the same logic functionality as the higher supply voltage devices, but on a lower voltage ranging from 1.0V to +3.6V.

**Table 7.5.** Selected low voltage CMOS (and BiCMOS *) family variants

| Low-Voltage CMOS Variant | Description | |
|---|---|---|
| 74LV | Low voltage CMOS. | Low speed operation, 1.0 – 3.6V power supply (some functions up to 5.5V power supply) |
| 74LVC | Low voltage CMOS. | Medium speed operation, 1.2 – 3.6V power supply (5V tolerant I/O) |
| 74LVT* | High-Speed, low voltage BiCMOS Technology. | High speed operation, 2.7 – 3.6V power supply (5V tolerant I/O) |
| 74ALVC | Advanced low voltage CMOS. | High speed operation, 1.2 – 3.6V power supply (5V tolerant I/O on bus hold types) |
| 74ALVT* | Advanced low voltage BiCMOS Technology. | Very high Speed operation, 2.3 – 3.6V power supply (5V tolerant I/O) |
| 74AVC | Advanced Very low voltage CMOS. | Very high speed operation, 1.2 – 3.6V power supply (3.6V tolerant I/O) |

The issues relate to the speed of operation, power supply voltage range and ability to interface to logic ICs with different voltage levels for both CMOS and BiCMOS technology devices.

## 7.3.6 Digital Cell Schematics

The following schematics, see Fig. 7.10, show the circuit schematics for representative circuit designs for input, output and bidirectional cells in CMOS logic.

# 7.4 Digital I/O Test

## 7.4.1 Introduction

Measurement of the parameters:

- Input cell voltage:     $V_{IL}, V_{IH}$
- Output cell voltage:   $V_{OL}, V_{OH}$
- Input cell current:     $I_{IL}, I_{IH}$ (also referred to as **leakage current**)
- Output cell current:   $I_{OL}, I_{OH}$

is undertaken using a source and measurement unit within the external ATE or other tester arrangement. Within the ATE, the **Precision Measurement Unit** [12] (PMU) would be used to create and measure accurate currents and voltages.

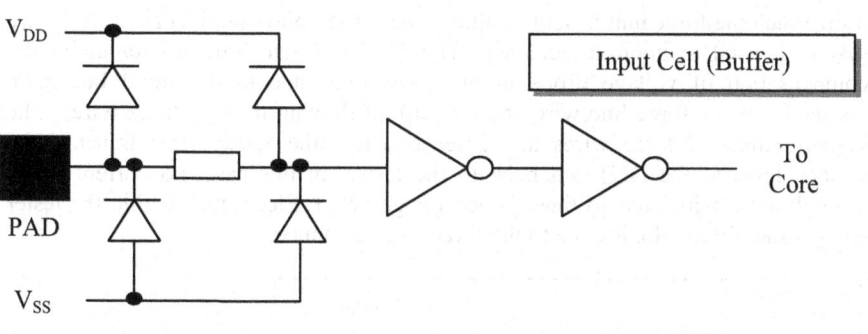

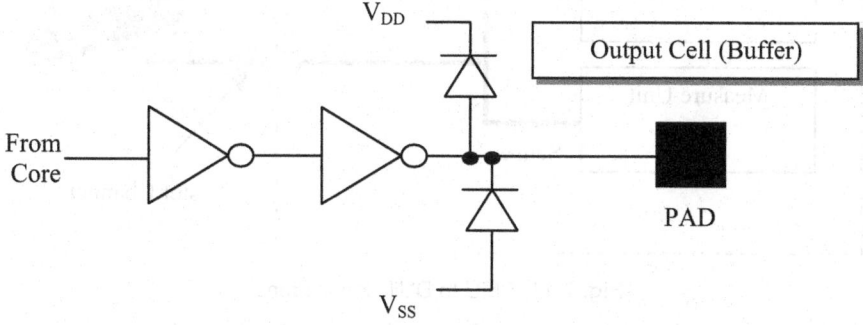

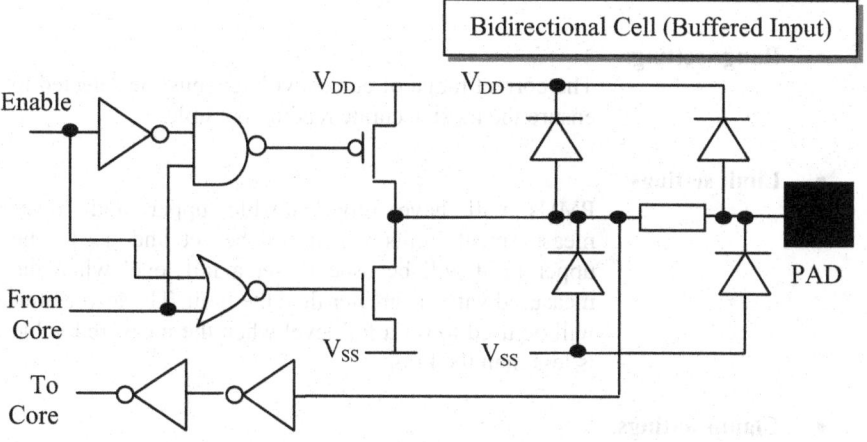

**Fig. 7.10.** Input, output and bidirectional cells in CMOS

The **PMU** is used to undertake accurate DC measurements. It can **force** (create) both currents and voltages, and **sense** (measure) currents and voltages. Testers would typically have one PMU that would be available on the various Input/Output channels available with the tester. The PMU, see Fig. 7.11, is programmable by the user. When the force unit is set to voltage, the sense unit is set to current. A 4-wire system is used to improve accuracy. The 4-wire system provides for automatic compensation of voltage drops in the force lines due to the small but finite resistance of the force line wire and the current flowing through these wires. The system utilizes 2-force wires and 2-sense wires (the sense wires transmit the voltage level at the DUT pin back to the tester. In this case, no current flows through these wires and so there is no voltage drop). Electronics within the tester compensate for any losses due to the force-line resistance.

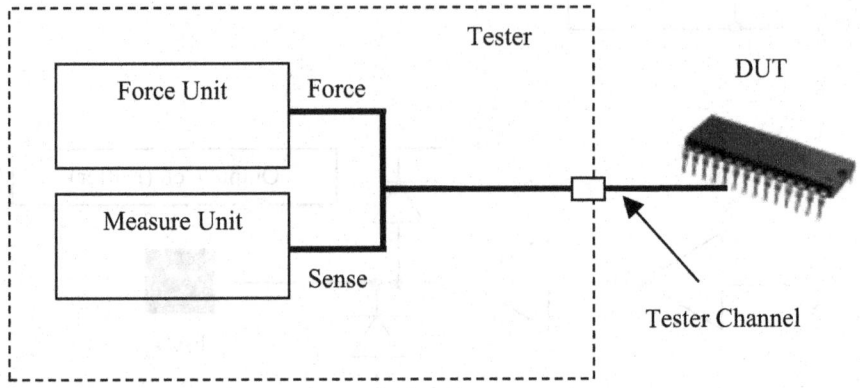

**Fig. 7.11.** PMU to DUT connections

When using this type of system, then there are a number of issues that need to be carefully addressed:

- **Range setting**:

  The correct range of current/voltage must be selected to ensure the most accurate reading possible

- **Limit settings**:

  PMUs will have programmable upper and lower measurement limits which can be set and used. The upper limit will be used to set a fail level when the measured value is greater than the limit. The lower limit will be used to set a fail level when the measured value is less than the limit

- **Clamp settings**:

  This sets the maximum value of current or voltage that is supplied by the PMU during the test. This would be required to protect the operator, the tester electronics and the device under test from excessive values

## 7.4.2 Measuring Input Cell Voltage and Current

Measurement of $V_{IH}$, $V_{IL}$, $I_{IH}$ and $I_{IL}$ can be undertaken using either static (DC) or dynamic (AC) measurements.

With the **static test**:

- To measure $I_{IH}$, the tester applies a high level voltage and the current flow into the device pin (input cell) is measured. All other device input pins are provided with a logic 0 input during the test. The measured value is then compared to the predefined current limit. A current clamp limit is to be set before undertaking the test.

- To measure $I_{IL}$, the tester applies a low level voltage and the current flow out of the device pin (input cell) is measured. All other device input pins are provided with a logic 1 input during the test. The measured value is then compared to the predefined current limit. A current clamp limit is to be set before undertaking the test.

- To measure $V_{IH}$ and $V_{IL}$, a functional test is to be run on the device with the input signal voltage levels set to the device specification levels for $V_{IH}$ and $V_{IL}$. It is then to be determined whether the device has passed or failed the test at these input signal levels.

The **leakage current** ($I_{IH}$ and $I_{IL}$) can be measured using either a **serial** or **parallel** measurement arrangement. In the **serial** measurement arrangement, each input cell leakage current is measured in turn. A value for each input pin is attained but this measurement would be time consuming, as the measurement would be repeated for each cell. In the **parallel** measurement arrangement, all input pins are connected together and a single current measurement is taken. This is faster than the serial arrangement as only one measurement is taken. However, there is no ability to diagnose the location of a fault should one exist.

## 7.4.3 Measuring Output Cell Voltage and Current

Measurement of $V_{OH}$, $V_{OL}$, $I_{OH}$ and $I_{OL}$ can be undertaken using either static (DC) or dynamic (AC) measurements.

With the **static test**:

- To measure $V_{OH}$, the device is set-up for the particular output cell to create a logic 1. All other device input pins are set to produce a logic 0 output. An $I_{OH}$ value current is applied to the device pin and the resulting voltage measured. This is compared to the predefined voltage limit for $V_{OH}$. A voltage clamp limit is to be set before undertaking the test.

- To measure $V_{OL}$, the device is setup for the particular output cell to create a logic 0. All other device input pins are set to produce a logic 1 output. An $I_{OL}$ value current is applied to the device pin and the resulting voltage measured. This is compared to the predefined voltage limit for $V_{OL}$. A voltage clamp limit is to be set before undertaking the test.

For output cells with tristate (high impedance) output, then the cell is placed in high impedance mode and the **high impedance leakage current** ($I_{OZ}$) is measured by applying a high and low voltage to the device pin, and measuring the resulting current flow. A current clamp limit is to be set before undertaking the test.

An additional current to be measured is the **output short circuit current** ($I_{OS}$). This is the current that the device supplies when the output applies a logic 1 and a zero-volt signal is applied to the pin. A current clamp limit is to be set before undertaking the test.

### 7.4.4 Dealing with Bidirectional Cells

Bidirectional cells have both input and output modes. The cell can be put into the relevant mode and treated as either an input or an output cell.

### 7.4.5 Dealing with Internal Pull-Ups and Pull-Downs

Some I/O cells with have integral pull-up or pull-down devices. These can be thought of as resistor structures connected to either $V_{DD}$ (pull-up), or $V_{SS}$ (pull-down) and will set the relevant input to either logic 1 or logic 0 when no external drive signal is applied. These are in many cases implemented using diode connected MOSFETs. If the cell is driven by a logic value, then dependent on the internal device structure and applied logic value, a current may flow through the device. It would be important not to confuse this with a small leakage current that would be tested for.

## 7.5 Analogue I/O Structures

For analogue I/O, then the requirement is to allow for analogue signals (current/voltage) to pass through the I/O cell with limited interference and known operation, whilst the I/O cell provides for EOS/ESD protection. A basic analogue I/O cell is shown in Fig. 7.12

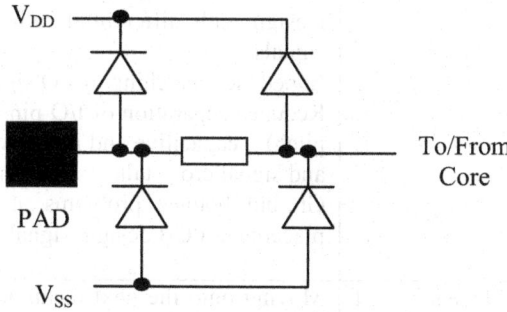

**Fig. 7.12.** Basic analogue I/O cell

## 7.6 Analogue I/O Test

Specific testing of analogue I/O is not discussed explicitly here, but is considered in the testing of the analogue core circuitry, and determination of the correct functionality of the core circuitry.

## 7.7 Future Directions

The future advances in the IC I/O test are based on the demands of the end-user. A summary is provided in Table 7.6.

**Table 7.6:** Future I/O test issues

| Technology driver | Description |
|---|---|
| Design styles | The use of multiple power supply voltage rails on a single IC<br>The development of electronic systems using ICs with varying power supply voltage operation<br>Increased data I/O rates<br>Move away from parallel transmission on and off-chip of data to high-speed serial bus architectures |
| Faster operating speeds | Higher signal frequencies<br>Driven by communications applications<br>Increased data I/O rates |

| | |
|---|---|
| | Move away from parallel transmission on and off-chip of data to high-speed serial bus architectures<br>Design with differential rather than single ended signals<br>Impedance matching of I/O signals<br>Reduced separation of I/O pins on the package (pin pitch) – capacitive and inductive coupling problems and signal cross-talk – signal integrity issues<br>Ground bounce problems at package pin – pin placement, PCB design, signal integrity issues |
| Higher levels of integration (higher density) | Moving onto the next technology node to improve performance of components and reduce interconnect delays<br>Newer failure mechanisms -- limitations of existing fault models and need for models more representative of process defects<br>New test issues emerging for SoC and SiP devices |
| Lower operating voltages | A need to improve device reliability by reducing electric field strength in dielectrics<br>Aim to lower device power consumption<br>Portable, battery operated circuits<br>Need for lower voltage IC to be tolerant to higher voltage IC operation<br>Higher operating currents and need for additional power supply pins and power supply decoupling |
| Test equipment and costs | Need for reduced test time and test costs<br>Need for reduced external ATE requirements and costs – simpler and less expensive ATE<br>More pins per device |

## 7.8 Summary

This chapter has discussed the structures of typical input-output cells and the technologies used for their implementation. From the early TTL (+5V) logic levels, the role of CMOS in the creation of I/O cells for operation at voltages down to and beyond +1V operation is key for the continued move for lower power supply voltage operation. The problem is added to by the need for higher operating frequencies and the development of electronic systems using ICs with varying power supply voltages. This chapter has concentrated on digital I/O cell design and test issues.

# 7.9 References

[1]     Kang S. and Leblebici Y., "CMOS Digital Integrated Circuits Analysis and Design", McGraw-Hill International Editions, Singapore, 1996, ISBN 0-07-114423-4

[2]     Bellaouar A. and Elmasry M., "Low-Power Digital VLSI Design Circuits and Systems", Kluwer Academic Publishers, The Netherlands, 1995, ISBN 0-7923-9587-5

[3]     Haskard M.R. and May I.C., "Analog VLSI Design nMOS and CMOS", Prentice Hall Silicon Systems Engineering Series, Australia, 1988, ISBN 0-7248-0027-1

[4]     Tocci R.J., Widmer N.S. and Moss G.LK., "Digital Systems 9th Edition", Pearson Education International, USA, 2004, ISBN 0-13-121931-6

[5]     Diaz C.H., Kang S. and Duvvury C., "Modeling of Electrical Overstress in Integrated Circuits", Kluwer Academic Publishers, 1994, ISBN 792395050

[6]     MIL-STD-833 Method 3015.7, "Electrostatic Discharge Sensitivity Classification"

[7]     ESD-STM5.1-1998, "ESD Association Standard Test Method for Electrostatic Discharge Sensitivity Testing: Human Body Model (HBM) – Component Level", ESD Association, 1998

[8]     EIA/JEDEC Test Method A114-A, "Electrostatic Discharge (ESD) Sensitivity Testing Human Body Model (HBM)", Electronic Industries Association, 1997

[9]     Diaz C., Kang S.M. and Duvvury C., "Electrical overstress and electrostatic discharge", IEEE Transactions on Reliability, Vol. 44, Issue 1, March 1995, pp2-5

[10]    Sadiku M.N.O. and Akujuobi C.M., "Electrostatic discharge (ESD)", IEEE Potentials, Vol. 22, Issue 5., December 2003 – January 2004, pp39-41

[11]    Sanjay Dabral S. and Maloney T., "Basic ESD and I/O Design", Wiley, 1999, ISBN 0-471-25359-6

[12]   Sunter S. and Nadeau-Dostie B., "Complete, Contactless I/O Testing – Reaching the Boundary in Minimizing Digital IC Testing Cost", Proceedings of the International Test Conference, 2002, pp446-455

[13]   Xilinx Inc., USA, http;//www.xilinx.com

[14]   "Dual Positive Edge Triggered D-Type Flip-Flops with Clear and Preset", datasheet, Texas Instruments Inc., USA

[15]   "Logic Selection Guide", Texas Instruments Inc., USA

# Chapter 8

# Design for Testability – Structured Test Approaches

*Over the last few years, the evolution of test engineering has brought it closer to design, bridging the "traditional" gap between design and test. Here, where once the design and test activities were separate and distant, the gap has been bridged by developing and adopting a unified "Design for Testability" (DfT) approach. For digital circuits and systems, DfT techniques have been embraced and are well supported. For analogue and mixed-signal circuits, the problems encountered due to the application specific nature of the circuits themselves has been a limiting factor in the widespread recognition and adoption of standardised analogue and mixed-signal DfT techniques.*

## 8.1 Introduction

It is widely accepted that the testing of Integrated Circuits (ICs) needs to be considered early in the design development process, due to increasing circuit complexities and greater demands on the design specification. This, coupled with the move towards lower device geometries, moving along the process technology nodes as identified in the International Technology Roadmap for Semiconductors (ITRS) [1-2], is making the ability to monitor internal nodes (the controllability and observability [3] of these nodes) increasingly difficult. In production test, the problem is added to by the need for faster and more complex ATE systems [4-5] in order to keep pace with new device designs. ATE systems with high speed, precision, memory and performance are increasingly expensive to purchase and maintain. Reducing ATE test times, applying simpler and faster ATE based tests and using less expensive test equipment are moves to reduce potentially the cost of test. This can be supported by moving some or all of the workload away from the ATE and placing self-test circuitry within the Device Under Test (DUT). In order to address the test problem, a Design for Testability (DfT) [6-8] approach is taken to addressing the controllability and observability problems associated with IC designs. Here, a suitable test strategy is developed during the design development stage and testability built into the design as and where required. Building in testability into the design can be achieved in both hardware and software (software for use in digital and mixed-signal processor based designs). Addressing the testability problem within the design is considered in two ways:

- **Design for Testability (DfT)**:

    Adding specific circuitry to access internal nodes for addressing the controllability and observability problem. This allows for enhanced external test access to the internal nodes via the IC package pins.

- **Built-In Self-Test (BIST)**:

    The BIST approach uses on-chip signal generation, results capture and analysis in order to provide local (on-chip) self-test capabilities. An external tester is used to initiate a self-test and to capture the self-test results. Although BIST is part of a DfT approach, it is usually considered as a subject in its own right, considered alongside DfT. This is a recognition of the size of the subject area. In this case, the subject of DfT is more concerned with providing test access to the device than providing a self-test mechanism.

DfT and BIST for digital circuits is commonplace and are well supported with software CAT (Computed Aided Test) tools for the development, insertion of, and test pattern generation for DfT and BIST structures. Software tool support links the test activities to the design activities and provides for automation of specific tasks – automation being a key to dealing with the problem of design complexity and the need to perform tasks in a minimal time and cost. Automation eases the burden on the design and test engineers, and many of the repetitive are dealt with "in the background", so freeing the engineer to undertake the high-level test strategy work.

## 8.2 Observability and Controllability

In order for a design to be tested, there is a requirement to ensure that the design is both controllable and observable:

- **Controllability**:

    The ability to control specific parts of a design in order to set particular values at specific points within the design

- **Observability**:

    The ability to observe the response of a circuit to a particular circuit stimulus

In a digital logic design, this would be a particular logic value. In an analogue circuit, this would be a particular voltage or current. However, this has to be achieved economically and with minimal impact on the design performance in normal operating mode. The economic issues would relate to the effort required to achieve the controllability and observability, along with the size of the circuit structures required – if the test access circuitry was comparable in size with, or even larger than the circuit under test, would this be considered economical? It would really depend on the design.

# 8.3 Digital DfT

## 8.3.1 Design Partitioning

If a large and complex design was to be analysed in full detail at the start of a test development process, then the complexity of the problem would quickly become unmanageable. In order to make the task more manageable, a design would be partitioned into specific blocks which may be tested as separate entities initially and which would then be tested together at a final test stage. This **divide and conquer** approach may be considered for circuit and systems ranging from circuits with a few logic gates to complex digital systems. The partition strategy would be considered based on the design architecture and would be made on a case by case basis. However, a starting point would be the circuit/system functional block diagram. This allows for the design partitioning for test purposes to be made at an early stage in the design process, prior to any detailed design work being undertaken, and any problematic parts for test identified at this early stage. For the more complex system designs, the design partitioning for test development would be made at two levels:

- **System Level**:

    This addresses the test access problem at a high level, typically the system block diagram and prior to any circuit level design. This level addresses the high-level test strategy issues.

- **Circuit Level**:

    This addresses the test access problem at a low-level, at the logic gate level and would be undertaken during the circuit design process. This level addresses logic gate count (test circuit) and timing (impact on the design performance and speed of test) issues.

## 8.3.2 Scan Path Test

Scan path testing is the main method to provide access for internal node controllability and observability of sequential logic circuits, that is, circuits that include bistable elements implementing counter and finite state machine (FSM) designs. In this method, the circuit is designed to allow for two modes of operation:

- **Normal Operating Mode**:
  The circuit is running as per its required end-user function.

- **Scan Test Mode**:
  The circuit is operating in a scan test mode, in which logic values are serially clocked into circuit bistable elements from an external signal source, and the results serially clocked out for external monitoring.

The incorporation of a scan path into a design will require the inclusion of inputs specifically used for the test procedure. These inputs, and the scan test circuitry, would not be used by the end user and so are transparent to the user. A single scan path within an IC will require the following additional I/O (primary I/O will be pins on the package):

**Primary Inputs**:
- **Scan Data Input** (SDI) – the data to clock serially into the circuit
- **Scan Enable** (SE) – enables the scan path mode

**Primary Output**:
- **Scan Data Out** (SDO) – the data (results) that is serially clocked out of the scan path for external monitoring

In practice, the scan path I/O can be dedicated pins on the package, or may be multiplexed pins in which specific pins will have different roles, dependent on the mode the device is in. The scan path is based on the use of the D-Type bistable and the serial-input, serial-output (SISO) shift register, see Fig. 8.1

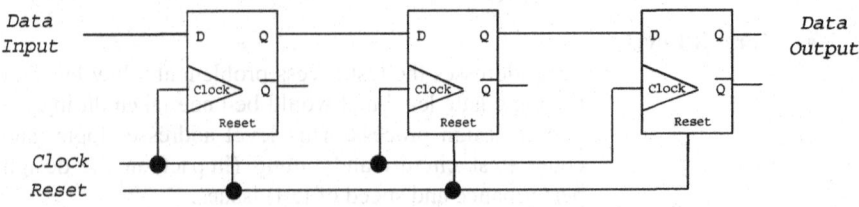

**Fig. 8.1.** SISO shift register

In this view, there are three D-Type bistables with the Q-output of one bistable connected to the D-input of the following bistable. With suitable application of input signals, the data input and clock signal nodes, a serial data stream at a primary input will eventually be monitored at the data output node, see Fig. 8.2. The **Clock** and **Reset** inputs are common to all bistables. With the three bistables in the circuit, the value placed on the data input node will be seen at the data output node after three clock cycles.

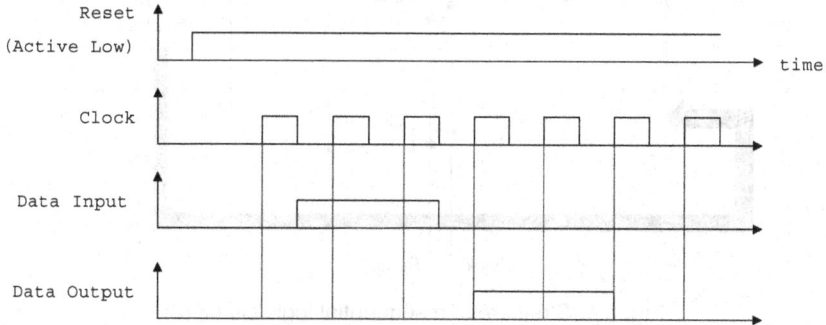

**Fig. 8.2.** Serial data stream

With the SISO shift register in mind, this will be fitted into a sequential logic circuit. With reference to the structure of a sequential logic circuit, see figure 8.3, the principle of operation can be identified. In this view, the design is separated into the sequential element block (the bistables) and two combinational logic blocks – one at the input and one at the output.

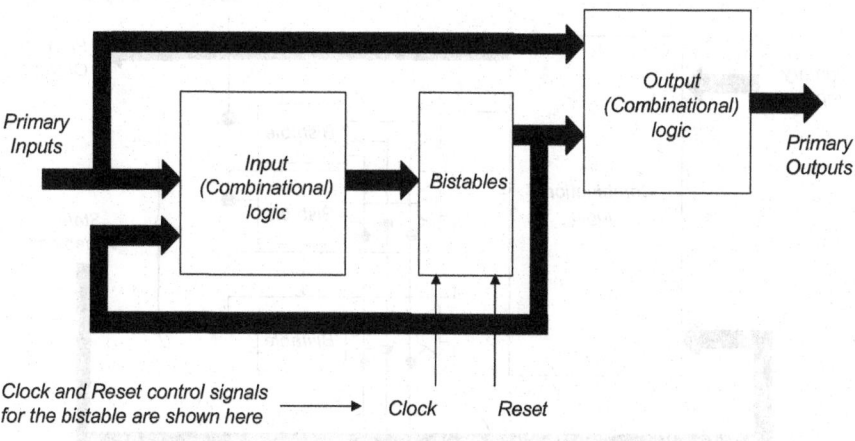

**Fig. 8.3.** Structure of a sequential logic circuit (1)

With a modification to Fig. 8.3, the design may be viewed as shown in Fig. 8.4. In this view, the combinational logic blocks have been combined into a single block of circuitry and each bistable is shown individually.

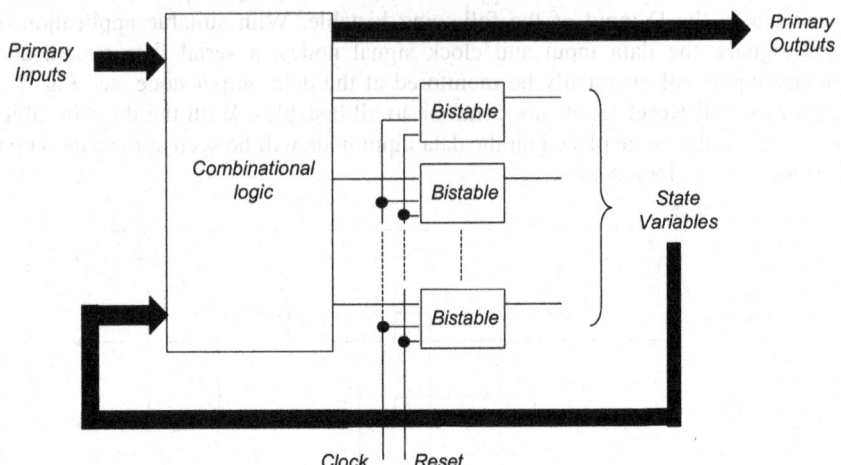

**Fig. 8.4.** Structure of a sequential logic circuit (2)

Now, by modification of the SISO shift register to allow for serial-input, serial output operation and also for parallel input (to the D-inputs) and parallel output (from the Q-outputs), this can then be integrated into the sequential logic circuit of Fig. 8.4. The resulting circuit is shown in Fig. 8.5 where the D-inputs to the bistables are multiplexed between the output of the combinational logic block and the Q-output of the previous bistable

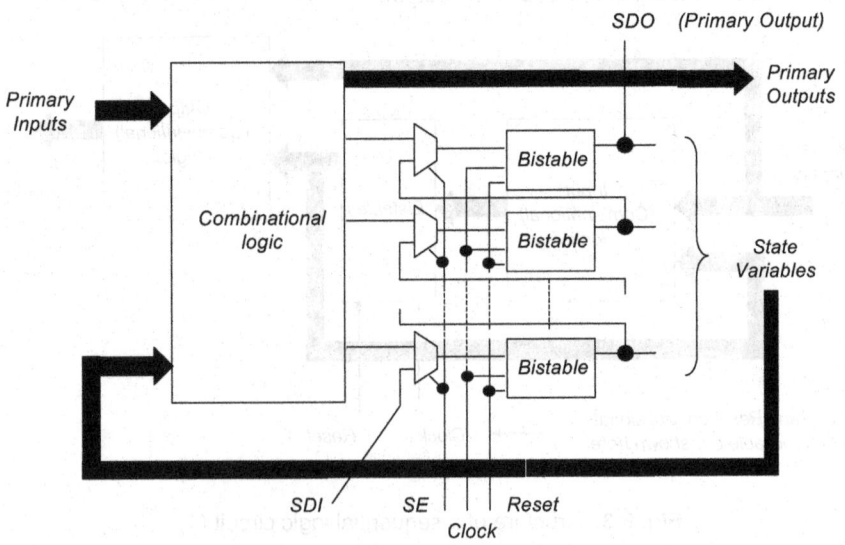

**Fig. 8.5.** Scan path insertion

An example operation of this scan path would be:

- Put the circuit into scan test mode (by control of the **SE** – scan enable input pin). Serially scan in a sequence of logic values to set the bistable Q-outputs to known values (*i.e.* put the circuit into a known, initial state) by applying data to the **SDI** (scan data input) pin.

- Put the circuit into its normal operating mode and operate the circuit for a set number of clock cycles (a simple approach would be to clock the circuit once into its next state).

- Put the circuit back into scan test mode. Serially scan out the values stored on the Q-outputs of the bistables, and monitor the **SDO** (scan data output) pin.

- Compare the received values with the expected values.

The adoption of the scan path techniques is widespread. However, there are a number of issues that would need to be addressed in order to effectively utilise the scan path:

- The amount of scan path circuitry to include. This would range from **no-scan** (for simple circuits where test access is not an issue or cannot be included for functionality or cost reasons) through **partial-scan** (some of the bistables in the design are included within the scan path) to **full-scan** (all bistables are included within a scan path).

- The number of scan paths that would be incorporated within the design. There can be a single scan path or multiple scan paths within the design, and the decision would be based on the length of the resulting scan paths (longer scan paths will require more clock pulses to get data through), the availability of primary I/O to support multiple scan paths, and the ability to control and observe multiple scan paths through the available ATE system.

- Speed of operation of the scan path. With delays within the circuit due to a combination of logic gate and interconnect delays, it may not be possible to operate the scan path at the device maximum clock frequency (at-speed). However, at-speed scan testing is increasing in importance [9].

- The ability to generate suitable test patterns through ATPG with one or more scan path designs.

- Clock domains -- simple devices may incorporate a single clock which is common to all bistables in the design. This would allow for direct operation of the scan path. In more complex designs, there would be multiple clock domains, with different parts of the design operating on different clock signals and also there may be gated clocks (the clock signal is passed through a logic gate before application to the bistable). In these more complex scenarios, the clock signals for the bistables would need to be designed so that in normal operating mode, the required clock functionality is provided, whilst in scan-test mode, a common clock is applied to all bistables within a particular scan path.

The diagram in Fig. 8.5 uses multiplexer and D-Type bistable circuit elements. These would normally be combined into a single **Scan D-Type bistable** circuit element, see Fig. 8.6. This would have the same logic functionality as a discrete bistable and multiplexer arrangement, but would be optimised for size and speed of operation. This would have two data inputs (D – normal data and SD – scan data), a scan enable (SE) control input to select between normal and scan test modes, in addition to the clock and reset (and/or set) inputs and Q/Q outputs.

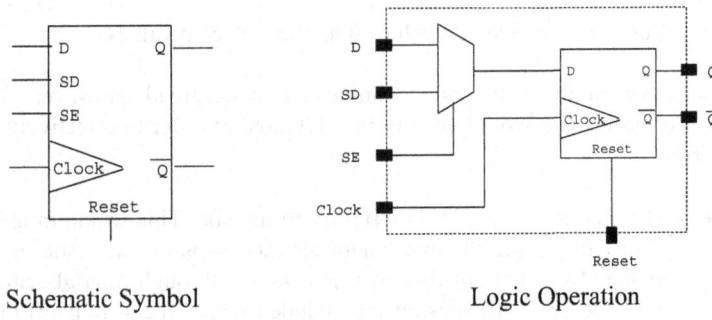

Schematic Symbol                    Logic Operation

**Fig. 8.6.** Scan D-type bistable

## 8.3.3 Built-In Self-Test (BIST)

The BIST approach uses on-chip signal generation and analysis, see Fig. 8.7, in order to provide local (on-chip) tester resources that would otherwise be required by external Automatic Test Equipment (ATE). Here, the circuit under test (CUT) is provided with either the normal system inputs, or test inputs generated on-chip. The circuit under test output is then monitored by an on-chip results analyser. Either a binary pass/fail signal or a test signature (identifying the cause(s)) of the failure) would be produced.

In digital logic, the test signature would be a sequence of 0s and 1s. The value of the code that is monitored would (ideally) be different between a fault-free and faulty circuit, and the faulty circuit output code value would be dependent on the particular fault. It may be possible to have a unique code for each fault, or a code may define a range of faults and it would be up to the user to investigate the failure in more detail and identify the particular fault. In the digital circuit, a simple results analyser would logically compare the output of the CUT with the ideal response generated by the signal generator.

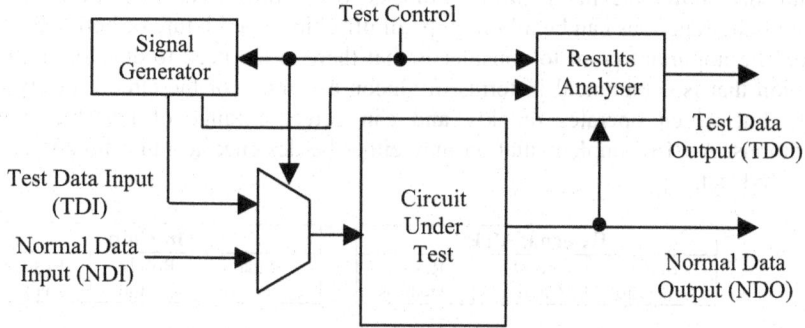

**Fig. 8.7.** BIST operation

In testing an IC without BIST, see Fig. 8.8, the external tester undertakes all of the signal generation, data capture and analysis operations.

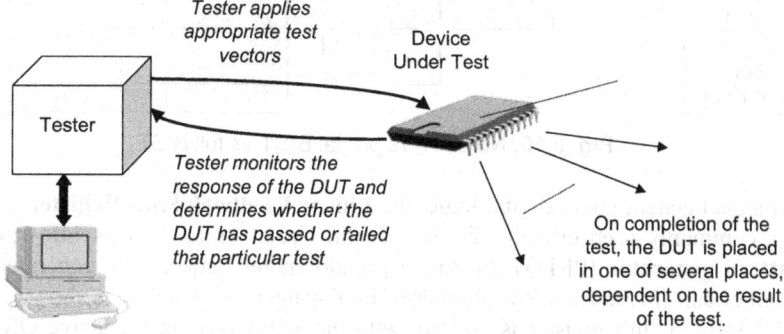

**Fig. 8.8.** Testing a Design without BIST

In testing a design with BIST, see Fig. 8.9, the role of the tester is then placed within the DUT. The external tester would then be used to initiate the self-test function and to be used to test the parts of the design not covered by the BIST.

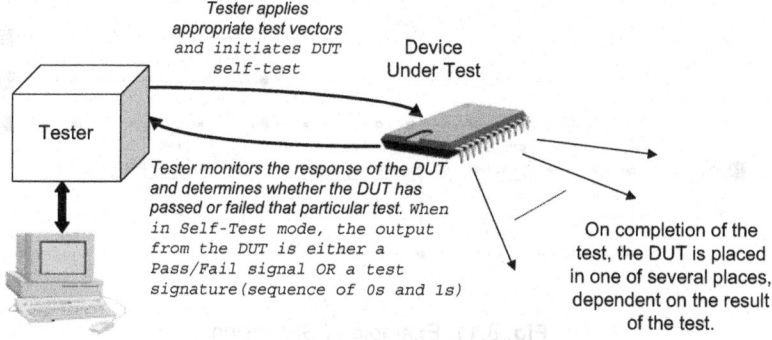

**Fig. 8.9.** Testing a design with BIST

Figure 8.7 shows the main building blocks to creating a self-test operation. In general, the functions can be all on-chip, all off-chip, or a mixture of both, see Fig. 8.10. The important point to consider is that there is the need to develop a BIST solution that is economical in terms of silicon area (size of the circuit and hence cost), but which operates quickly and can detect a range of possible faulty conditions. A BIST implementation may either be designed to run a functional or structural test.

| | External ATE | | | On-Chip | | |
|---|---|---|---|---|---|---|
| | Signal Generation | Results Capture | Results Analysis | Signal Generation | Results Capture | Results Analysis |
| No BIST | ▓ | ▓ | ▓ | | | |
| Partial BIST | ▓ | ▓ | | | | ▓ |
| Partial BIST | ▓ | ▓ | | | ▓ | ▓ |
| Partial BIST | ▓ | ▓ | ▓ | ▓ | | |
| Partial BIST | | | ▓ | ▓ | ▓ | |
| Full BIST | | | | ▓ | ▓ | ▓ |

**Fig. 8.10.** No BIST vs partial BIST vs full BIST

For signal generation in digital logic, the **Linear Feedback Shift Register** (LFSR) is a commonly used circuit. This is a useful circuit to act as a **pseudo-random pattern generator** (PRPG). In this, a pseudorandom sequence is generated in a small circuit (*i.e.* using a low number of logic gates). An example is shown in Fig. 8.11. Here, a shift register is created, with the addition of an Exclusive OR (EX-OR) gate as a feedback element. In general, the position of the feedback element (and number of elements) can be determined from the LFSR requirements.

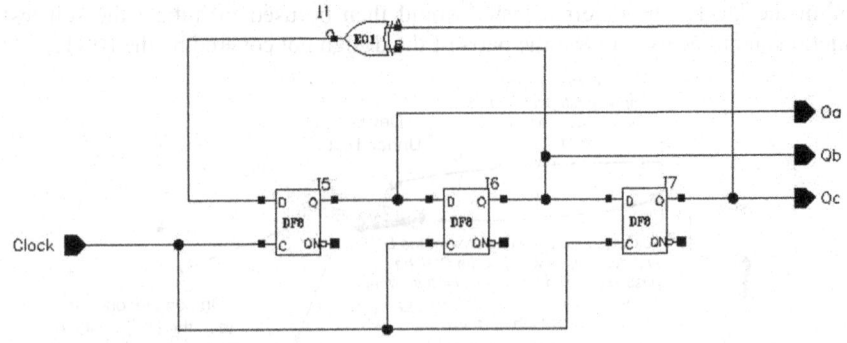

**Fig. 8.11.** Example LFSR design

When analysing the operation of the LFSR, the output code generated will on first inspection appears to be random. However, it does repeat itself after a set number of clock cycles, hence the output is actually pseudorandom. For an n-bit LFSR (where n is the number of bistables used), the LFSR will produce a maximum of $(2^n-1)$ codes – this is a **maximal length sequence**. There will be one forbidden state (all 0s code) as if the LFSR was to initialise or enter into this state, it would remain there indefinitely. If required, the LFSR can be designed to have a number of states less than the maximal length sequence.

The LFSR can also be modified to implement a **signature analyser**. By incorporating additional inputs (the outputs from the CUT) and applying these signals with feedback terms via EX-OR gates within the LFSR circuit, a **Single-Input Signature Register** (SISR) where the CUT has one output, or a **Multiple-Input Signature Register** (MISR) where the CUT has multiple outputs, can be generated. Figure 8.12 shows the principle of operation, when the CUT is a combinational logic circuit.

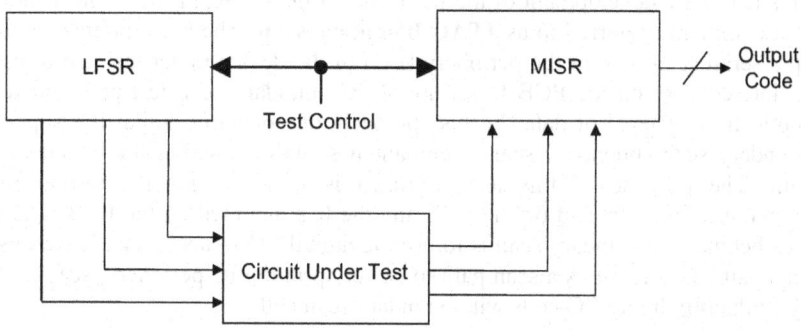

**Fig. 8.12.** BIST implementation using an LFSR and MISR

Once the SISR or MISR has been clocked a number of times, then the output code is monitored. This output code will identify whether the circuit is considered to be fault-free or faulty (and ideally, the location of the fault). There is however a potential problem with this approach. It is assumed that the signature from a faulty circuit will be different from a fault-free circuit. There is a small probability that the signatures will be the same and so a fault will not be detected. This is referred to as **aliasing** or **error masking**. The LFSR, SISR and MISR circuits also require additional logic within the IC in order to implement the self-test signal generation and results analysis. This is a circuit overhead that needs to be minimised if the costs are to be kept as low as possible.

The **Built-In Logic-Block Observer** (BILBO) [10] is also utilised, and is a circuit that implements the required self-test operations, but which uses the same circuitry in normal and self-test modes. The circuit therefore has dual purpose, in a similar way as the scan path uses the same D-Type bistables in normal and scan test modes.

## 8.3.4 1149.1 Boundary Scan

Scan path testing was developed to allow for test access to the core of the IC via the circuit bistable elements, allowing for data to be serially clocked into, and out of, the device under test via the circuit primary I/O (*e.g.* the device package pins). This allows for the device to be effectively tested during production and away from its final application. With the increase in the complexity of digital ICs, the reduction in package dimensions, and the increased complexity of the printed circuit board (PCB) designs that the ICs are placed on, the ability to test the ICs at the PCB level was identified as becoming increasingly difficult to achieve. The traditional way in which to test the populated PCB, for example by using **flying probe** and **in-circuit test** (ICT) equipment that physically probe the PCB in order to measure electrical values at predefined locations on the PCB, was identified as becoming difficult to implement due to the reduced package sizes and higher packing densities of the ICs on the PCB. In order to address the problem, the **Joint Test Action Group** (JTAG) was established to identify a solution to this problem. This led to the development of the IEEE Std 1149.1-1990[11]-15]. The standard is also commonly referred to as **JTAG boundary scan**. The basic premise is to set-up a serial scan path in the periphery (I/O) of the device under test and to use the IC interconnect on the PCB to set-up the IC boundary scan test path and for the application/retrieval of data. In this approach, one or more digital ICs supporting boundary scan (boundary scan compliant ICs) will be within the boundary scan path. The purpose of this test approach is to allow for the testing of the interconnections (tracks) on the PCB and the ICs mounted on the PCB. The basic idea behind the boundary scan is to use the digital I/O to access the IC via a serial scan path. This requires a scan path to be set up in the IC periphery, see Fig. 8.13, by **replacing** digital I/O cells with boundary scan cells.

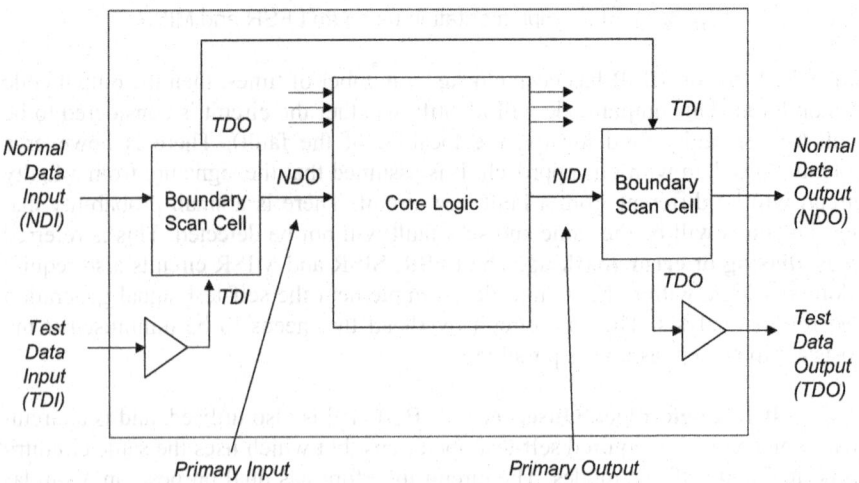

**Fig. 8.13.** Use of boundary scan cells

With this arrangement, the IC will have the normal data I/O, but also the ability to clock-in and clock-out test data serially. The ICs can be connected in a serial chain once mounted on the target application PCB, see Fig. 8.14. The selection of the mode of operation for each IC is arranged through the use of a **Test Access Port** (TAP) placed within the IC. In Fig. 8.14, there is one boundary scan path on the PCB, with the following connections on each IC:

- **TDI**   Test Data Input:
        This input is used for providing the IC with both data and instructions

- **TCK**   Test Clock:
        This input is the test clock

- **TDO**   Test Data Output:
        This output is used for providing with both data and instruction outputs from the IC

- **TMS**   Test Mode Select:
        This input, together with the TCK input, is used to control the TAP controller, a 16-state finite state machine used as the controller of the boundary scan circuitry on the IC

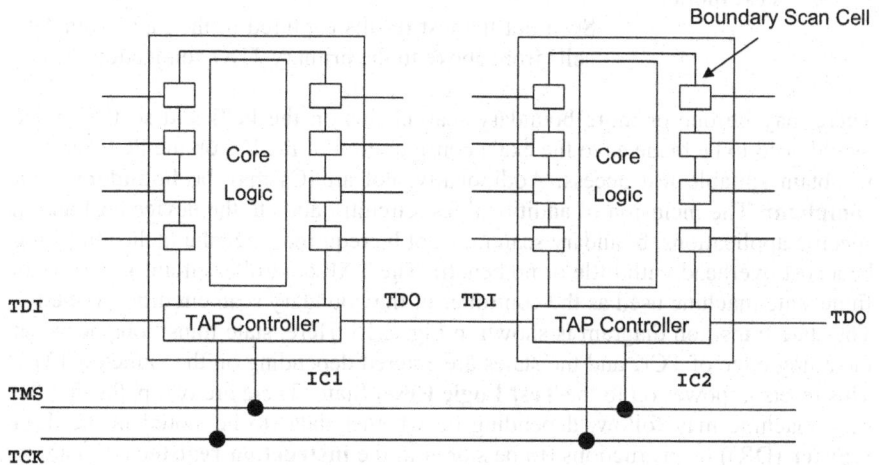

**Fig. 8.14.** Two ICs on a PCB

An optional test reset (TRST) input can also be added to the circuit. The device is put into either a normal operating mode, or scan test mode. In these two modes, there will be a seven possible operations for the boundary scan for IC and PCB interconnect testing:

- **Normal operating mode:**

  The boundary scan cell is transparent – the circuit operates in its normal operating mode.

- **Test mode:**

  Test data (on TDI) is serially scanned in to each input boundary cell (as in an internal scan path) and the data are used as inputs to the core logic of the IC.

- **Test mode:**

  The test data that has been scanned in are used as inputs to operate the core logic, and the outputs from the core logic stored in the output boundary cells.

- **Test mode:**

  The test results are serially scanned out from the IC through TDO.

- **Test mode:**

  The operation here is to scan in test data serially and scan out values to act as test inputs to the following ICs (produce a scan path for all ICs within a particular boundary scan path).

- **Test mode:**

  Use the test data from above in order to control surrounding interconnect and devices and store results using the inputs of specific ICs.

- **Test mode:**

  Scan out the test results captured in the input boundary cells from above to the primary TDO connection.

There may be one or more boundary scan chains on the PCB and so the choice would need to be made as to the exact configuration of the ICs on the PCB in order to obtain suitable test access. Additionally, not all ICs may be **boundary scan compliant**. The inclusion of additional test circuitry adds to the device cost and in specific applications, boundary scan may not be required since the inclusion would be a cost overhead with little or no benefit. The TAP controller circuit is a 16-state finite state machine used as the controller of the boundary scan circuitry on the IC. The state transition diagram is shown in Fig. 8.15. Here, state transitions occur on the rising edge of TCK and the states are entered depending on the value on TMS. This resets at power on to the **Test Logic Reset** State. There are two paths that the state machine may follow, depending on whether data (to be stored in the **data register** (DR)) or instructions (to be stored in the **instruction register** (IR)) are to be loaded into the IC. The 1149.1 standard is provided alongside the following other standards:

- IEEE Std 1149.4-1999   IEEE standard for a mixed-signal test bus
- IEEE Std 1149.5-1995   IEEE standard for Module Test and Maintenance Bus (MTM-Bus) protocol

- IEEE Std 1149.6-2003     Augment to IEEE Std 1149.1: improving the ability for testing differential and/or ac-coupled interconnections between ICs on circuit boards and systems
- IEEE Std 1532-2002     IEEE Standard for In-System Configuration of Programmable Devices

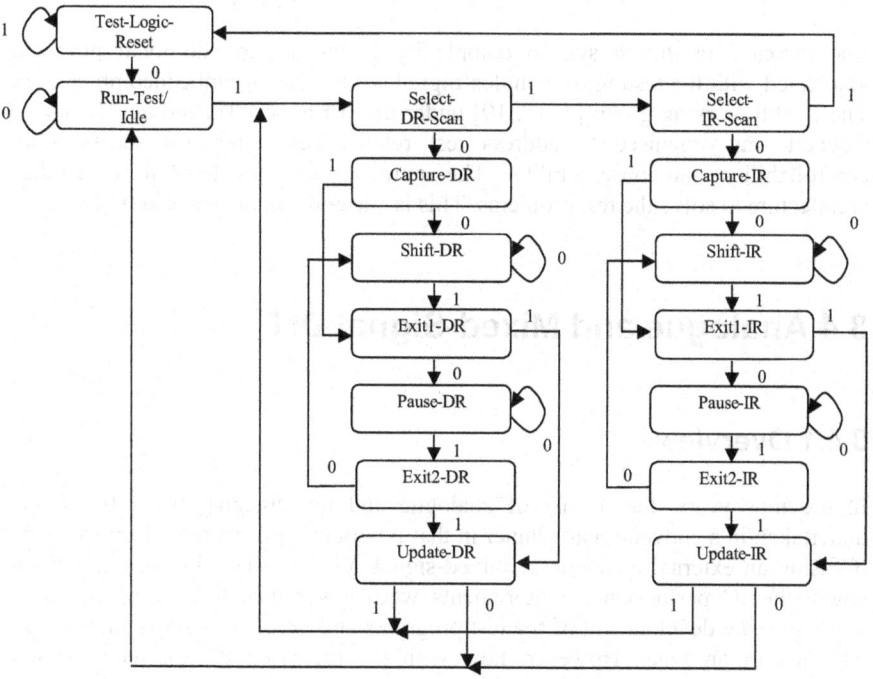

**Fig. 8.15.** TAP controller state transition diagram

The standard provides for a number of registers in the design that have not been described here. For example. the standard provides for **bypass** and **identification registers**. Today, boundary scan is used in the following scenarios:

- Testing of digital ICs in the final application (on the populated PCB)

- Testing of the IC itself away from the final application (*e.g.* during production test)

- Programming of programmable logic devices (PLDs) – CPLDs (Complex PLDs) and FPGAs (Field Programmable Gate Arrays). This is supported through the IEEE Std 1532-2002 [16]

The above discussion introduced the hardware aspects of the standard. In order to use this however, there will be the need for software support. 1149.1 is supported with the use of the **Boundary Scan Description Language** (BSDL) [17]. This is a subset of VHDL. The full boundary scan operation is described in the IEEE Std 1149.2001 document.

### 8.3.5 P1500 Core Test Standard Development

The increase in digital system complexity is leading to additional problems associated with the testing of complex digital cores, such as embedded processors. The P1500 working group [4, 18, 19] under the IEEE Test Technology Technical Council was organised to address test related issues for test access, node controllability and observability. The purpose was to develop a standard architecture to solve the test problems. This is currently an on-going activity.

# 8.4 Analogue and Mixed-Signal DfT

## 8.4.1 Overview

In previous years, the testing of analogue and mixed-signal ICs [20-25] was undertaken in a conventional manner in that functional tests were performed on the IC using an external analogue or mixed-signal ATE system. This was acceptable where the IC performance requirements were lower than today, and the costs relating to the development of the test programs and the ATE systems themselves, was less of an issue. However, now with the increased IC test requirements, relating to improvements in the device performance and the integration of analogue and mixed-signal functionality into System on a Chip (SoC) designs, the test problem is increasing. DfT and BIST for digital circuits is commonplace and are well supported with software tools for the development, insertion of, and test pattern generation for DfT and BIST structures. Whilst the uptake of analogue and mixed-signal DfT techniques have met with less success, the basic idea is the same as for digital circuits in the generation (on-chip) of test signals, the control of the circuit between normal and one or more test modes, the capture of test results and analysis.

Much of the reticence encountered by the design community in adopting DfT lies with circuit performance degradation and cost when the additional test circuitry is included. In the design, essentially, the placement of any additional circuitry in the path of signals on-chip will degrade signal quality. This may or may not be acceptable, depending on the circuit final application requirements. Today, there is still a lack of standardisation in DfT and BIST for analogue and mixed-signal ICs, and solutions will tend to be ad-hoc, on a device-by-device basis. Realistically, the main areas of research have been limited to the main mixed-signal building blocks such as data converters (A/D and D/A) and phase locked loops (PLLs) [26, 27]. However, if a complete BIST solution for an IC is possible, then the benefits would be that an external tester would then only need to provide a small number of specialised signals such as high-speed clock signals, test-control signals and power. In analogue and mixed-signal designs, the problems associated with DfT and BIST are:

- **Stimulus generation**:
  Required to generate the necessary waveforms to extract circuit performance information. Typical waveforms include DC, ramp, triangular, sawtooth and sine (single tone and multi-tone). For example, signal generation based on sigma-delta ($\Sigma\Delta$) techniques [28-33] have been well reported. Here, a pulse density modulated (PDM) signal is generated using digital circuitry (immune to process variations) and when this PDM signal is passed through a low-pass analogue filter to remove high-frequency noise, high quality signals can be produced on- or off-chip. This principle uses oversampling and noise shaping techniques in the signal generation process.

- **Results access and monitoring**:
  Required to access the circuit output for analysis.

- **Results analysis**:
  Required to extract circuit performance information. A set of circuit performance criterion is required to achieve this.

- **System level test development**:
  Required to ensure that the above test aspects complement the overall system test requirements.

A number of problems occur with attempting to build-in DfT into the types of circuits encountered. This complicates the ability to develop and adopt a standardised approach for the board range of circuits encountered due to the application specific nature of these circuits. An example of an early implementation of a standardised DfT approach developed was based on the **Hybrid Built-In Self-Test** for Analogue/Digital Integrated Circuits (HBIST) [34, 35]. Recently, the **sigma-delta** ($\Sigma\Delta$) approach to signal generation has been developed as it lends itself well to mixed-signal circuits in that it uses digital circuitry to generate the PDM signal representing an analogue signal. The original analogue signal (required to be generated in the application) would have been initially encoded from the continuous time signal into a PDM bitstream. This can then be stored (*e.g.* in memory) and provided as a stimulus to a circuit under test. By passing the bitstream through an analogue low-pass filter, the original signal is restored, see Fig. 8.16. Enabling test access will require the ability to switch in and out analogue signals (voltages / currents) – both the normal circuit signals and the test signals. In an ideal case, the switch will have infinite impedance when open and zero impedance when closed. On-chip switches using transistors (*e.g.* an analogue switch in CMOS) will have non-ideal (*e.g.* a non-linear operation and frequency dependence) electrical performance that must be accounted for. This may limit the ability to switch signals in high-performance circuits.

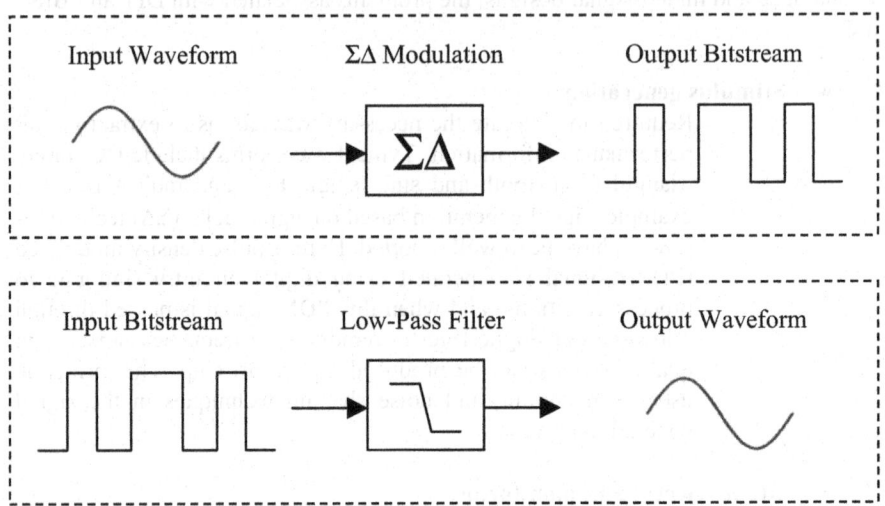

**Fig. 8.16.** Sigma-delta encoding (*top*) and signal recovery (*bottom*)

An alternative to making the decision to creating an on-chip stimulus generator is based on **oscillation based test** (OBT) [36, 37]. In this method, reported for both continuous-time and discrete-time circuits such as filters and A/D converters, the circuit is configured to produce an oscillation at the output. Here, in normal operating mode, the circuit under test transfer function is determined and the poles and zeros of the system identified. In test mode, the normal input is disconnected and the output of the circuit is fed back to the input via a feedback circuit, see Fig. 8.17. The feedback circuit is used to modify the overall system transfer function so that oscillation is produced. This oscillation can then be monitored, and suitable circuit parameters extracted.

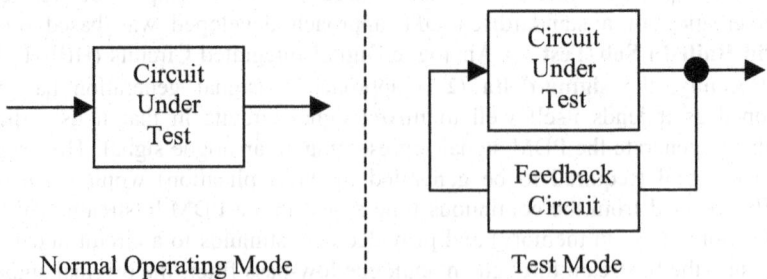

**Fig. 8.17.** Oscillation based test approach

## 8.4.2 1149.4 Mixed-Signal Test Bus

An attempt to enable test access for analogue and mixed-signal circuits resulted in the development of the mixed-signal test bus standard, **IEEE Std 1149.4-1999**. In this, the 1149.1-1990 standard was extended to incorporate on-chip analogue test capability. The standard relates to the **Mixed-Signal Test Bus** [38-41], although it is sometimes referred to as **Analog Boundary Scan**. This was proposed at the International Test Conference in 1992 and then led to the proposed architecture in 1993. The digital boundary scan capability is the same as before, but now internal switching of analogue signals within the IC and access at the primary I/O is achieved. The architecture for an IC incorporating the mixed-signal test bus is shown in Fig. 8.18.

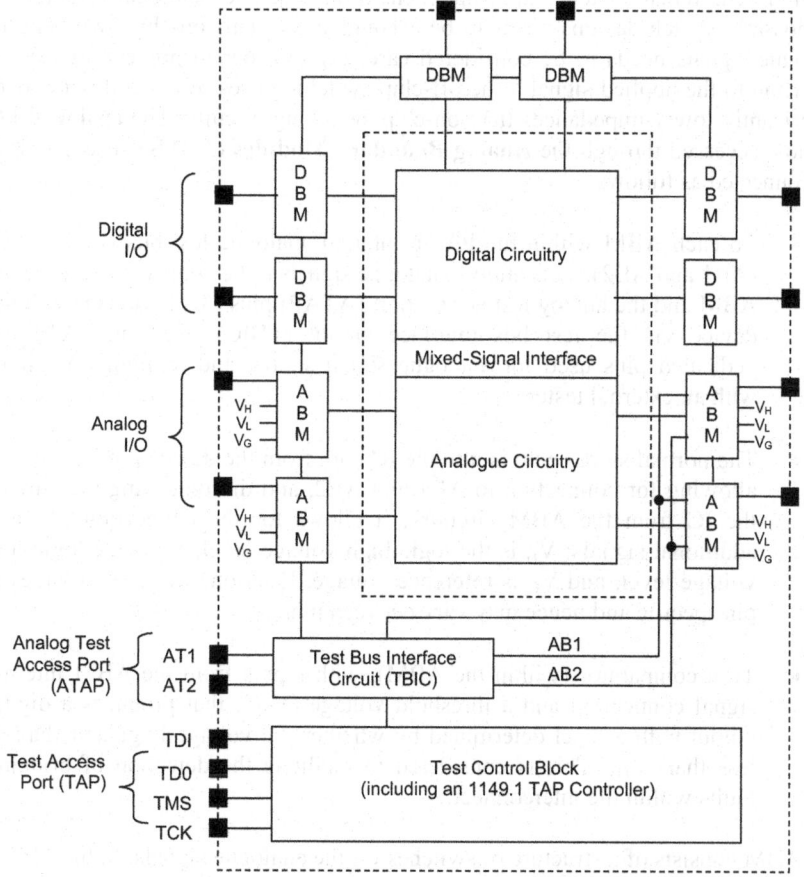

**Fig. 8.18.** Mixed-signal test bus compliant IC architecture

In this view, the digital and analogue circuits within the core are separated and there is a mixed-signal interface in the core. The digital and analog(ue) input and output cells have been replaced by **boundary modules**. For the digital cells, these are referred to as **Digital Boundary Modules** (DBMs). For the digital cells, these are the boundary scan cells as defined in the 1149.1 standard and are accessed through the test control block (also as in the 1149.1 standard).

The standard is not concerned with capturing analogue voltages and scanning them out in the same manner as in digital boundary scan, but rather with enabling analogue measurement and multiplexing/switching to be moved from the ATE resource requirements and onto the silicon. The silicon resources are physically smaller than those required by the ATE and this may reduce the tester costs. However, on-chip switching needs careful consideration. Within the ATE, switching is achieved through the use of relays with the relay contact resistance being a high resistance open-circuit when open and a low resistance short-circuit when closed. When switches are implemented on-chip, the non-ideal response of the possible switch designs needs to be accounted for. This on-chip switching of analogue signals needs to be considered carefully in order to prevent significant distortion to the applied signal – the off-chip switching uses relays and wires with significantly lower impedances than on-chip switching circuits. The analogue I/O are now accessed through the **Analog Boundary Modules** (ABMs). These ABMs are connected as follows:

- To each ABM within the IC, an internal analogue test bus is connected (AB1 and AB2). This allows analogue signals to be connected between an ABM and the analog test access port (ATAP) pins (AT1 and AT2) on the device via the test bus interface circuit (TBIC). AT1 and AT2 are dedicated pins used for analogue signal source and monitor operations with an external tester.

- The normal signal path is into the IC core from the device pin. As well as allowing for connection to AB1 and AB2, and disconnecting the core of the IC from the ABM circuitry, it allows for the connection of three additional signals: $V_H$ is the logic high voltage level, $V_L$ is the logic low voltage level, and $V_G$ is reference voltage. Each one of these voltages is pin specific and hence may vary between pins.

- To a comparator (within the ABM), with inputs from the ABM internal signal connection and a threshold voltage ($V_{TH}$), that produces a digital output with a level determined by whether the voltage is greater than or less than $V_{TH}$. This is considered to facilitate the detection of bridging faults within the interconnect.

The ABM consists of a structure of switches for the analogue signals. In the 1149.4 standard, these are identified as **conceptual switches** rather than physical switches that may be implemented, for example, using CMOS switches or buffers. The standard does not specify the structure of the switches.

When considering the **interconnect** on the **PCB**, for digital signals, this will take the form of metal tracks on the PCB forming a direct connection between IC pins.

For analogue signals, this **interconnect** between ICs may be considered as **simple** or **extended**, see Fig. 8.19. In this, a simple interconnect is a direct connection between IC pins. In extended interconnect, there will be a component (*e.g.* resistor) in the path. Additionally, the analogue signal may be either single ended or differential. This would need to be considered in the overall implementation on an IC. The full mixed-signal test bus operation is described in the IEEE Std 1149.4-1999 document.

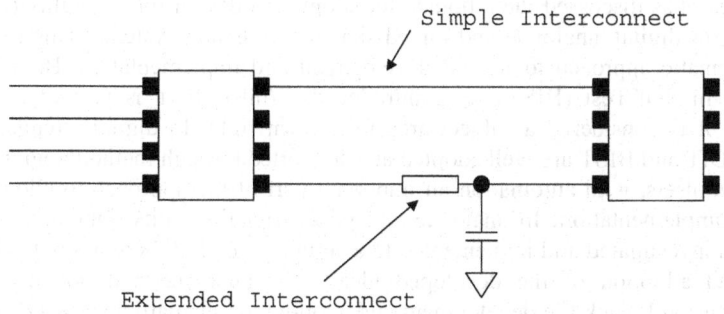

**Fig. 8.19.** Simple and extended interconnect

# 8.5 Future Directions for DfT and BIST

The inclusion of both DfT and BIST have become commonplace in digital circuits and systems, although they have only had a limited up-take in the analogue and mixed-signal domains. Future directions for DfT and BIST will be:

- Higher device complexities and greater demands on the testing phase of an IC product development.

- An increased use and standardisation of DfT and BIST for digital circuits and systems, with an emphasis on complex digital SoCs. A need to address increased design operating frequencies.

- An increased use and standardisation of DfT and BIST for analogue and mixed-signal circuits.

- Development of complex System in Package (SiP) designs with multi-domain (electrical and non-electrical) parts. These will include Micro-Electro-Mechanical devices (MEMs).

- Providing a cost benefit through reduced tester (ATE) requirements, costs and test times.

- Development of test methods to test effectively high-performance mixed-signal circuits and systems on digital only testers.

## 8.6 Summary

This chapter has discussed the rationale for adopting a design for testability (DfT) approach for digital, analogue and mixed-signal circuits and systems. This is part of a systematic approach to IC test development and implementation. Both DfT and Built-In Self-Test (BIST) were introduced. Whilst BIST is part of a DfT approach, it is considered a subject area in its own right. In digital circuits and systems, DfT and BIST are well adopted and supported through methodologies and software toolsets, with automation an important part of the approach to allow for effective implementation. In analogue and mixed-signal circuits, DfT and BIST have been investigated and implemented to a certain level, but there has not been a widespread adoption of the developed ideas. The analogue and mixed-signal domains currently lack the development and adoption of standard approaches with the support of automation.

## 8.7 References

[1]     International Technology Roadmap for Semiconductors, 2003 Edition, "Executive Summary"

[2]     International Technology Roadmap for Semiconductors, 2003 Edition, "Test and Test Equipment"

[3]     Hurst S., "VLSI Testing digital and mixed analogue/digital techniques", IEE, 1998, ISBN 0-85296-901-5

[4]     Rajsuman, R., "System-on-a-Chip Design and Test", Artech House Publishers, USA, 2000, ISBN 1-58053-107-5

[5]     Rivoir J., "Lowering cost of test: parallel test or low-cost ATE?" Proceedings of the 12th Asian Test Symposium, 2003, pp360 – 363

[6]     Bushnell M. and Agrawal V., "Essentials of Electronic Testing for Digital, Memory & Mixed-Signal VLSI Circuits", Kluwer Academic Publishers, 2000, ISBN 0-7923-7991-8

[7]     Needham W., "Designer's Guide to Testable ASIC Devices", Van Nostrand Reinhold, 1991, ISBN 0-442-00221-1

[8]     Smith M., "Application Specific Integrated Circuits", Addison-Wesley, 1999, ISBN 0-201-50022-1

[9]     Xijiang L. et al., "High-Frequency, At-Speed Scan Testing", IEEE Design and Test of Computers, September-October 2003, pp17-25

[10]    Zwolinski M., "Digital System Design with VHDL", Pearson Education Limited, 2000, England, ISBN 0-201-36063-2

[11]    IEEE standard test access port and boundary - scan architecture, IEEE Std 1149.1-2001, IEEE, USA

[12]    Van treuren B. and Miranda J., "Embedded Boundary Scan", IEEE Design and Test of Computers, March-April 2003, pp20-25

[13]    Parker K., "The Boundary-Scan Handbook, Analog and Digital", 2nd Edition, Kluwer Academic Publishers, USA, 2000, ISBN 0-7923-8277-3

[14]    Eklow B., Barnhart C. and Parker K., "IEEE 1149.6: A Boundary-Scan Standard for Advanced Digital Networks", IEEE Design and Test of Computers, September-October 2003, pp76-80

[15]    Bennetts B., "Status of IEEE testability standards 1149.4, 1532 and 1149.6", Proceedings of the Design, Automation and Test in Europe Conference and Exhibition, 2004, Vol. 2 , 2004, pp1184 – 1189

[16]    IEEE Std 1532-2002, IEEE Standard for In-System Configuration of Programmable Devices, IEEE, USA

[17]    Oakland S., "Considerations for implementing IEEE 1149.1 on system-on-a-chip integrated circuits", Proceedings of the International Test Conference, 2000, pp628-637

[18]    Zorian Y., "Test requirements for embedded core-based systems and IEEE P1500", Proceedings of the International Test Conference, 1997, pp191 – 199

[19]    P1500 Working Group, http://www.grouper.ieee.group/1500

[20]    Roberts G., "Metrics, Techniques and Recent Developments in Mixed-Signal Testing," Proceedings of the IEEE/ACM International Conference on Computer Aided Design, USA, 1996, pp. 514-521

[21]    Kuijstermans, F., Sachdev, M. and Thijssen, A., "Defect-oriented test methodology for complex mixed-signal circuits", Proceedings of the European Design and Test Conference, 1995, pp18 – 23

[22]   Russell G. and Learmouth D., "Systematic approaches to testing embedded analogue circuit functions", Microelectronics Journal, No. 25, 1994, pp133-138

[23]   Dufort B. and Roberts G., "On-Chip Analog Signal Generation for Mixed-Signal Built-In Self-Test", IEEE Journal of Solid-State Circuits, Vol. 34, No. 3, March 1999, pp318-330

[24]   Majernik, D. et al., "Using simulation to improve fault coverage of analog and mixed-signal test program sets", Proceedings of the IEEE Autotestcon, 1997, pp371-375

[25]   Zorian Y., "Testing semiconductor chips: trends and solutions", Proceedings of the XII Symposium on Integrated Circuits and Systems Design, 1999, pp226-233

[26]   Stoffels R., "Cost effective frequency measurement for production testing: new approaches on PLL testing", Proceedings of the International Test Conference, 1996, pp708-716

[27]   Burbidge M. et al., "Motivations towards BIST and DfT for embedded charge-pump phase-locked loop frequency synthesisers", IEE Proceedings on Circuits, Devices and Systems, Vol. 151, Issue 4, August 2004, pp337-348

[28]   Roberts G. and Lu A., "Analog Signal Generation for Built-In Self-Test of Mixed-Signal Integrated Circuits", Kluwer Academic Publishers, USA, 1995, ISBN 0-7923-9564-6

[29]   Lubaszewski, M., et al., "A built-in multi-mode stimuli generator for analogue and mixed-signal testing", Proceedings of the XI Brazilian Symposium on Integrated Circuit Design, 1998, pp175 – 178

[30]   Cassol L. et al., "The $\Sigma\Delta$-BIST Method Applied to Analog Filters", Journal of Electronic Testing: Theory and Applications, No. 19, 2003, pp13-20

[31]   Provost B. and Sanchez-Sinencio E., "On-Chip Ramp Generators for Mixed-Signal BIST and ADC Self-Test", IEEE Journal of Solid-State Circuits," Vol. 38, No. 2, February 2003, pp263-273

[32]   Guanglin W. et al., "Implementation of a BIST scheme for ADC test", Proceedings of the 5th International Conference on ASIC, 2003, Vol. 2, 2003, pp1128-1131

[33]   de Vries R. et al., "Built-in self-test methodology for A/D converters", Proc. of the European Design and Test Conference, 1997, pp353-358

[34]    Ohletz M., "Hybrid Built-In Self-Test (HBIST) for Mixed Analogue/Digital Integrated Circuits", Proceedings of the European Test Conference, 1991, pp307-316

[35]    Hoffmann C. and Ohletz M., "Feasibility Study for the Hybrid-Built-In Self-Test (HBIST) for Mixed-Signal Integrated Circuits", IEEE Design & Test of Computers, July-September 2000, pp106-115

[36]    Huertas G. et al., "Practical Oscillation-Based Test of Integrated Filters", IEEE Design & Test of Computers, November-December 2002, pp64-72

[37]    Huertas G. et al., "Testing Mixed-Signal Cores: A Practical Oscillation-Based Test in an Analog Macro", IEEE Design & Test of Computers, November-December 2002, pp73-82

[38]    IEEE Standard for a Mixed-Signal Test Bus, IEEE Std 1149.4-1999, IEEE, USA

[39]    Sunter S., "Cost/benefit analysis of the P1149.4 mixed-signal test bus" IEE Proceedings on Circuits, Devices and Systems, Vol. 143, No. 6, December 1996, pp393-398

[40]    Acevedo G. and Ramirez-Angulo J., "Built-in self-test scheme for on-chip diagnosis, compliant with the IEEE 1149.4 mixed-signal test bus standard", Proceedings of the IEEE International Symposium on Circuits and Systems, Vol. 1, 2002, pp I-149 - I-152

[41]    Uros K. et al., "Extending IEEE Std. 1149.4 Analog Boundary Modules to Enhance Mixed-Signal Test", IEEE Design & Test of Computers, March-April 2003, pp32-39

# Exercises

*The following questions will require access to the PC based tester arrangement identified in Appendix E.*

## Question 1

For the following Boolean expression for a circuit under test (CUT):

$$Z = (A . B) + (\overline{(A + B)} . (A + C))$$

- Develop a maximal length LFSR to produce a signal generator for this design
- Develop an SISR to act as a results analyser
- Implement the CUT, LFSR and SISR within the tester CPLD and operate the design in both normal operating and self-test modes
- For every stuck-at-fault possible in the CUT, implement the LFSR, faulty CUT and SISR within the tester CPLD. Identify the stuck-at-faults that are detected and the point in the self-test operation the fault can be detected
- For each detected fault, is there a unique output from the SISR that identifies the location of the fault? Comment on the results

## Question 2

For the gray-code counter CUT in Chap. 3, Question 7, develop a scan path using scan D-Type bistables (or discrete D-Type bistables and multiplexers) and implement within the tester CPLD. For each possible stuck-at-fault within the CUT, identify the test vectors that would stimulate the faults and show how these can be applied and the fault effect detected via the scan path.

## Question 3

For the TAP controller state machine for the 1149.1 standard, design a TAP controller circuit. Implement within the tester CPLD and perform a functional test on the circuit. For each possible stuck-at-fault, identify the test vectors that would stimulate the faults and show how these can be applied and the fault effect detected directly and via the scan path.

*The following questions will require access to an analogue circuit simulator and a digital logic simulator. In the following questions, the HSPICE® circuit simulator and Verilog-XL® logic simulator are considered, although with suitable modifications, a suitably available simulator could be used.*

## Question 4

Repeat the requirement of Question 1 via digital logic simulation.

## Question 5

Repeat the requirement of Question 1 via analogue circuit simulation.

## Question 6

Repeat the requirement of Question 2 via digital logic simulation.

## Question 7

Repeat the requirement of Question 2 via analogue circuit simulation.

## Question 8

Repeat the requirement of Question 3 via digital logic simulation.

## Question 9

Repeat the requirement of Question 3 via analogue circuit simulation

# Chapter 9

# System on a Chip (SoC) Test

*The ability to integrate complex digital circuits and systems on a single circuit die has led to the ability to incorporate the functionality that was once manufactured as a discrete chip-set on a printed circuit board within the single IC itself. A system once composed of multiple ICs can now be realised within a single IC, providing for physical size reduction, and leading to increased operating speed and portability for mobile applications. The system on a chip (SoC) is testament to the recent advances in design methods and fabrication processes. However, this leads to increased problems for test that need to be resolved.*

## 9.1 Introduction

The advances in digital IC design, fabrication and test are driven by the end user requirements for:

- Increased device functionality (more circuitry per $mm^2$ of silicon area and higher operating frequencies).
- Reduced physical size (more circuitry is a smaller package to support miniaturisation of the product and aid portability for mobile applications).
- Lower cost (the need to sell products with higher performance for less cost to the customer).

This has been in the main driven for increased system complexities by the PC, Communications [1, 2] and Multimedia markets. A system can however be defined in a number of ways. Here, the following definition is used:

*"A system is a group or combination of interrelated, interdependent, or interacting elements forming a collective entity"*
*(Collins Concise English Dictionary).*

225

Within an SoC [3], the types of interrelated elements are the electronic system *sub-systems* forming specific functions. Typical SoC designs incorporate functions such as:

- Processor core (*i.e.* microprocessor (µP), microcontroller (µC) or digital signal processor (DSP) cores)
- Embedded memory (RAM and ROM)
- Dedicated graphics hardware for fast graphics related operations
- Dedicated arithmetic hardware (*e.g.* adder, multiplier) for high-speed computation
- Bus control circuitry. Internal to the SoC will be systems buses for data, addresses and control signals between the main functional blocks
- Serial and parallel I/O connecting the device to its environment
- Glue logic. Miscellaneous logic for sub-system interfacing purposes
- Data Converters - ADC and DAC
- PLL – phase-locked-loop (for internal clock generation circuitry)

A generic SoC architecture can be viewed as a system block diagram. Figure 9.1 identifies a floorplan of an *imaginary* SoC that contains the key building blocks for a typical system. The design is separated into the digital logic and analogue/mixed-signal parts.

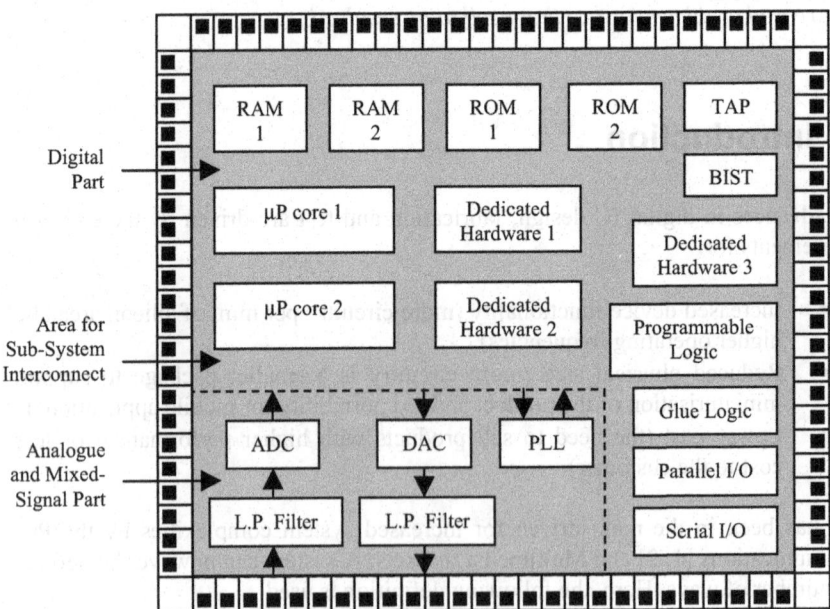

**Fig. 9.1.** Example SoC floorplan.

In this view, the main part of the device is undertaken within the digital logic and memory. In many SoC designs today, there may be more area dedicated to logic than memory – producing a **logic dominant** design. However, with greater demands on fast data storage and processing operations requiring a substantial amount of memory, SoC designs would dedicate more area to memory than logic – producing a **memory dominant** design.

The example design in Fig. 9.1 identifies two microprocessor (μP) cores (with a RISC (Reduced Instruction Set Computer) or CISC (Complex Instruction Set Computer) architecture) and four associated memory blocks (two RAM (for temporary data storage) and two ROM (for program code)). For high-speed operations such as digital multiplication, three dedicated hardware blocks are present. These would be under the control of the microprocessors and would allow for operations (*e.g.* mathematical) to be undertake in a minimum number of clock cycles and in parallel with the main microprocessor code operation. Two dedicated digital I/O circuits (parallel and serial) are provided to allow for high-speed communications to the external environment. Where parallel communications is considered, this will require a large number of pins on the device package. However, high-speed serial communications [4, 5] have become popular due to the increases in data transfer rates possible and the low device pin-count. Both single-ended and differential serial communications interfaces are utilised depending on the application requirements. Analogue I/O is supported through the use of suitable A/D and D/A converters with the analogue signal I/O passing through on-chip low-pass (LP) filters. For test [6] purposes, adequate controllability and observability is essential. The IEEE 1149.1[7-9] standard is utilised and on-chip, there is a Test Access Port (TAP) circuit included to facilitate this. Additionally, scan path test and BIST is incorporated into the design. The BIST here is set-up for the device to self-test the memory macros and the dedicated (high-speed) hardware macros. In order to operate in its intended environment, the system needs to be clocked at a high frequency [10]. Internal clock generation is achieved using an on-chip phase-locked loop (PLL). With the size of the design, and the delays in moving data around the die with the long interconnect, the logic gate and interconnect delays need to be minimised to achieve this. The final block of circuitry is a programmable logic block. This might be an FPGA (Field Programmable Gate Array) or CPLD (Complex Programmable Logic Array) architecture which is controlled via the microprocessor (or external programmer), and is available for implementing digital logic functions that may need to change either predefined by the user before purchase, or **on-the-fly** [11] in the final application. The design itself can be created using a combination of custom design (designed in-house using Hardware Description Langauge (HDL) design entry (both VHDL and Verilog®-HDL) and synthesis methods), in-house schematic capture, by purchasing IP (Intellectual Property) blocks from third party vendors and by synthesising C-code [12] descriptions into the final hardware. The target fabrication process would currently be in the range of a 90nm technology node CMOS process with a high number of interconnect layers (probably using copper interconnect and low-K (low dielectric constant) dielectric in order to reduce interconnect delays).

SoC designs are considered to be single die solutions. However, where such a single die solution is not possible for technological or cost reasons, a multi-die solution may be realisable, so producing the multi-chip module (MCM) structure. The MCM is in-line with the ITRS definition for a System in Package (SiP) device. The ITRS definition for the SiP is:

*"any combination of semiconductors, passives, and interconnects integrated into a single package"*

However, SiP designs [13-15] extend the concept of the MCM.

## 9.2 Examples of SoC Devices

A number of examples of SoC devices, their application areas and technical specifications have been published. Table 9.1 provides examples of these devices for further reading.

**Table 9.1.** Examples of SoC devices

| Example | Brief Description | Reference |
|---|---|---|
| 1 | Intel® Pentium® 4 microprocessor | [16] |
| 2 | Intel® Celeron® microprocessor | [17] |
| 3 | AMD Athlon™ microprocessor | [18] |
| 4 | AMD Duron™ microprocessor | [19] |
| 5 | Xpipes: Network-on-Chip Architecture | [2] |
| 6 | 100GOPS Programmable Processor for Vehicle Vision Systems | [20] |
| 7 | Sun Microsystems Inc., "Niagara" microprocessor | [21] |
| 8 | Infineon Technologies, Hard-disk controller SoC design | [22] |
| 9 | IBM, PowerPC 603™ microprocessor | [23] |
| 10 | Xilinx Inc., Virtex®-II Pro FPGA | [24] |

## 9.3 Test Complexity and Additional Problems

Due to the complexity of the test problem, the testing of an SoC is non-trivial and will require a **structured** (as opposed to an **ad-hoc**) *Design for Testability* (DfT) approach to be adopted right from the conception of the design [25-29]. A number of the key test related issues include:

- Faster on-chip operating speeds leading to increased I/O bandwidth problems [30] - the internal circuitry will operate at a higher operating frequency than the I/O. For testing, this can cause problems with both at-speed testing of the core, and the required test times – limited I/O bandwidth will require the need for longer test times.

- Very deep sub-micron (VDSM) fabrication processes leading to problems with $I_{DDQ}$ testing – higher static currents may mask faults than cause an increase in $I_{DDQ}$.

- VDSM fabrication processes leading to process variability problems between batches of wafers, across a wafer and across a die.

- Signal integrity problems [31]. Higher signal frequencies on-chip and the reduced interconnect pitch (physically closer interconnect) leads to noise issues such as signal cross-talk and power supply noise (power and ground).

- A problem with testing of large digital circuits and systems which may not immediately be obvious as an issue relates to the power consumption of the device during test [32, 33]. The application of test vectors may lead to a **higher switching activity** within the device during the testing phase than would be encountered during normal operation. This high activity may be the result of efficient test vectors, but can also lead to excessively high current flow during switching and resulting in heating of the device. If this exceeds the maximum limits of the device, then decreased reliability or immediate failure may result due to the testing phase.

- New fault mechanisms – *e.g.* soft errors due to noise sensitivity of the circuits at reduced power supply voltages with $V_{DD}$ moving towards and beyond 1V operation. Additionally, high currents in the power and ground lines may cause additional noise and also resistive voltage drop on these lines and that can result in otherwise good devices failing. This would lead to yield loss.

- The increase in the size of the circuit (number of transistors and the physical size of the die) may result in small parts of the device failing a particular test. If this cannot be accounted for, then the whole device would fail. However, designs may be created to incorporate **Built-In Self-Repair (BISR)** [34]. Here, if a part of the design fails, then it can be isolated and the device essentially configured to account for this. BISR is commonly applied to memory [35] where particular addresses can be isolated.

# 9.4 P1500 Core Test Standard Development

The increase in digital system complexity is leading to additional problems associated with the testing of complex digital cores, such as embedded processors. The P1500 working group [36-38] under the IEEE Technical Council on Test Technology was organised to address test related issues for test access, node controllability and observability. The purpose was to develop a standard architecture to solve the test problems. This is currently an on-going activity.

# 9.5 Future Directions for SoC Test

The future advances in SoC test are essentially the same as identified in chapter 3 (digital test). A summary is provided in table 9.2.

**Table 9.2.** Future SoC test issues

| Technology driver | Description |
|---|---|
| Design styles | Increasing use of hardware description languages and synthesis techniques<br>Increasing complexity of designs<br>Behavioural level synthesis and issues for test insertion<br>Increased integration – SoC and SiP test issues<br>The use of multiple power supply voltage rails on a single IC<br>Use of designs with multiple threshold voltage ($V_T$) MOS transistors<br>Move away from parallel transmission on and off-chip of data to high-speed serial bus architectures<br>Move to mixed-technologies incorporating logic, memory, mixed-signal and RF circuit elements<br>Integrated optical and MEMs devices alongside the circuit |
| Faster operating speeds | Higher signal frequencies<br>Driven by communications applications<br>Reduced device geometries (gate length and gate oxide thickness)<br>Reduced separation of interconnect tracks (track pitch) – capacitive and inductive coupling problems and signal cross-talk – signal integrity issues<br>Moving onto the next technology node to improve performance of components and reduce interconnect delays<br>Move into the nanotechnology domain |
| Higher levels of integration (higher density) | Moving onto the next technology node to improve performance of components and reduce interconnect delays<br>Newer failure mechanisms - limitations of existing fault models and need for models more representative |

| | |
|---|---|
| | of process defects<br>Higher fault-free $I_{DDQ}$ levels and limitations of $I_{DDQ}$ testing with the newer fabrication processes<br>Higher logic to pin ratio making it increasingly difficult to provide adequate controllability and observability<br>New test issues emerging for SoC and SiP devices<br>Move into the nanotechnology domain |
| Lower operating voltages | A need to improve device reliability by reducing electric field strength in dielectrics<br>Aim to lower device power consumption<br>Portable, battery operated circuits<br>Multi-voltage power supply device operation and interfacing |
| Test equipment and costs | Need for reduced test time and test costs<br>Need for reduced test pattern generation time and maintenance<br>More formal approaches to test program generation – test re-use<br>Problem of increasingly long test pattern generation and application times<br>Increasing requirements for test data storage in ATE<br>Increasingly difficult to perform at-speed testing of devices using current ATE<br>Need for reduced external ATE requirements and costs – simpler and less expensive ATE<br>Multi-Site test<br>Increased use of BIST<br>Increased use of BISR |

# 9.6 Summary

This chapter has introduced the role of test for SoC designs and the challenges that complex circuits and systems provide in the development of suitable and cost-effective test programs. The area of SoC design, fabrication and test is relatively new and is still developing. As such, many of the challenges have yet to be resolved, or even yet identified. Whilst this chapter has aimed to provide an introduction, it has not been comprehensive in its discussions. Further information can be found in the provided references.

# 9.7 References

[1]     Liu J. and Lin X., "Equalization in High-Speed Communication Systems", IEEE Circuits and Systems Magazine, Vo.4, No. 2, 2004, pp4-17

[2]     Bertozzi D. and Benini L., "Xpipes: A Network-on-Chip Architecture for Gigascale Systems-on-Chip", IEEE Circuits and Systems Magazine, Vo.4, No. 2, 2004, pp18-31

[3]     Rajsuman, R., "System-on-a-Chip Design and Test", Artech House Publishers, USA, 2000, ISBN 1-58053-107-5

[4]     Sauer C. et al. "Developing a Flexible Interface for RapidIO, Hypertransport, and PCI-Express", Proceedings of the International Conference on Parallel Computing in Electrical Engineering, 2004, pp129-134

[5]     Baosheng W. et al., "Yield, overall test environment timing accuracy, and defect level trade-offs for high-speed interconnect device testing", Proceedings of the 12th Asian Test Symposium, 2003, pp348-353

[6]     Zorian Y., "Testing Semiconductor Chips: Trends and Solutions", Proceedings of the XII Symposium on Integrated Circuits and Systems Design, 1999, pp226 - 233

[7]     IEEE standard test access port and boundary - scan architecture, IEEE Std 1149.1-2001, IEEE, USA

[8]     IEEE Standard for a Mixed-Signal Test Bus, IEEE Std 1149.4-1999, IEEE, USA

[9]     Bennetts B., "Status of IEEE testability standards 1149.4, 1532 and 1149.6", Proceedings of the Design, Automation and Test in Europe Conference and Exhibition, 2004, Vol. 2 , 2004, pp1184 – 1189

[10]    Edwards C., "Speeding up is hard to do", IEE Review, September 2004, pp44-46

[11]    Flaherty N., "In the chip or on the fly", IEE Review, Sept. 2004, pp48-51

[12]    Takach A., "Turning C into hardware", IEE Electronics Systems and Software, December/January 2004/05, pp20-23

[13]    Miettinen, J., Mantysalo, M., Kaija, K. and Ristolainen, E.O. "System design issues for 3D system-in-package (SiP)", Proceedings of the Electronic Components and Technology Conference (ECTC), 2004, Vol. 1, pp610-614

[14]   Tai K.L., "System-In-Package (SIP): challenges and opportunities", Proceedings of the Asia and South Pacific Design Automation Conference, 2000, pp191-196

[15]   Song Y. et al., "The reliability issues on ASIC/memory integration by SiP (system-in-package) technology", Proceedings of the IEEE International SOC Conference, 2003, pp7-10

[16]   Intel Corporation, USA, "Intel® Pentium® 4 Processor 660, 650, 640, and 630, and Intel® Pentium® 4 Processor Extreme Edition Datasheet"

[17]   Intel Corporation, USA, "Intel® Celeron® M Processor Datasheet"

[18]   Advanced Micro Devices, Inc., USA, "AMD Athlon™ 64 FX Product Data Sheet"

[19]   Advanced Micro Devices, Inc., USA, "AMD Duron™ Processor Model 8 Data Sheet"

[20]   Raab W. et al., "A 100GOPS Programmable Processor for Vehicle Vision Systems", IEEE Design & Test of Computers, Jan-Feb 2003, pp8-16

[21]   Geppert L., "Sun's Big Splash", IEEE Spectrum magazine, January 2005, pp50-54

[22]   Schrader M. and McConnell R., "SoC Design and Test Considerations", Proceedings of the Design, Automation and Test in Europe Conference and Exhibition, 2003, pp202-207

[23]   Hunter C. Vida-Torku E. and LeBlanc J., "Balancing Structured and Ad-hoc Design for Test: Testing of the PowerPC 603™ Microprocessor", Proceedings of the International Test Conference, 1994, pp76-83

[24]   Xilinx Inc. http://www.xilinx.com

[25]   Li J. et al., "A hierarchical test methodology for systems on chip", IEEE Micro (Special Issue on Design and Test of Systems on Chip), Vol. 22, Issue 5, September-October 2002, pp69-81

[26]   Marinissen E. et al., "Towards a standard for embedded core test: an example", Proceedings of the International Test Conference, 1999, pp616-627

[27]   Dervisoglu B., "A unified DFT architecture for use with IEEE 1149.1 and VSIA/IEEE P1500 compliant test access controllers", Proceedings of the Design Automation Conference, 2001, pp53-58

[28]   Koranne S., "Design of reconfigurable access wrappers for embedded core based SoC test", IEEE Transactions on Very Large Scale Integration (VLSI) Systems, Vol. 11, Issue 5, October 2003, pp955-960

[29]    Goel S. and Marinissen E., "Effective and efficient test architecture design for SOCs", Proceedings of the International Test Conference, 2002, pp529-538

[30]    Ahmed N., Tehranipour, M and Nourani, M, "Extending JTAG for testing signal integrity in SoCs", Proceeding of the Design, Automation and Test in Europe Conference and Exhibition, 2003, pp218-223

[31]    Maroufi W. et al., "Solving the I/O bandwidth problem in system on a chip testing", Proceedings of the 13th Symposium on Integrated Circuits and Systems Design, 2000, pp 9-14

[32]    Rosinger P. et al., "Analysing trade-offs in scan power and test data compression for systems-on-a-chip", IEE Proceedings on Computers and Digital Techniques, Vol. 149, No. 4, July 2002, pp188-196

[33]    Wang S. and Gupta S., "ATPG for heat dissipation minimization during test application", IEEE Transactions on Computers, 1998, Vol. 47, No.2, pp256-262

[34]    Nicolaidis, M, Achouri, N. and Anghel, L., "Memory built-in self-repair for nanotechnologies", Proceedings of the 9th IEEE On-Line Testing Symposium, 2003, pp94-98

[35]    Lu S. "Built-in self-repair techniques for embedded RAMs", IEE Proceedings on Computers and Digital Techniques, Vol. 150, Issue 4, July 2003, pp201-208

[36]    P1500 Working Group, http://www.grouper.ieee.group/1500

[37]    Zorian Y. et al, "Testing embedded-core based system chips", Proceedings of the International Test Conference, 1998, pp130-143

[38]    Zorian Y., "Test requirements for embedded core-based systems and IEEE P1500", Proceedings of the International Test Conference, 1997, pp191-199

[39]    Chakrabarty K., Iyengar V. and Krasniewski M., "Test Planning for Modular Testing of Hierarchical SOCs", IEEE Transactions on Computer-Aided Design of Integrated Circuits and Systems, Vol. 24, No. 3, March 2005, pp435-448

[40]    Psarakis M. Gizopoulos D. and Paschalis A., "Built-In Sequential Fault Self-Testing of Array Multipliers", IEEE Transactions on Computer-Aided Design of Integrated Circuits and Systems, Vol. 24, No. 3, March 2005, pp449-460

# Chapter 10

# Test Pattern Generation and Fault Simulation

*Structural test programs are used in production test in order to reduce the time taken to test the fabricated IC when compared to an exhaustive functional test. Structural tests are based on the development of test vectors to detect specific faults that are considered to exist in a circuit due to process defects. The generation of the necessary test vectors is undertaken using test pattern generation and fault simulation techniques and tools.*

## 10.1 Introduction

Testing [1, 2] of an integrated circuit (IC) is undertaken to consider three types of test: functional, structural or parametric:

- A **functional test** will exercise the design in such a way as to exercise the operation of the design through the various functional operations that it would be designed to undertake. For complex digital circuits and systems, then this can be extremely time consuming and hence costly.

- A **structural test** will exercise the design in such a way as to stimulate faults that may exist in the design due to fabrication defects. The idea is to apply suitable digital vectors that will sensitise the fault such that the faulty circuit will produce a different result at the primary output from a fault-free circuit. This requires suitable fault models to be created to model fabrication defects, and for these models to be simulated in the design to identify the right set of digital vectors to apply to the actual fabricated circuit.

- A **parametric test** will determine specific parameters such as current and voltage levels. **Parametric tests** are undertaken to measure specific

235

electrical characteristics of a device. **DC parametric** tests measure voltages and currents, along with open/short circuit tests. These are not time dependent. **AC parameter** tests measure time dependent characteristics such as delays and rise and fall times of signals.

The **structural test** approach is based on the development of test vectors to detect specific faults that may exist in a fabricated circuit. Fault models are created for specific fabrication process defects and inserted into a model of the circuit. A requirement in the creation of structural test programs is the detection of faults. The idea is then to generate a minimal set of test vectors that detect a maximum number of faults, or a specific number of faults required to be detected according to the required quality levels. In the ideal case, 100% of faults will be detected. However, it may not necessarily be possible to detect 100% of faults in the circuit. Typically, the fault coverage would be required to be in excess of 95% and those parts of the circuit that may contain faults not covered by purely structural testing would require additional functional testing to be undertaken. The fault coverage is calculated as:

$$FC\ (\%)\ =\ \frac{Number\ of\ faults\ detected}{Total\ number\ of\ faults\ considered}\ x\ 100\ (\%)$$

It should be noted that the fault coverage figure would need to be considered with care. The figure will relate only to those faults considered. Faults that are not considered may have importance, but their effect will not be included. If care is not taken in the faults considered, then a fault coverage figure of 100% may be attained for specific faults, but faults of more relevance to the particular design and fabrication process might not actually be detected. Additionally, all faults have equal weighting. This means that each fault is considered to be of equal importance and there is no information relating to the importance of the fault. **Fault weighting** can be used however to reduce the size of the fault list to those faults which may be considered more likely to occur.

Structural tests will be based on the detection of specific fault represented by their own set of fault models. The fault models will be considered to be either **logical fault models** or **defect oriented fault models**. Fault models will be based on electrical faults caused by process defects within a circuit, and:

- **Logical fault models** are translations of electrical faults into logic level models. The models are based on logic levels (and possibly timing).

- **Defect oriented fault models** are electrical faults based on the properties of the defect that created the fault. These are not simple digital (logical) models, but consider the electrical operation (voltage and current) of the fault in terms of analogue circuit primitives (resistance, capacitance, *etc.*).

Examples of electrical faults within a circuit include:

- Process variations outside the normal process spread (*e.g.* excessive change in transistor threshold (MOS) voltage value).
- Open circuits in metal interconnect. These will be resistive open circuits, the value of the resistance dependent on the physical nature of the open.
- Short-circuits (bridges) between metal interconnect tracks. These will be resistive short circuits, the value of the resistance dependent on the physical nature of the short.
- Transistor stuck-open and stuck-short faults where the transistor is considered as a switch.
- Excessive steady-state (quiescent) power supply current.
- Transistor (MOS) open/short circuits (defects in the gate oxide and resistive opens/shorts between nodes in the transistor).

A number of fault models have been developed for IC test purposes. The key models used for digital circuits are:

**Logical fault models** including:

- Stuck-At-Fault
- Bridging fault (Wired-AND and Wired-OR)
- Delay fault (the delay fault might also be considered as a defect oriented fault)
- Memory fault (logical faults considered)

**Defect-Oriented fault models** including:

- Bridging fault (resistive)
- Memory fault (non-logical faults considered)
- Open-circuit fault in interconnect metal
- Stuck-open and stuck-short faults
- Transistor (MOS) open/short faults (used for analogue circuit analysis)
- $I_{DDQ}$ fault

Usually, a **single fault assumption** is made in that a faulty circuit is to contain only one fault at any one time. However, **multiple faults** may exist and can be considered. The process of generating the fault list, determining the test vectors that detect the faults considered, and identification of the minimum number of test vectors required to detect the maximum number of faults (or to create a required fault coverage figure) can be undertaken either manually (time consuming, expensive and likely to suffer from human-induced errors), or by using a suitable software tools which partially or completely automates the process (faster and more reliable than a purely manual approach).

The fault model is of major concern to the generation of the test patterns. Models will need to be simple and quick to use, but must adequately reflect the process defect that they represent. The relevance of a particular fault model can be a cause of major debate, particularly for analogue circuits. In **digital** logic, the stuck-at-fault has been the major fault model utilised and is well supported with Automatic Test Pattern Generation (ATPG) tools. With the lower geometry fabrication processes, interconnect faults are becoming dominant for designs with a large number of interconnects, and the relevance of the simple stuck-at-fault is increasingly questioned. In **analogue** circuits, the fault models that have been developed and demonstrated have encountered resistance to their universal adoption. In analogue circuits, the major concern is the applicability of structural testing to replace functional testing. The complex and application specific behaviour of analogue circuits requires a range of tests which will be particular to a circuit design and if, by replacing such functional tests with structural tests, the ability to identify specific device characteristics may not be possible, this situation would need a great deal of careful consideration. A structural test would then not necessarily be able to guarantee the fault-free circuit operation within specific boundaries of operation (due to the process variations).

## 10.2 Test Pattern Generation

**Test pattern generation** [2, 3] (exhaustive, pseudo-random or algorithmic) and **fault simulation** techniques are used in the test vector generation process. **Test pattern generation** (TPG) is the process of generating the test vectors required to stimulate a circuit at the primary inputs so that effect of the considered fault (the fault effect) is propagated to the primary outputs. A difference between the fault-free and faulty circuit can then be detected. It is common, and sensible, to derive a minimal set of test vectors as this will reduce the overall test set size and hence test time. **Manual** or **automatic** TPG can be undertaken. Where **exhaustive**, **algorithmic** or **pseudo-random** methods would be employed, the basic steps that are required to be undertaken in TPG are:

1. Generate a model of the circuit under test (CUT) in a suitable format (usually in either VHDL [4] or Verilog®-HDL [5]).

2. Generate a fault list of considered faults based on the circuit netlist and fault types.

3. For all faults in the fault list, the following sequence will be repeated until either all faults have been considered, or a target fault coverage has been met:

   a) Select the fault from the fault list.
   b) Insert the fault into the circuit, apply a test vector and determine whether the fault has been detected or not with that vector. This would be undertaken for all considered test vectors.
   c) Remove the fault from the circuit.

d) Determine the fault coverage (%).
e) Remove the fault from the fault list.
f) Repeat from (a) until all faults have been considered, or the target fault coverage has been met

**ATPG** (Automatic Test Pattern Generation) for combinational digital logic is well supported, although for sequential logic circuits, the problems associated with the sequential nature of the circuits becomes more problematic for automation tools. Additionally, for an ATPG tool to be able to interpret and work with a particular circuit design, the design will need to adhere to specific **design rules**. These rules may exist either to allow the tool to operate without **error**, or to enable the tool to run more **efficiently**. For combinational logic circuits, an ATPG tool will use a particular method (exhaustive, pseudo-random or algorithmic) in the development of the required set of test vectors (the test set). The methods used are:

**Exhaustive**:
In this approach, the complete set of primary input combinations is applied.

**Pseudo-Random**:
In this approach, a pseudo-random pattern generator is used to create patterns. These are evaluated to identify their effectiveness as to the fault detection capabilities.

**Algorithmic:**
Most algorithmic TPGs are based on identifying a sensitized path in the design. The path from the fault location to the primary output is sensitized so that the effect of the fault can be captured at the primary output (the value of the signal at the output varies between the fault-free and faulty conditions). In terms of digital logic, this would be the logic value (0/1) and/or timing. The need then is to identify the primary inputs required to set-up the sensitised path from the primary inputs through to the primary outputs and to ensure that this condition can be attained. The best known algorithms are the simple, D-algorithm, PODEM and FAN:

> **Simple**: This algorithm selects a fault to consider at a particular node in the design and setup the primary inputs in order to monitor the value of the faulty node. The detection or non-detection of the fault is then determined. This simple procedure can encounter problems with circuit fan-out and reconvergence.
>
> **D-Algorithm**: An improvement on the simple algorithm which utilises five-value algebra $(0, 1, X, D, \overline{D})$ to represent values at nodes in the circuit under fault-free and faulty conditions.
>
> **PODEM**: This algorithm (**P**ath **O**riented **DE**cision **M**aking) attempts to reduce the time taken for decision making within the algorithm when compared to the D-Algorithm.

**FAN**: This algorithm (**FAN**-out-oriented test-generation algorithm) attempts to reduce the time taken when compared to PODEM and has been quoted as to being three to five times faster than the PODEM algorithm.

The output from the ATPG process will be a test set which will require to fulfil a number of requirements including:

The time to create the test set – the time required for the ATPG tool to run. The higher the fault coverage figure required, the longer the tool is required to run. For example, it may be quick and easy to reach, say, a 90% fault coverage figure, but increasingly difficult and time consuming to move beyond this fault coverage figure.

The size of the test set – there may be a limit as to how many test vectors can be dealt with either by the software tools used, or by the ATE.

The time that would be required to load the test patterns into the ATE during production test – the more patterns, the longer it takes the ATE to load the vectors and so the longer the overall test time.

ATPG for sequential logic circuits requires the taking into account the combinational logic parts of the circuit, the states of the circuit (current and next states), the ability to set the required state in order to access the faulty node behaviour, the need to detect the faulty behaviour at the primary output and the operation of the circuit under fault-free and fault conditions. An example of a test pattern generation tool is the TetraMAX® [6] ATPG tool from Synopsys Inc. [7]. This ATPG supports sequential circuits incorporating full-scan and partial-scan test access. It supports the generation of patterns for testing of a range of test methologies and is linked to the Synopsys DFT Compiler® [8] tool.

## 10.3 Digital Fault Simulation

Fault simulation [1, 2] is based on the insertion of a fault model into a circuit design, and the simulation of the design under the fault conditions. Fault simulators will be available for either digital logic or analogue circuits. They will be either dedicated fault simulators, or will be "add-ons" to existing logic/circuit simulation tools.

Whatever the type of circuit, the resulting output of the faulty circuit is then compared to the output from a fault-free circuit. If the results differ at the points in time that the outputs are monitored, then the fault is detected. If no difference is detected, then the fault is not detected.

For a given fault simulation study (circuit design, test vector, fault model and fault detection criterion), faults may therefore be considered to fall into one of three categories:

- **Detected**:
    A difference at the monitored output between the fault-free circuit and faulty-circuit is detected.

- **Not detected**:
    In the study, with the given test stimulus and output monitoring points defined. The fault may however be detected if different stimulus is applied and/or different monitor points (circuit outputs and/or time points) are used.

- **Not detectable**:
    The fault cannot be detected under any input stimulus and/or monitor points.

Fault simulation can be undertaken using a number of techniques. The common techniques are:

- **Serial fault simulation**:
    This is the simplest approach in which a single simulation is run at any one time. Each fault is inserted into the circuit and the circuit operation simulated. The total number of simulation runs is equal to the total number of faults plus one (the initial simulation of the fault-free circuit).

- **Parallel fault simulation**:
    Parallel fault simulation uses the word width of the computer on which it is run in order to simulate faulty circuits in parallel. Bit-wise logical operations within the computer are used to simulate the logical operation of the gates in the circuit.

- **Concurrent fault simulation**:
    Concurrent fault simulation is more efficient than parallel fault simulation in terms of simulator run time, but not necessarily so in terms of its memory usage requirements. Only a single pass through the circuit is needed during fault simulation. Concurrent fault simulation maintains a linked list of the "faulty circuits". Only the circuits that do not agree, in terms of their input and output relationships, are explicitly simulated.

- **Deductive fault simulation**:
  Deductive fault simulation [9] is also more efficient than parallel fault simulation in terms of simulator run time, but not so in terms of its memory usage requirements. It is a "one-pass" simulation.

Digital fault simulation can be undertaken with either a specific fault simulator, or it might be possible to utilise an existing logic simulator. In the case of using a logic simulator, then serial fault simulation can be performed by either modification to the circuit description (schematic or HDL description), or by the addition of suitable statements within the testbench/testfixture. For example, in the case of Verilog®-HDL [5], the use of the **force** (force a node to a set logic value – *i.e.* insert the fault) and **release** (remove the force – *i.e.* remove the fault) commands at suitable points in the testfixture operation can allow the performing of serial fault simulation. This can be a manual (time consuming) or automatic (using a suitable program (*e.g.* C-based)) to identify the node names, to create the fault list, to undertake the fault insertion, simulation, fault retraction, application of fault detection criterion, updating of fault list (removing the fault from the fault list), fault coverage calculation and continue/stop fault simulation run operations.

An example of a fault simulator is the Verifault-XL® [10] fault simulator from Cadence Systems Inc. [11]. This tool works along with the Verilog-XL® [12] logic simulator. A *gate-and-switch* or *structural* description of the circuit is used and the fault simulator allows for the user to identify the test vectors that will detect the faults considered. It does not generate the test vectors required, rather the user must define the vectors to apply and the tool will identify the effectiveness of the test vectors to detect the considered faults. The types of faults considered are the stuck-at, bridging and open faults. Verifault-XL® is used by generating a Verilog®-HDL description of the circuit and a Verilog®-HDL testfixture for use with Verilog-XL®. Specific Verifault-XL® tasks are added to the design and testfixture files, and Verifault-XL® is run. Figure 10.1 shows an example of a Verilog®-HDL design (*design.v* file - this is the Verilog-XL® design file with additional Verifault-XL® statements) and the Verifault-XL® testfixture file (*testfixture.v* - this is the Verilog-XL® testfixture file with additional Verifault-XL® statements). The simulator user guide provides a comprehensive description of the fault simulator and commands, along with interpreting the results statistics. In the testfixture file of Fig. 10.1, the design has one 2-input AND gate and so there are 3 nodes with 6 possible stuck-at-faults. Here, only SAFs are to be considered. With this design, it is possible to detect 100% of faults with 3 out of the possible 4 vectors (the vectors are (A/B): 0/1, 1/0, 1/1). With such a small circuit, it would actually be faster and simpler to manually derive the required vectors (through a fault matrix), but once the number of gates and nodes increases, the manual approach is very limited. Using Verifault-XL®, the simulator commands require to be executed by the user only once, and will require a short time to execute. For larger designs, the execution time may run into hours (or days), depending on the design complexity and structure of the testfixture. It is however a requirement for the user to define the test vectors, the output monitoring points (and time), and to interpret the fault simulation results.

design.v file                                          testfixture.v file

```
/////////////////////////          //////////////////////////////////////////////
//                                 `timescale 1ns/1ps
// Module description              //////////////////////////////////////////////
// of logic (2-input               module testfixture;
// AND gate)
// for Verifault-XL                reg in1, in2;
// fault simulation study          wire out;
/////////////////////////          //////////////////////////////////////////////
//                                 and_gate cut(out, in1, in2);

`enable_portfaults                 initial

module and_gate(Z, A, B);          begin
                                   $write("\t\tTime\tin1\tin2\tout\n");
output Z;
input A, B;                        in1 = 1'b0;
                                   in2 = 1'b0;
and #2 (Z, A, B);
                                   #5  $write($time, "\t",in1,"\t",in2,"\t",out,"\n");
endmodule                          #10 $write($time, "\t",in1,"\t",in2,"\t",out,"\n");
                                   #10 $write($time, "\t",in1,"\t",in2,"\t",out,"\n");
`disable_portfaults                #10 $write($time, "\t",in1,"\t",in2,"\t",out,"\n");
                                   #5  $finish;
                                   end
                                   //////////////////////////////////////////////
                                   always #10 in1 = ~in1;
                                   always #20 in2 = ~in2;
                                   //////////////////////////////////////////////
                                   initial
                                   begin
                                           $fs_options ("one_level_only = no");
                                           $fs_model ("exclude;");
                                           $fs_model ("include;");
                                           $fs_add;
                                           $fs_dictionary;
                                           $fs_inject;
                                       #5  $fs_strobe(out);
                                       #10 $fs_strobe(out);
                                       #10 $fs_strobe(out);
                                       #10 $fs_strobe(out);
                                   end
                                   //////////////////////////////////////////////
                                   endmodule
```

**Fig. 10.1.** Fault simulation example files

It is important to identify that the use of fault simulation can require substantial computing power (time and memory) and would require careful placement within the test development process in order to minimise costs whilst maximising the benefits. The specific details of the particular fault simulator used need to be considered and the results obtained readily interpreted.

# 10.4 Analogue Fault Simulation

Digital circuits are simplified representations of the analogue nature of the circuit behaviour and so in simulation, information relating to the circuit currents and voltages are not modelled. When it comes to analogue circuits and the analogue parts of mixed-signal circuits, then the current and voltage information is required along with analogue behaviour of the circuit elements (transistors, resistors, capacitors, *etc.*). For larger designs, this would require substantial run time and computing power that may not be practical. A number of analogue fault simulation tools [13-15] have been developed to investigate the problem. The simplest approach is to undertake serial fault simulation on a single machine. This can be readily automated. There has been much debate into the relevance of fault simulation [16] of analogue circuits as it is a feed into the development of structural test programs. Whilst digital designs lend themselves readily to structural testing, the nature of the analogue signals and variations in signal values between "good" devices, means that a window of fault free behaviour needs to be defined, as shown in Fig. 10.2.

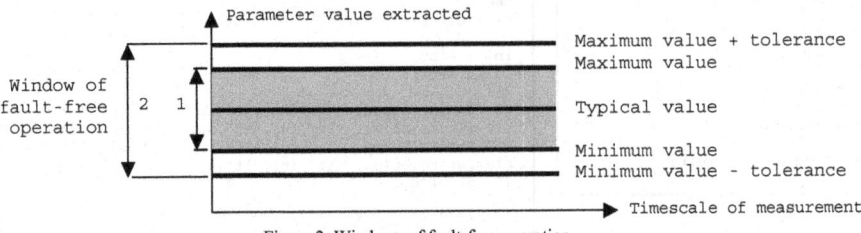

Figure 2: Windows of fault-free operation

**Fig. 10.2.** Window of fault-free behaviour

Here, the window of fault-free behaviour exists between an upper and lower limit. These limits would need to take into account fabrication process variations (to define the minimum and maximum values (1)) and a tolerance (to account for measurement equipment tolerances (2)). An example of an approach to undertaking a fault simulation study is shown in Fig. 10.3. Here, the circuit simulation model and test stimulus are derived and the fault-free behaviour determined. The considered faults are then inserted one at a time into the circuit, the simulation run repeated and the detection/non-detection of the fault (specific parameters defined by the user) determined. After all faults considered have been simulated, the results are post-processed and presented to the user. Additionally, the relevance of the fault model to the fabrication process [19-21] and how accurately the model reflects the physical defect needs to be questioned. Typical analogue component fault models are shown in Fig. 10.4. Here, the models are based around resistive open/short faults on the nodes of specific circuit components, and rely on representative values (minimum/maximum values of fault resistance) to be used. The fault must be detectable at these limits. Interconnect faults will also be considered as resistive open-short faults. However, obtaining up-to-date and accurate process defect information can be difficult to acquire, and models may need to be developed based on public-domain (published) material unless access to specific foundry information can be obtained. This however may not be possible due to commercial requirements and certain company specific information may not be accessible outside the company. Additionally, the rapid progress in fabrication processes moving along the technology roadmap introduces additional problems given the introduction of new defect mechanisms that may not have existed in the *coarser* fabrication processes.

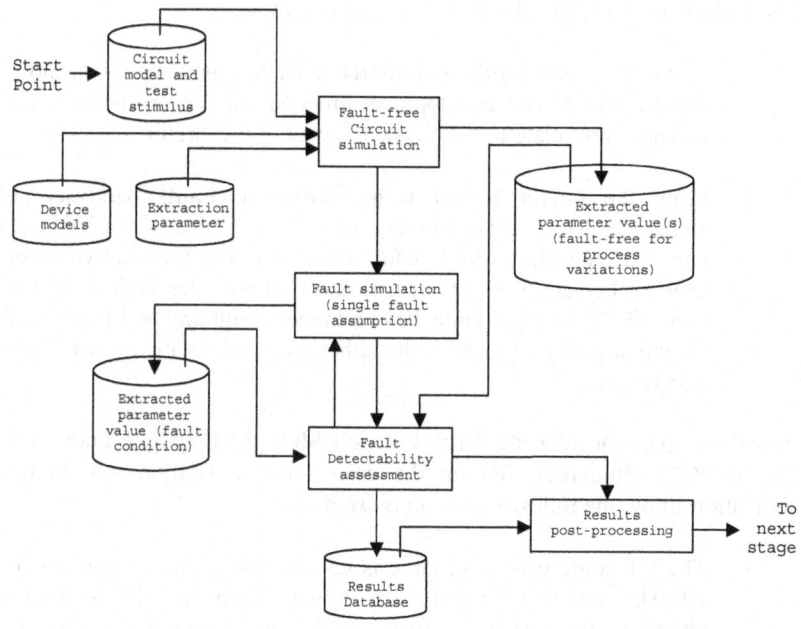

**Fig. 10.3.** Example analogue circuit fault simulation study approach

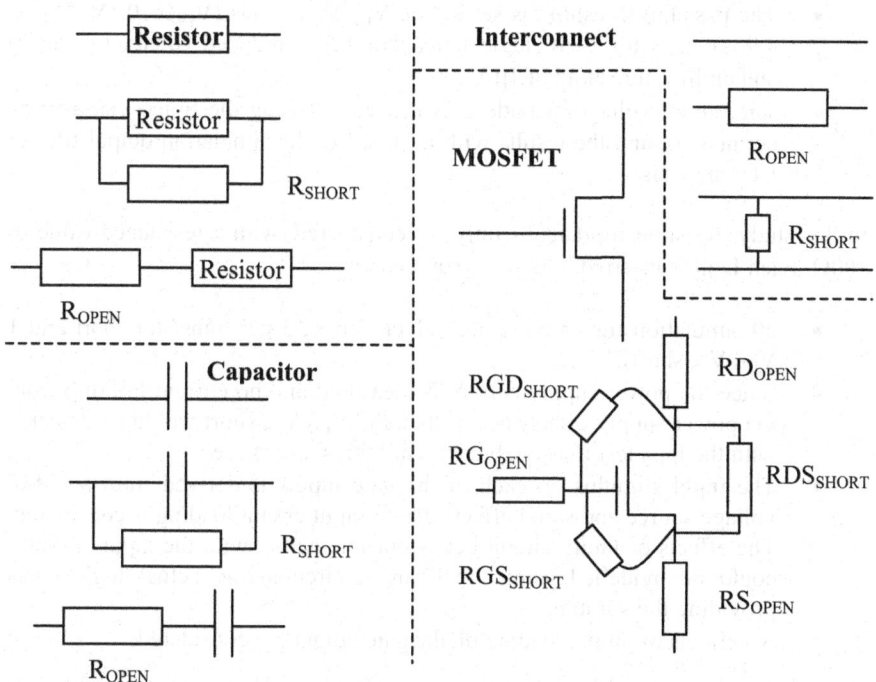

**Fig. 10.4.** Example analogue fault models (open/short fault model)

The faults considered may be derived in one of two ways:

- From the circuit **netlist/schematic**. This can be achieved at any time in the design and test development process, but can create large fault lists and does not relate to the physical layout of the circuit.

- From the circuit **layout** using **Inductive Fault Analysis** [17, 18] techniques. Here, either Monte Carlo [19] or Critical Area [20] techniques are used. This allows for a reduced fault list to be created (when compared to the schematic based fault list) and also provides fault weighting – the more likely to occur faults have a higher weighting and this would allow for the targeting of these faults rather than faults with a low probability of occurrence.

As an example, consider the 2-input static CMOS AND gate. Figure 10.5 shows the HSPICE$^®$ simulation file for the design and simulation run. In this fault simulation study, the following set-up is arranged:

- The full logic gate truth-table is tested with an input application rate of 20MHz. The input signals are derived from an ideal voltage source, although the drive circuitry could be added to realise a more "representative" input circuit.
- The output voltage is monitored 25ns after the signal is applied.
- The pass/fail threshold is set by the $V_{OL}/V_{OH}$ levels ($V_{OL} = 0.1V$, $V_{OH} = 4.9$ – e.g. as for 74HC logic series, but with a reduced voltage tolerance) and an $I_{DDQ}$ threshold of 1μA.
- The output voltage at node Z is measured (using the Spice .measure command), and the results will be stored in the simulation output file for later analysis.

In this study, resistive short faults only are considered, with a resistance value of 100Ω. Each fault is inserted into the circuit netlist:

- 20 simulation runs are required (1 fault-free, 3 per transistor short and 1 $V_{DD}/V_{SS}$ short).
- Since the power supply current is measured and no current limiting from the power supply effects occur, then the $V_{DD}/V_{SS}$ short would be detected with the $I_{DDQ}$ test if a suitable pass/fail threshold is used.
- The input stimulus on each of the gate inputs is derived from an ideal voltage source and so no effects due to input circuit loading is considered. The effects of faulty circuit behaviour interacting with the input circuitry could be modelled using "real" input circuits (*i.e.* actual logic gates providing the stimulus).
- No effects of output loading on the gate output are considered.

```
*2-input Static CMOS AND Gate Fault Simulation Study

.OPTIONS POST
.TEMP 25

Vdd  vdd  0  5V
Va   A   0   PWL(0,0 50n,0 50.1n,5 100n,5 100.1n,0
+ 150n,0 150.1n,5
Vb   B   0  PWL(0,0 100,0 100.1,5)

Mn1  x1   A    x2   0    MODN    W=4um    L=0.6um
Mn2  x2   B    0    0    MODN    W=4um    L=0.6um
Mn3  Z    x1   0    0    MODN    W=4um    L=0.6um
Mp1  x1   A    vdd  vdd  MODP    W=5um    L=0.6um
Mp2  x1   B    vdd  vdd  MODP    W=5um    L=0.6um
Mp3  Z    x1   vdd  vdd  MODP    W=5um    L=0.6um

.model MODN   nMOS  (LEVEL=1 VTO=0.7V KP=25e-6 LAMBDA=0)
.model MODP   pMOS  (LEVEL=1 VTO=-0.7V KP=10e-6 LAMBDA=0)

.measure point1 V(Z) 25n
.measure point2 V(Z) 75n
.measure point3 V(Z) 125n
.measure point4 V(Z) 175n

.TRAN 1ns 200ns

.end
```

**Fig. 10.5.** Fault simulation study

# 10.5 Mixed-Signal Fault Simulation

The problems associated with either digital only fault simulation or analogue only fault simulation come together with the need to run true mixed-signal fault simulation. The availability of suitable simulation tools, coupled with the effort required to learn the simulator/language and to develop suitable fault simulation studies, cannot be underestimated. The approaches that can be undertaken to attempt mixed-signal simulation are:

- Use a digital logic simulator and behavioural model the analogue parts using the available variables types and creating a digital equivalent model. For example in VHDL, use of the *real* types can aid this.

- Use an analogue circuit simulator and behavioural model the digital parts using the available macros. For example, Spice has a limited capability, but the use of a simulator such as Spectre® [10] from Cadence Design Systems Inc. with the ability for behavioural modelling using Verilog-A [11], may be advantageous.

- Use a mixed-signal language and simulator. Over the last number of years, the development of mixed-signal [12] and mixed-technology simulation capabilities has led to a number of moves for the development of standardised languages. The support for analogue and mixed-signal extensions to the VHDL and Verilog®-HDL languages has led to the introduction of VHDL-AMS [13] (VHDL – Analogue and Mixed-Signal) and Verilog-AMS [14]. VHDL-AMS is documented in IEEE Standard 1076.1-1999.

# 10.6 Issues with Fault Simulation

In order to undertake fault simulation as part of the test pattern generation process of an IC product, then the effort required must be considered as an integral part of the overall design, test and fabrication tasks, and suitably built into the overall product development process. If it is considered as an "add-on", which is not a *necessity*, it would be considered a task that would not be suitably resourced and would never be able to reach its full potential. The efforts required would include:

- Project management support.

- CAD and CAT tool support. These include the development and utilisation of realistic fault models and, where possible, to utilise input from the fabrication process and failure analysis in order to retain the relevance of the fault models.

- Identification of the procedures required in order to perform effectively fault simulation studies.

- Identification of the points during the design and test development tasks at which fault simulation studies will be undertaken. For example, such studies could be undertaken during the design development, and would be part of the test development process.

- Integration of the fault simulation outputs (results) into the overall test pattern generation activities.

- Identification of the cost benefits that may be provided.

- Identification of the strengths and weaknesses associated with a particular approach adopted.

- Training of designers, test engineers and DfT engineers in performing fault simulation activities.

## 10.7 Circuit vs Behavioural Level Fault Simulation

Depending on the level of detail required during simulation, a circuit/system description may range from a high-level behavioural description to a low-level structural description. Using these different levels of description [25-33], and a mixture of levels, is commonplace in design. For use in fault simulation, care would need to be taken to ensure that relevant information is incorporated into the design description and important information is not lost due to simplifications in the model. For example, in analogue circuit simulation at the basic transistor and passive component level, this will identify the currents and voltages that exist within the circuit. If a behavioural model is written for part or all of the design, this may be written to include or exclude this level of detail. Simpler models will simulate faster and so results will be obtained in a shorter time. This is important at an early stage to obtain an understanding of a design operation. However, where behavioural models exclude performance affecting issues such as input and output impedance, non-linear effects, saturation effects and power supply issues, then the results obtained would need to be used with care.

## 10.8 Future Directions

The need for fault simulation in the future can be considered in relation to the following issues:

- Easier application and formalisation of fault simulation approaches.

- More realistic fault models with a closer relevance to the process defects that they model.

- Formalisation of the models and universal recognition, in particular for analogue circuit faults.

- Reduction in fault simulation resource requirements – both computing hardware, software and application time.

- Mixed-technology fault simulation – where electrical and non-electrical (mixed-technology) simulation is undertaken. This is particularly important for MEMS devices.

## 10.9 Summary

This chapter has discussed the role of test pattern generation and fault simulation in the generation of the test vectors for structural testing of ICs. Structural tests, as opposed to functional tests, aim to stimulate the operation of the circuit so that any faults that may be considered to exist in the design, due to fabrication defects, can be detected. The need to consider the availability of the necessary software tools to support ATPG and fault simulation, along with the effort required to integrate and operate these methods for TPG, require careful consideration. The need also to identify and utilise realistic fault models within the process is problematic, in particular for analogue circuits. The lack of universally agreed analogue fault models and the need to consider the ability for structural testing to partially or completely replace functional testing has to date limited the uptake of analogue fault simulation.

## 10.10 References

[1]   Bushnell M. and Agrawal V., "Essentials of Electronic Testing for Digital, Memory & Mixed-Signal VLSI Circuits", Kluwer Academic Publishers, 2000, ISBN 0-7923-7991-8

[2]   Hurst S., "VLSI Testing digital and mixed analogue/digital techniques", IEE, 1998, ISBN 0-85296-901-5

[3]   Zwolinski M., "Digital System Design with VHDL", Pearson Education Limited, 2000, England, ISBN 0-201-36063

[4]   IEEE Std 1076-2002, IEEE Standard VHDL Language Reference Manual, IEEE, USA

[5]   IEEE 1364-1995, IEEE Standard Verilog® Hardware Description Language, IEEE, USA

[6]   TetraMAX® ATPG, Synopsys Inc., USA

[7]   Synopsys Inc., USA, http://www.synopsys.com

[8]   DFT Compiler®, Synopsys Inc., USA

[9]   Miara A. and Giambiasi N., "Dynamic and deductive fault simulation", Proceedings of the 15th conference on Design Automation Conference, USA, 1978, pp439-443

[10]    Verifault-XL® Fault Simulator User Guide, Cadence Design Systems Inc., USA

[11]    Cadence Design Systems Inc., USA, http:://www.cadence.com

[12]    Verilog-XL® User Guide, Cadence Design Systems Inc., USA

[13]    Grout I., "An Analogue and Mixed-Signal Fault Simulation Tool based on Tcl/Tk and HSpice", Proceedings of the Iberchip 2002 Workshop, Mexico, 2002

[14]    "Antics Analogue Fault Simulation Software", University of Hull, UK, http://www.eng.hull.ac.uk/research/ee_vlsi/antics.htm

[15]    Mir S. et al., "SWITTEST: Automatic Switch-level Fault Simulation and Test Evaluation of Switched-Capacitor Systems2, Proceedings of the 34th Design Automation Conference, 1997, pp281-286

[16]    Straube B., Vermeiren W., Müller B., "Using an Analogue Fault Simulator for Microsystem Fault Simulation", Proceedings of the 4th IEEE International Mixed Signal Testing Workshop, The Netherlands, 1998

[17]    Olbrich T. et al, "A New Quality Estimation Methodology for Mixed-Signal and Analogue ICs", Proceedings of the European Design & Test Conference, France, 1997, pp573-580

[18]    Olbrich T. et al., "Defect-Oriented Vs Schematic-Level Based Fault Simulation for Mixed-Signal ICs", Proceedings of International Test Conference USA, 1996, pp 511-520

[19]    Walker H. and Stephen W., "VLASIC: A Catastrophic Fault Yield Simulator for Integrated Circuits", IEEE Transactions on Computer-Aided Design, Vol.CAD-5, No. 4, October 1986, pp541-556

[20]    Ferguson F. and Shen J., "A CMOS Fault Extractor for Inductive Fault Analysis", IEEE Transactions on Computer-Aided Design, Vol. 7, No. 11, November 1988, pp1181-1194

[21]    Hawkins C. Soden J., Righter A. and Ferguson F., "Defect Classes – An Overdue Paradigm for CMOS IC Testing", Proceedings of International Test Conference, USA, 1994, pp 413-425

[22]    Hawkins C. and Segura J., "Failure Modes in Nanometer Technologies", Tutorial D2, Design and Automation in Europe Conference (DATE), 2003

[23]     Gaitonde D. and Walker D., "Hierarchical Mapping of Spot Defects to Catastrophic Faults – Design and Applications", IEEE Transactions on Semiconductor Manufacturing, Vol. 8, No. 2, May 1995, pp167-177

[24]     Chess B., Roth C. and Larrabee T., "On Evaluating Competing Bridge Fault Models for CMOS ICs", Proceedings of the 12th IEEE VLSI Test Symposium, 1994, pp446-451

[25]     MacMillen D. et al., "An Industrial View of Electronic Design Automation", IEEE Transactions on Computer Aided Design of Integrated Circuits and Systems, Vol. 19, No. 12, December 2000, pp1428-1448

[26]     IEEE standard VHDL analog and mixed-signal extensions, Std 1076.1-1999, IEEE, USA

[27]     Open Verilog International (OVI), Verilog Analog Mixed-Signal Group, http://www.eda.org/verilog-ams/

[28]     Fang L., Kerkhoff H. and Gronthoud G., "Reducing Analogue Fault-Simulation Time by Using High-Level Modelling in Dotss for an Industrial Design", Proceedings of the IEEE European Test Workshop (ETW'01), 2001, pp61-67

[29]     Kilic Y. and Zwolinski M., "Speed-up Techniques for Fault-based Analogue Fault Simulation", Proceedings of the European Test Workshop, 2001

[30]     Perkins A. et al., "Fault Modeling And Simulation Using VHDL-AMS", Analog Integrated Circuits and Signal Processing, Vol. 16, No. 2, 1998, pp141-155.

[31]     Grout I. and Santana J., "Mechatonic System Fault Simulation study with Spectre and Verilog-A", Proceedings of the Mechatronics Forum Conference, The Netherlands, 2002

[32]     Soma M. et al., "Hierarchical ATPG for Analog Circuits and Systems", IEEE Design and Test of Computers, January-February 2001, pp72-81

[33]     Godambe N. and Shi C., "Behavioral Level Noise Modeling and Jitter Simulation of Phase-Locked Loops with Faults using VHDL-AMS", Proceedings of the 15th IEEE VLSI Test Symposium, 1997, pp177-182

# Exercises

### Question 1

Identify the range of ATPG tools available for digital combinational logic and their creators/vendors. For each fault simulator, identify the fault simulation principles on which they are based and the fault models utilised.

### Question 2

Identify the range of ATPG tools available for digital sequential logic and their creators/vendors. For each fault simulator, identify the fault simulation principles on which they are based and the fault models utilised.

### Question 3

Identify the range of ATPG tools available for analogue circuits and their creators/vendors. For each fault simulator, identify the fault simulation principles on which they are based and the fault models utilised.

### Question 4

Identify the range of digital fault simulators available and their creators/vendors. For each fault simulator, identify the fault simulation principles on which they are based and the fault models utilised.

### Question 5

Identify the range of analogue fault simulators available and their creators/vendors. For each fault simulator, identify the fault simulation principles on which they are based and the fault models utilised.

### Question 6

For the fault models identified in Question 5, identify from published material, the range of analogue circuit component values used for modelling analogue circuit faults. How do these relate to the process defects that they are attempting to model?

*The following questions will require access to a fault simulator for digital combinational. In the following questions, the Verifault-XL® is considered, although with suitable modifications, a suitably available logic simulator or fault simulator could be used.*

## Question 7

Consider the circuit in Fig. 3.15 (Chap. 3).

1. Create a gate-level Verilog®-HDL module and testfixture for this design and simulate using Verilog-XL® for all possible input combinations.

2. Modify the circuit description in order to run a fault simulation study using Verifault-XL® and determine the response of the circuit to the inserted faults. What faults were considered and what was the fault coverage figure?

3. What is the minimum set of test vectors and patterns required to detect the maximum number of faults?

## Question 8

Repeat Question 7, except now use the circuit in Fig. 3.26 (Chap. 3).

## Question 9

Repeat Question 7, except now consider the following circuit:

The circuit is a combinational logic circuit with 4 inputs (X1, X2, X3 and X4, where X1 is the least significant bit). The circuit has 7 outputs (a – g) to drive a 7-segment display (see below), where a logic 1 output indicates a segment on the display to be ON. The display is to represent the hexadecimal equivalent of the 4-bit binary input.

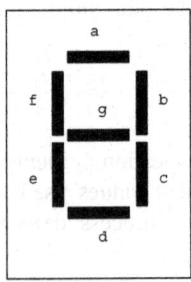

*The following questions will require access to an analogue circuit simulator or a suitable analogue circuit fault simulator. In the following questions, the HSPICE® circuit simulator is considered, although with suitable modifications, a suitably available simulator could be used.*

## Question 10

Repeat Question 7 except now create the circuit design in Spice (transistor level). Identify the assumptions made in the creation of the circuit description and simulation arrangement. How does the effort required to set-up and run the fault simulation study compare with the digital fault simulation study?

## Question 11

Repeat Question 8 except now create the circuit design in Spice (transistor level). Identify the assumptions made in the creation of the circuit description and simulation arrangement. How does the effort required to set-up and run the fault simulation study compare with the digital fault simulation study?

## Question 12

Repeat Question 9 except now create the circuit design in Spice (transistor level). Identify the assumptions made in the creation of the circuit description and simulation arrangement. How does the effort required to set-up and run the fault simulation study compare with the digital fault simulation study?

# Chapter 11
# Automatic Test Equipment (ATE) and Production Test

*Once an IC is in production, the need to test for fabrication defects is essential to providing the customer with working devices and to guarantee low defect levels. The tests undertaken in production must be suitably comprehensive, but at the lowest cost possible. The role of the physical test equipment is essential to achieving this. During production test, Automatic Test Equipment (ATE) is used to reduce the test times by automating as much of the test process as practical.*

## 11.1 Introduction

It could be said that given a perfect design and a perfect fabrication process, then there would be no need for test. However, given that new design ideas are not usually fully functional first time and the issues relating to the fabrication processes (spot defects and process variability), then the role of test is ever more important. An Integrated Circuit (IC) undergoes a range of tests from the initial design specification through to a full-scale production run, and the role of test at the specific identified different points in time reflects the need for obtaining information on the device for a range of reasons. The testing of a design is undertaken for a number of reasons [1]:

- **Design uncertainty**:
  Required to ensure that the design works according to the required specification. During the initial design development and before any device is fabricated, then this is undertaken in simulation. Once prototyped devices have been fabricated, then device characterisation (design debug/functional verification, and determination of parametric (DC and AC) specifications), is undertaken.

- **Fabrication/manufacturing (Production Test)**:

  Once the design has been accepted for production, then production testing (identifying defective devices due to fabrication problems) is undertaken on a device that is considered to be correct in design. Hard errors (local spot defects), soft errors (process variations) and reliability hazards are tested for. This is automated as far as possible as machines can undertake the operations required faster than a person and will also readily repeat procedures.

- **Regulations**:

  It may be necessary to demonstrate that specific regulatory requirements are adhered to, such as EMC/EMI and safety.

- **Maintenance**:

  Once the device is in service, located in its final application, then routine tasks such as checking for correct device operation and periodic circuit calibration are undertaken.

- **Contracts**:

  Specific customers may require particular tests to be undertaken on the devices to match their particular requirements.

Additionally, devices will need to work within specific environmental conditions. Devices will be classified as for **commercial, industrial** or **military** applications.

## 11.2 Production Test Flow

Testing of the device will be undertaken both at the wafer level (prior to packaging) and also on the final packaged devices, see Fig. 11.1. The key point is to identify failures quickly after a major fabrication step so that these failures are not passed on to the next stage in fabrication. The initial stage is to test the wafer to ensure that the electrical properties (*e.g.* material resistivity, MOS transistor threshold voltage, *etc.*) are within the process spread. Special "drop-in" circuits with defined circuit test structures are placed at key points on the wafer for this test. If the wafer is within limits, then each die on the circuit is tested. Parametric and functional tests would be applied. Only those dies that pass the test are then used. Faulty dies are marked (either with ink dots or electronically within an electronic database). The wafer is diced and the individual (working) dies are packaged. The packaged device is tested (**1ˢᵗ level package test**). Specific parametric, functional and structural tests would be undertaken. After packaging, an optional step will be to perform **burn-in**. Here, the device is tested for a set time (*e.g.* 24 h) at an elevated temperature (*e.g.* 125°C) and at an elevated power supply voltage (*e.g.* 1.4 x $V_{DD}$) [3]. This is undertaken to accelerate circuit failure that would be caused by any circuit defects that may exist within the device. These faults may not necessarily occur during production test, but would cause the device to fail early within the final application. After burn-in, a final, **2$^{nd}$ level, package test** is undertaken. Again, specific parametric, functional and structural tests would be undertaken. The cause of a failure will also require analysis and the test results will be fed-back to the process engineer (if the failure was process induced) or the design engineer (if the failure was due to the design).

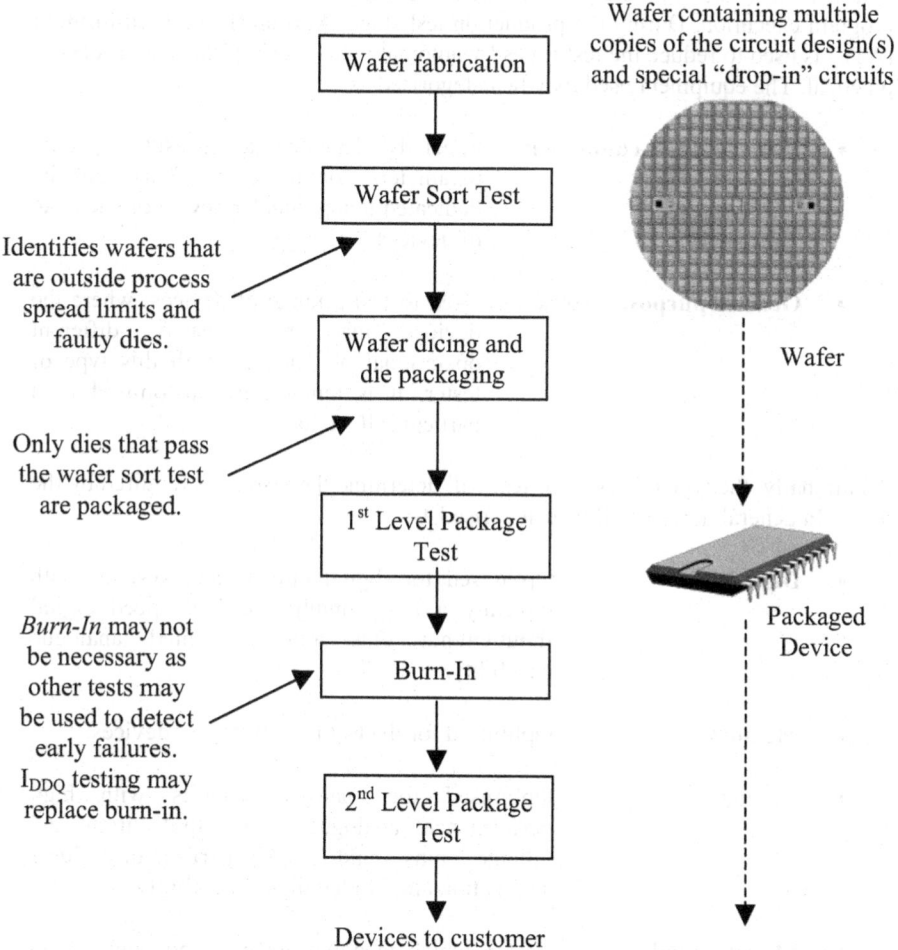

**Fig. 11.1.** Production test steps

It may not be necessary to undertake all the identified stages in Fig. 11.1. For devices requiring a less demanding production test, then a wafer sort test followed by a single package level test would be undertaken.

# 11.3 ATE Systems

In production test, the need is for tests to be undertaken that are suitably comprehensive for the particular product and product application area, but at the lowest cost possible. This means using the most cost-effective equipment possible and minimising the test time per device. The role of the test equipment [2-5] used is essential to achieving this. In general, a (software) test program is run on a tester to implement the required test procedure. The semiconductor testers used require both hardware and software parts in order to set-up and control the tester and test program execution. During the production test stage, **Automatic Test Equipment** (ATE) is used to reduce the test times by automating as much of the test process as practical. The equipment used may be categorised as:

- **Dedicated test equipment**   Specially designed to measure specific parameters for a device. This will be dedicated to a particular device or small set of devices.

- **General purpose testers**    Used to test a range of devices, where the devices may have vastly different operational parameters. With this type of tester, it is temporarily customised to a particular IC to test.

Additionally, the type of device to test will determine the resources required by the tester. In general, testers will be categorised as:

- **Digital**          Optimised for digital circuits and systems with typically a large number of high-speed digital Input/Output pins and a limited analogue capability.

- **Memory**           Optimised for the testing of memory devices.

- **Analogue**         Optimised for analogue circuits with high performance analogue Input/Output current and voltage pins and high performance data acquisition and limited digital capability.

- **Mixed-Signal**     Testers which need to provide a good level of both digital and analogue Input/Output capabilities, but which may not necessarily reach the performance levels attained by digital or analogue testers.

In addition, testers to support **System on a Chip** (SoC) devices and testers to support devices with specific **Design for Testability** (DfT) structures are available from the major ATE vendors. Figure 11.2 shows an example of an ATE system. The *Catalyst* from Teradyne Inc. is a mixed-signal tester system that can be utilised for production testing of devices including SoC, wireless and RF. It is a generic tester that is customised to a particular Device Under Test (DUT) through the test program and a Device Interface Board (DIB).

**Fig. 11.2.** Teradyne Catalyst ATE (courtesy of Teradyne Inc.)

This shows the key features of an ATE system. Figure 11.3 identifies these key components for the ATE in Fig. 11.2.

Dual Monitors    Mainframe    Manipulator

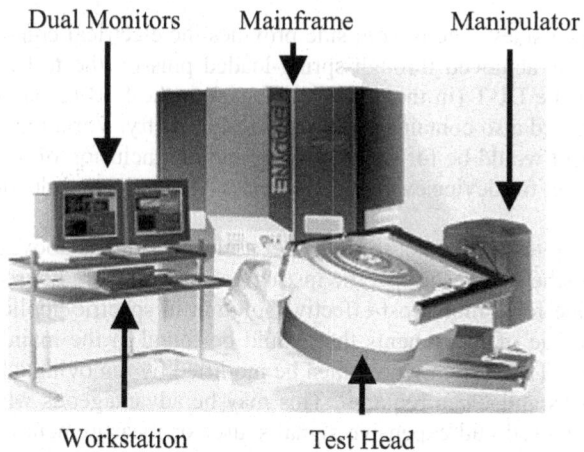

Workstation    Test Head

**Fig. 11.3.** Main components of the ATE system

The workstation acts as the user interface, test program control, data collection and results analysis. It is usual for the workstation to incorporate a single user monitor: however, here a dual monitor system is used to provide the ability to view more information at the same time. The mainframe contains the tester electronics and interfaces between the workstation and the test head. The test head is the physical structure that the **wafer** or individual **Device Under Test** (DUT) is placed on during the test procedure. This test head is a movable structure (mounted on a manipulator) that is connected to the tester mainframe through a cable loom. The test head can therefore be placed some distance from the mainframe. The test head contains additional electronics, close to the DUT, and allows for the DUT to be tested through an intermediate PCB – the **Device Interface Board** (DIB). The DIB is a custom PCB design for a particular DUT and provides the mechanical and electrical interface to the generic tester. Figure 11.4 is an example of a DIB for the Catalyst tester.

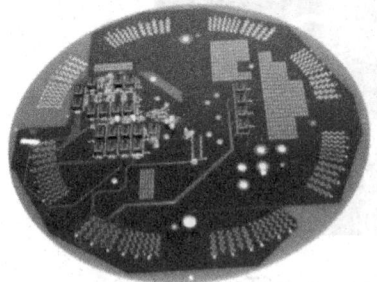 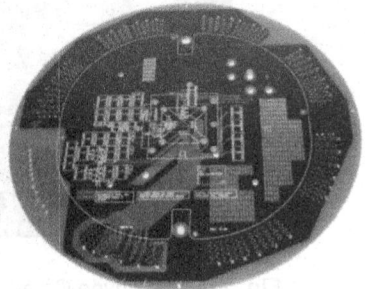

Bottom: Tester side                              Top: DUT side

**Fig. 11.4.** Device Interface Board (courtesy of Teradyne Inc.)

The DIB has two sides. The bottom side provides the electrical connections to the test head. This is achieved through spring-loaded pins on the test head. The top side allows for the DUT (in this case) to connect to the DIB (centre) via pads on the DIB. The board also contains additional local circuitry. Variations on this basic DIB arrangement would be for wafer probing and the inclusion of a zero insertion force (ZIF) socket for device evaluation purposes (rather than production test).

The ATE electronics can be housed in either a **mainframe** (as above), or by using a **rack system** where only the specific instruments required are housed in the rack. This can provide for a more cost-effective solution in specific applications where the complete range of instruments that would be found in the mainframe system are not required. The system would then be modified by removing/adding specific tester modules as and when required. This may be advantageous where the DUT requires a specialised and expensive signal source or measurement capability that would be particular to the particular device or small range of devices – this capability might not be required by the majority of users. Packaged devices will be offered to the tester using a **device handler**. Here, the untested device will be automatically taken from a **handler storage bin** and mated with the DIB. The test procedure is then performed and depending on the results of the test, the device is then placed in one of several **bins** – separating good devices from the devices that have failed particular tests. Good devices may also be **speed rated** – for example, microprocessors that operate correctly at different speeds can be passed and rated according to their maximum operating frequency. Tests may be performed at room temperature, or at specified temperatures. An additional part of the tester will be a **temperature forcing unit**. This can locally control the temperature of the DUT, to be either higher or lower than room temperature.

Test equipment can be sourced from a number of vendors. Table 11.1 provides a list of a number of the major test equipment vendors.

**Table 11.1.** Test equipment and support vendors

| Vendor | Internet resource |
| --- | --- |
| Advantest Corporation | http://www.advantest.com |
| Agilent Technologies | http://www.agilent.com |
| Credence Systems Corporation | http://www.credence.com |
| CTS Division, Analog Devices Inc. | http://www.analog.com |
| Fluence Technology Inc. | http://www.fluence.com |
| LogicVision | http://www.logicvision.com |
| LTX Corporation | http://www.ltx.com |
| Teradyne Inc. | http://www.teradyne.com |

A number of issues relating to ATE systems need to be addressed. These include:

- Number of I/Os and power supplies required by the DUT and available from the tester [6]
- Quality of signals capable to be generated by the tester
- Data capture and analysis capabilities
- Operational frequency limitations due to length of cable loom to the test head
- Test time per IC
- ATE costs:
  > Purchase cost
  > Running costs
  > Operator training costs
  > Upgrade costs

In traditional ATE systems, the test procedures were developed to test one IC at a time (**single site test**). However, there has been increased interest and work undertaken to provide the ability to test multiple ICs in parallel. This **multi-site test** [7, 8] capability has the potential to reduce the test times and so reduce costs. However, the capability to undertake multi-site testing requires the infrastructure to support this:

- ATE hardware support
- Software support for generating the programs to support the parallel testing of ICs

- Operator training and support
- The test economics issues relating to the cost effectiveness of adopting a multi-site test approach

Additionally, the trend has been towards the testers becoming increasingly expensive and newer, lower cost, tester platforms becoming increasingly desirable to use.

# 11.4 Future Directions for ATE Systems

The future direction for ATE[9] systems lies in the ability to provide cost-effective solutions for production test activities. Table 11.2 provides a summary of the main issues.

**Table 11.2.** Future ATE system issues

| Technology driver | Description |
| --- | --- |
| Faster operating speeds | Higher device signal frequencies<br>Driven by communications applications<br>Requirement for ATE to support devices into the GHz frequencies<br>Moving onto the next technology node to improve performance of components and reduce interconnect delays<br>Move into the nanotechnology domain |
| Higher levels of integration (higher density) | Reduced package dimensions<br>More pins on a package<br>Move into the nanotechnology domain<br>SoC and SiP devices<br>SoC devices moving away from mainly logic devices (with some memory) to memory oriented devices (with some logic)<br>Multiple dies and passive components within packages<br>Integration of MEMs (Micro-Electromechanical) devices (*e.g.* sensor integration)<br>Process integration – analogue, digital and memory on the same die |
| Lower operating voltages | A need to improve device reliability by reducing electric field strength in dielectrics<br>Aim to lower device power consumption |

| | Portable, battery operated circuits<br>Devices with multiple power supply voltage levels |
|---|---|
| ATE architectures | Reduced test time<br>Devices with increased complexity<br>Reduced tester costs<br>More flexible ATE architectures [10]<br>Standards to be adopted (open architecture?) [11])<br>Requirement for ATE to support devices into the GHz frequencies<br>Support for multi-site testing<br>Support for DfT within the DUT<br>Support for remote access (Internet based access) for globally distributed DfT teams |

## 11.5 Summary

This chapter has discussed the role of Automatic Test Equipment (ATE) in production test. The role of test equipment and examples were provided, along with considerations for the use of ATE. The future role of test and ATE systems was also provided. Currently, ATE systems from different vendors will have different architectures. The move towards standardisation, and potential open architecture approach, to ATE system design is considered. STIL (IEEE Standard 1450-1999 – Standard Test Interface Language) and the work of the Semiconductor Test Consortium [12] are moves in the right direction for standardisation.

## 11.6 References

[1]    O'Connor P., "Test Engineering, A Concise Guide to Cost-effective Design, Development and Manufacture", Wiley, England, 2001, ISBN 0-471-49882-3

[2]    Bushnell M. and Agrawal V., "Essentials of Electronic Testing for Digital, Memory & Mixed-Signal VLSI Circuits", Kluwer Academic Publishers, 2000, ISBN 0-7923-7991-8

[3]     Rajsuman, R., "System-on-a-Chip Design and Test", Artech House Publishers, USA, 2000, ISBN 1-58053-107-5

[4]     Hurst S., "VLSI Testing digital and mixed analogue/digital techniques", IEE, 1998, ISBN 0-85296-901-5

[5]     Sunter S. and Nadeau-Dostie B., "Complete, Contactless I/O Testing – Reaching the Boundary in Minimizing Digital IC Testing Cost", Proceedings of the International Test Conference, 2002, pp446-455

[6]     Burns M. and Roberts G.W., "An Introduction to Mixed-Signal IC Test and Measurement", Oxford University Press, New York, 2001, ISBN 0-19-514016-8

[7]     Jannesari S., "HP94000 Multi_site Mixed Signal Test Development Considerations", Proceedings of the 16th IEEE Instrumentation and Measurement Technology Conference, Vol. 3, 1999, pp1563 - 1568

[8]     Rivoir J., "Lowering cost of test: parallel test or low-cost ATE?" Proceedings of the 12th Asian Test Symposium, 2003, pp360-363

[9]     International Technology Roadmap for Semiconductors, 2003 Edition, "Test and Test Equipment"

[10]    West B., "Highly Reconfigurable ATE for Rapidly Evolving Test Requirements", Proceedings of the 9th IEEE European Test Symposium, 2004, pp87-93

[11]    Parnas B., Pramanick A., Elston M. and Adachi T., "Software Development for an Open Architecture Test System", Proceedings of the 9th IEEE European Test Symposium, 2004, pp81-86

[12]    The Semiconductor Test Consortium (STC), http://www.semitest.org

# Chapter 12

# Test Economics

*Given that test activities must be undertaken, the cost to test is one of the factors that will determine the commercial viability of an IC. In order to identify the cost to test, the different parts that go into the creation and operation of a production test program must be identified, and a value associated with that part. A test economics model can be created which will allow for these costs to be formally defined and analysed.*

## 12.1 Introduction

The development of a test program for an IC can be a complicated task given the number of tasks that have to be undertaken from test program concept through to test program execution, and the linkages of test to design and fabrication. Throughout the process of the test program creation, there are a number of issues that arise and trade-offs that need to be made in order to arrive at a workable solution. The amount of testing required would be determined by issues including the type of device (digital, analogue, mixed-signal), the complexity of the design, the final application area of the device, and the fabrication process utilised. At one end of the test complexity spectrum, no testing would be undertaken (it could be said that given a perfect design and a perfect fabrication process, then there would be no need for test). In this case, the cost to test would be zero. However, given that design ideas are not usually fully functional first time (the initial design requiring debugging and verification of correct operation), and the issues relating to the fabrication processes (spot defects and process variability), then the role of test is ever more important. At the other end of the test complexity spectrum, the device is tested for every possible functional operation and for every possible fault that may occur in production. This would be extremely time consuming and in this case, the cost to test would be excessive. This then could lead to the overall product cost being prohibitively expensive, and hence uneconomic.

A compromise position must be undertaken between the two scenario extremes, with this position based on the economics of testing the device. This can be formalised by creating a suitable **test economics model**. This would be arranged to allow for the quality of the test to be maintained whilst allowing for the costs to be minimized [1-3]. The purpose of testing a device is to identify failures before they have the chance of being used by the customer. Testing will be undertaken on a circuit for a number of reasons, and at different points in time from design concept through to full device fabrication, see Table 12.1.

**Table 12.1.** Testing the design

| | |
|---|---|
| Simulation | During design development, the model of the design within the CAD tools is simulated in order to verify the correctness of the design. At this time, the test program development would commence and the program operation be simulated |
| Characterisation (device debug/verification) | Undertaken on fabricated samples of a device prior to the production phase. This is undertaken to verify the operation of the design and also to undertake exhaustive functional and parametric (DC and AC) tests. At this time, it is possible for the production test program operation to be verified |
| Production test | Once the design has been fully prototyped, the device will then go into full scale production (fabrication). The production test will need to be undertaken using Automatic Test Equipment (ATE) and each device must be tested in the shortest time possible without compromising the quality of the test |

Ideally, the scenario would exist where the manufacturer would provide the customer with devices knowing that 100% of all devices supplied are fully functional. However, the ability to test for every possible fault would be prohibitively expensive, and may not necessarily be possible to implement. In digital logic, for example, the stuck-at-fault is commonly used for structural test program development despite its known limitations. Despite every care possible being taken to identify device failures and prevent the customer from receiving these failures, the supply of 100% of fully functional devices working is unlikely to happen. The goal is to reduce the number of faulty devices to an acceptable level. The **Defect Level** (DL) is a measure of the test quality and is expressed as the number of faulty ICs that exist within a group of ICs that has passed the test. This is normally expressed in terms of **parts per million** (ppm). Typically, a defect level of 100 ppm would be considered to identify a high quality test. The defect level is determined from analysis of the returned faulty devices. These devices may have failed the customer **incoming acceptance test** (supplied devices are tested by the customer prior to use), may have failed an **in-system test** (the device has passed an incoming acceptance test but fails in the customer application), or may fail in the **in-field test** (either due to a system failure when in normal operation, or during routine maintenance test).

## 12.2 Purpose of a Test Economics Model

The purpose of the test economics model is to identify the cost to test and to formalise these costs within a suitable model. From the analysis of this model, a test with the right quality but at a minimal cost can be determined. The costs [4] may be considered as:

- **Fixed costs**:

  These costs would exist but will not vary with the production volume. For example, the initial purchase cost of the test equipment.

- **Variable costs**:

  These costs would exist and vary with the production volume. For example, the test equipment energy use.

- **Total cost**:

  The sum of the fixed and variable costs.

- **Average cost**:

  The total costs divided by the number of items produced.

A number of models [5] have been considered with the aim to formalise the cost modelling. The particular economics model for a product must take into account the following activities:

- Design

- Fabrication

- Test

- Other (*e.g.* management, marketing)

It is important to identify the costs and also their significance within the overall model. In economic terms, it would be pointless to spend time and effort to reduce costs that are insignificant to the overall cost whilst ignoring the more significant cost issues. However, considering a holistic approach to the overall product development and production problem, insignificant cost issues may be significant in the engineering or other activities that have not been factored into the model. It will depend on the activity the individual person is active in as to their own set of priorities. The increasing importance placed on test in the development and production of an IC product is leading to the increased importance of knowing the **cost to test**. Where once the cost to test was only a small portion of the overall costs of the device, it now presents an increasing percentage of the overall costs. Depending on the device, the test cost has been reported to be as high as 30% - 50% of the overall product (IC) cost. This is particularly so in mixed-signal devices where, when even though compared to digital only devices, the size of the circuitry is relatively small, the types of tests required for the mixed-signal parts can lead to long test times and the need for high performance test equipment. The production test issues and the need for high-cost, high–performance

Automatic Test Equipment (ATE) issues dominate many discussions in relation to reducing the test costs. The need for reduced test times and reduced tester costs are key examples of significant cost items. Traditional ATE systems, with high-performance but closed architectures that perform all tests external to the device under test (DUT), are now considered along with alternative, lower-cost testers with open architectures that support device Built-In Self-Test (BIST) operation.

The ATE costs that need to be considered would include:

- Initial purchase
- Tester maintenance
- ATE vendor support
- Device handler maintenance
- Tester depreciation
- Lifetime of the tester (for example, will it be capable of testing the future generation devices which will anticipated)
- Speed of test (the DUT throughput capabilities – the number of devices that can be tested in a particular time period)
- Upgrade cost
- Operator training
- Power consumption
- Flexibility in operation – multi-tasking a single ATE for different test development and production test activities

The problem is not however solved by simply replacing a high-cost tester with a lower-cost tester. The high-cost ATE systems available have in many cases been optimised for particular production test scenarios and are well supported by the vendors. They have the capability of high DUT throughput that allow for the test program to be operated in a specific time and at high speed – perhaps not at the full device speed, but will incorporate specific high-speed circuitry for fast signal generation and data capture/analysis. Of course, the meaning of "high-cost" and "lower-cost" is relative to the particular test scenario and does need to be used with care. If the choice of ATE is not carefully considered, the use of a lower-cost tester, whilst initially may seem to be a sensible approach to adopt, could result in higher test costs: they may not necessarily be capable of high throughput (therefore longer test times) and maintenance, vendor support and upgrade potentials may be less mature. As an addition to considering the performance (and performance limitations) of the ATE, design for testability (DfT) and BIST are considered to enable adequate levels of device controllability and observability, and to take workload away from the ATE into the device itself in order to test specific parts of the design. Part of the economics model will need to identify the level of DfT and BIST required and to enable a justification of the additional design effort and size of circuitry.

The time taken for particular activities to be undertaken impacts on the economics of a product. Identifying ways in which to obtain the required results in a shorter time are essential to cost minimisation. The need to control the **"time-to-quality"** (time taken to obtain the required quality levels), **"time-to-yield"** (time taken to obtain the required process yield) and **"time-to-market"** (time taken to get the product to the customer) are important challenges. The **process yield** is defined as the percentage of acceptable parts fabricated in the total number of parts fabricated. In a new fabrication process, the yield will be low as there will be the need to embed and commission the production equipment, and to adjust the process in order to obtain the required characteristics. The yield will need to be improved to the acceptable levels before it may be used in production.

The design activity has been successful in time reduction with the use of advanced computer aided design (CAD) tools, design automation (design compilers and HDL synthesis) tools and design reuse to improve the designer efficiency. Test activities are becoming more formalised with support for computer aided test (CAT), formal methods to test program generation and test program reuse.

## 12.3 Test Economics Model Development

The development of a test economics model [5, 6] is not trivial and the identification of the significant vs insignificant costs can be difficult to identify. It will require an understanding of the particular company operations and internal costing procedures, along with the external factors.

In many scenarios, an IC design development (from concept through to production) is undertaken by a product team, and this structure of the team may be used as the basis of the test economics model creation since all necessary company activities would be involved in the team. For example, the model must include the following activities and specific aspects that are internal to the activity, or cross-boundaries between activities:

- Design:

    Design engineer time; engineer training; purchase, maintenance and upgrade of CAD tools; the incorporation of DfT and BIST into the design.

- Fabrication:

    Process engineer time; engineer training; process set-up, maintenance and enhancement costs; Size of circuit die (cost per $mm^2$ of circuitry); packaging costs; raw material costs.

- Test:

    Test engineer time; engineer training; purchase, maintenance and upgrade of CAT tools; purchase, maintenance, upgrade and depreciation of test equipment costs; the incorporation of DfT and BIST into the design.

- Other:

    Project management costs; product marketing costs; company overheads.

Given that a suitable model can be created and incorporates the necessary information, the model is then created and may be used. The structure of the model [7, 8] must be carefully considered, with issues to be addressed that include:

- The choice of the model implementation (paper or software):

    The software based approach would in many situations be the preferred choice, so then a suitable tool would be required.

- The model should be user friendly:

    The usefulness of the model would be based on its' usability. A difficult to use model would not likely be successful as it would be too complicated and awkward to use.

- The model should be intuitive to use.

- It should support coarse and detailed modelling:

    As information relating to the model parts may not necessarily be available at the creation of the model, or only simplistic information available, the model must be capable of working with a range of information and information detail from little information (coarse modelling) through to detailed information (detailed modelling).

- It should clearly identify the significant and insignificant costs in order to allow for particular cost issues to be addressed.

- It should readily allow for "what if" scenarios to be undertaken in the model analysis.

- It should readily support the capability for exporting information out of the model, and for documentation generation.

- It should easily support model maintenance (a modular approach to model development would support this) and upgrading. Such an approach would allow for parts or all of the model to also be reusable.

- It should allow for the latest information (process parameter forecasting, process trend analysis, market information, *etc.*) to be used.

- It should allow for the different stages of test from test development though characterisation/debug to production to be modelled, and the impact of specific tests (and the complexity of the tests undertaken) to be considered, with trade-offs in cost, time and the amount of useful information gathered.

- It should allow for trade-offs between tester and DfT/BIST techniques to be considered.

## 12.4 Summary

This chapter has introduced the idea for test economics modelling. The requirement to minimise the test to cost whilst providing a test program with the right quality, is a key factor driving the decisions made in the creation and implementation of suitable test programs. In the test economics model, the various costs and their significance are considered. This will include design, fabrication, test and other project related issues that impact on the test program requirements and costs. Aspects relating to the development of a test economics model were identified and discussed.

## 12.5 References

[1]     International Technology Roadmap for Semiconductors, 2003 Edition, "Executive Summary"

[2]     International Technology Roadmap for Semiconductors, 2003 Edition, "Test and Test Equipment"

[3]     Volkerink E., Khoche A., Kamas L., Rivoir J. and Kerkhoff H., "Tackling Test Trade-offs from Design, Manufacturing to Market using Economic Modeling", Proceedings of the International Test Conference, 2001, pp1098-1107

[4]     Bushnell M. and Agrawal V., "Essentials of Electronic Testing for Digital, Memory & Mixed-Signal VLSI Circuits", Kluwer Academic Publishers, 2000, ISBN 0-7923-7991-8

[5]     Dance D., "Modeling Test Cost Ownership", Proceedings of the $3^{rd}$ International Conference on the Economics of Design, Test and Manufacturing, 1994, pp13-17

[6]     Burns M. and Roberts G.W., "An Introduction to Mixed-Signal IC Test and Measurement", Oxford University Press, New York, 2001, ISBN 0-19-514016-8

[7]     Volkerink E., Khoche A., Rivoir J. and Hilliges K., "Test Economics for Multi-site Test with Modern Cost Reduction Techniques", Proceedings of the $20^{th}$ IEEE VLSI Test Symposium, 2002

[8]     Iyengar V., Goel S., Marinissen E. and Chakrabarty K., "Test Resource Optimization for Multi-Site Testing of SOCs Under ATE Memory Depth Constraints", Proceedings of the International Test Conference, 2002, pp115-1168

# Appendix A

# Introduction to VHDL

## A.1 Introduction

VHDL is a **Hardware Description Language (HDL)** used to describe digital circuit and system operation. It is one of the two IEEE standard HDLs in use today:

- Verilog®-HDL          IEEE Standard 1364-2001
- VHDL                  IEEE Standard 1076-2002

The idea for VHDL (VHSIC HDL – Very High Speed Integrated Circuit Hardware Description Language) [1-4] began in 1980 under a US DoD (Department of Defense) requirement for the design of digital circuits following a common design methodology, providing the ability for self-documentation and reuse with new technologies.

The development of VHDL commenced in 1983 and the language became an IEEE standard in 1987 (IEEE Std 1076-1987). The language has been revised since then in 1993, 2000 and 2002, the latest release being 1076-2002. VHDL also has a number of associated standards relating to modeling and synthesis.

The choice of whether to use Verilog®-HDL or VHDL is not simple, with each language providing a range of benefits and limitations in its ability for use. A number of the key aspects in the choice of language to use include:

- The availability of suitable Electronic Design Automation (EDA) tools to support the language use (including design management capabilities and availability of tool use within a specific IC design project).

- Previous knowledge and personal preferences.

- Availability of simulation models and synthesis capabilities.

VHDL allows for a digital circuit or system to be described at a number of levels of abstraction starting at the technology independent high-level **behavioural** description and working through **RTL** (Register Transfer Level) to a **structural** (logic gate and interconnect) level. Both VHDL and Verilog®-HDL provide for the ability for design descriptions to be synthesised, usually from the RTL level, into a technology specific **structural** description. Examples of VHDL code are provided in Appendix F.

## A.2 Entities and Architectures

A design is described in VHDL as an **entity declaration** and an associated **architecture body**. An entity may have one or more associated architectures, but only one architecture would be used at any one time:

- The **entity declaration** defines the design interface and is used at higher levels of hierarchy, where a hierarchical design is created.

- The **architecture body** describes the operation of the entity as:

   The circuit behaviour
   The interconnection to other design entities
   A mixture of circuit behaviour and interconnection to other design entities

An example of a VHDL entity/architecture pair describes the circuit identified in Fig. A.1.

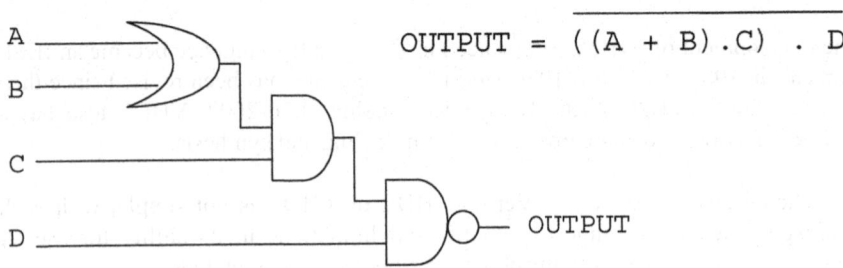

$$\text{OUTPUT} = \overline{((A + B).C) . D}$$

**Fig. A.1.** Combinational logic circuit

An example VHDL description for this circuit is provided in Fig. A.2.

```
library IEEE;
use IEEE.STD_LOGIC_1164.ALL;
use IEEE.STD_LOGIC_ARITH.ALL;
use IEEE.STD_LOGIC_UNSIGNED.ALL;


entity example is
    Port ( A : in std_logic;
           B : in std_logic;
           C : in std_logic;
           D : in std_logic;
           OUTPUT : out std_logic);
end entity example;


architecture Behavioural of example is

begin

        OUTPUT <= not (((A or B) and C) and D);

end architecture Behavioural;
```

**Fig. A.2.** VHDL entity/architecture pair

# A.3 Libraries, Packages and Configurations

VHDL provides the ability to define subprograms and to provide configurations in order to set-up specific scenarios. **Libraries** provide a means for design management. These contain mechanisms for accessing design entities, configurations and subprograms. These can be **reference** (read only) libraries or a user **work** (both read and write capabilities) library. **Packages** are used to group sub-programs and design entities. **Configurations** are used for customising design entities (**generics** can be used within entity declarations which are then customised for a particular use), and associates a particular architecture body to a particular entity declaration.

# A.4 VHDL Testbench

The testbench is VHDL is an entity/architecture pair that specifies the stimulus to apply to the design (circuit under test). It contains an instance of the top-level design along with the required stimulus. An example testbench for the description in Fig. A.2 is shown in Fig. A.3. This applies all possible binary input codes, with each code change every 10ns.

```
LIBRARY ieee;
USE ieee.std_logic_1164.ALL;
USE ieee.numeric_std.ALL;


ENTITY testbench IS
END testbench;


ARCHITECTURE behavior OF testbench IS

        COMPONENT example
        PORT(
                A, B, C, D : IN  std_logic;
                OUTPUT :     OUT std_logic);
        END COMPONENT;

        SIGNAL A, B, C, D, OUTPUT :  std_logic;

BEGIN

uut: example PORT MAP(A => A, B => B, C => C, D => D, OUTPUT => OUTPUT);

testbench : PROCESS
BEGIN
            wait for  0 ns; A <= '0'; B <= '0'; C <= '0'; D <= '0';
            wait for 10 ns; A <= '0'; B <= '0'; C <= '0'; D <= '1';
            wait for 10 ns; A <= '0'; B <= '0'; C <= '1'; D <= '0';
            wait for 10 ns; A <= '0'; B <= '0'; C <= '1'; D <= '1';
            wait for 10 ns; A <= '0'; B <= '1'; C <= '0'; D <= '0';
            wait for 10 ns; A <= '0'; B <= '1'; C <= '0'; D <= '1';
            wait for 10 ns; A <= '0'; B <= '1'; C <= '1'; D <= '0';
            wait for 10 ns; A <= '0'; B <= '1'; C <= '1'; D <= '1';
            wait for 10 ns; A <= '1'; B <= '0'; C <= '0'; D <= '0';
            wait for 10 ns; A <= '1'; B <= '0'; C <= '0'; D <= '1';
            wait for 10 ns; A <= '1'; B <= '0'; C <= '1'; D <= '0';
            wait for 10 ns; A <= '1'; B <= '0'; C <= '1'; D <= '1';
            wait for 10 ns; A <= '1'; B <= '1'; C <= '0'; D <= '0';
            wait for 10 ns; A <= '1'; B <= '1'; C <= '0'; D <= '1';
            wait for 10 ns; A <= '1'; B <= '1'; C <= '1'; D <= '0';
            wait for 10 ns; A <= '1'; B <= '1'; C <= '1'; D <= '1';
            wait for 10 ns;

END PROCESS testbench;

END;
```

**Fig. A.3.** VHDL testbench

# A.5 References

[1]     IEEE Standard VHDL Language Reference Manual, IEEE Std 1076-2002, IEEE

[2]     Smith D., "HDL Chip Design", Doone Publications, USA, 1996, ISBN 0-9651934-3-8

[3]     Navabi Z., "VHDL Analysis and Modeling of Digital Systems", McGraw-Hill International Editions, 1993, ISBN 0-07-112732-1

[4]     Zwolinski M., "Digital System Design with VHDL", Pearson Education Limited, 2000, England, ISBN 0-201-36063-2

# Appendix B

# Introduction to Verilog®-HDL

## B.1 Introduction

Verilog®-HDL is a **Hardware Description Language (HDL)** [1, 2] used to describe digital circuit and system operation. It is one of the two IEEE standard HDLs in use today:

- Verilog®-HDL         IEEE Standard 1364-2001
- VHDL                IEEE Standard 1076-2002

Verilog®-HDL [3] was released in 1983 by Gateway Design System Corporation, together with a Verilog®-HDL simulator. In 1985, the language and simulator were enhanced – with the introduction of the Verilog-XL® simulator.

In 1989, Cadence Design Systems, Inc. brought the Gateway Design System Corporation. In early 1990, Verilog®-HDL and Verilog-XL® were separated to become two separate products. Verilog®-HDL, until then a proprietary language, was released into the public domain. This was to facilitate the dissemination of knowledge relating to Verilog®-HDL and to compete with VHDL, with VHDL already existing as a non-proprietary language. Additionally, in 1990, Open Verilog International (OVI) was formed as an Industry consortium, consisting of Computer Aided Engineering (CAE) vendors and Verilog®-HDL users, in order to control the language specification.

In 1995, Verilog®-HDL was reviewed and then adopted as an IEEE standard – IEEE Standard 1364 (becoming IEEE Std 1364-1995). In 2001, the standard was reviewed, the latest version of the standard now being IEEE Std 1364-2001.

The choice of whether to use Verilog®-HDL or VHDL is not simple, with each language providing a range of benefits and limitations in its' ability for use. A number of the key aspects in the choice of language to use include:

- The availability of suitable Electronic Design Automation (EDA) tools to support the language use (including design management capabilities and availability of tool use within a project).
- Previous knowledge and personal preferences.
- Availability of simulation models and synthesis capabilities.

# B.2 Verilog®-HDL Modules

Within Verilog®-HDL, the module describes the design. All functions and procedures are contained within the module statement for the design. A design will consist of one or more interconnected modules, and each of these modules has an interface port to other modules in order to describe how they are interconnected.

The modules are contained within ASCII text files. For design maintenance and ease of understanding, it is common to place one design module per file, although this is not a requirement. The **top-level module** specifies the overall system in which other modules are used. Where modules are contained in separate text files, then these are included in the top-level design, or next level up in multi-level hierarchical designs.

# B.3 Verilog®-HDL Testfixture

The testfixture is a Verilog®-HDL module that specifies the stimulus to apply to the design (circuit under test) and simulator commands. It contains an instance of the top-level design, along with the required stimulus.

# B.4 Data Types

Verilog®-HDL uses **wire** and **reg** data types. Net data types are **wire**, whilst register data types are **reg**. A **wire** has a circuit equivalent of an electrical wire (net) that connects circuit elements. A **reg** is updated under the control of a procedural flow – a **reg** type variable stores the last value that was procedurally assigned to it.

The **reg** and **wire** data objects may have the following possible values:

| | |
|---|---|
| 0 | logical zero or false. |
| 1 | logical one or true. |
| x | unknown logical value. |
| z | high impedance output of a tristate gate. |

The **reg** variables are initialised to **x** at the start of the simulation. Any unconnected **wire** variable has an **x** value. The size of a wire or register (single bit or multi-bit) can be specified in the module declaration. For example:

```
reg [7:0]  C;
reg [7:0]  A, B;
wire [3:0] DataOut;
wire DataIn;
```

specifies registers A,  B and C to be 8-bits wide, with the most significant bit being the 7$^{th}$ bit. The wire DataOut is 4-bits wide and DataIn is 1-bit wide.

# B.5 Lexical Conventions

Verilog®-HDL is close to the C++ programming language. The language is case sensitive and comments are created on a single line by starting the line with a //, or by /* to */ where comments span across several lines.

A number of keywords, *e.g.* **module,** are defined and typed in all lower case letters. Spaces are also important in the language.

Numbers can be specified in the traditional form of a series of digits (with or without a sign), but also in the following form:

```
<size><base format><number>
```

where <size> contains *decimal* digits that specify the size of the number in the number of *bits*. The <size> is optional. The <base format> is the single character ' followed by one of the following characters b, d, o and h, which stand for binary, decimal, octal and hex, respectively. The <number> part contains digits that are legal for the <base format>. For example:

```
549         // decimal number.
'h8FF       // hex number.
'o752       // octal number.
4'b0101     // 4-bit binary number "0101".
3'b11x      // 3-bit binary number with least
            // significant bit unknown.
5'd3        // 5-bit decimal number.
```

A string is a sequence of characters enclosed in double quotes:

```
"this is a string"
```

# B.6 Example

Consider the following combinational logic circuit, see Fig. B.1. The Verilog®-HDL design module is shown in Fig. B.2. The testfixture file and simulation results are shown in Figs. B.3 and B.4, and the design module uses in-built Verilog®-HDL logic gate models.

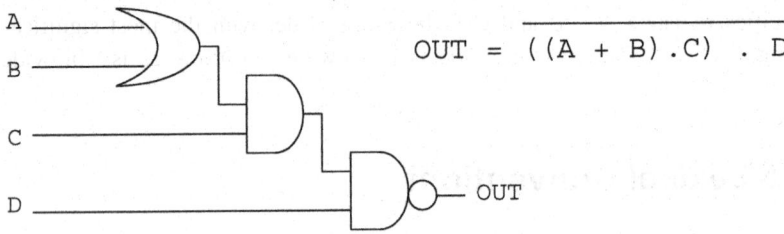

**Fig. B.1.** Combinational logic circuit

Two internal nodes (X1 and X2) are created for the OR and AND gate outputs.

```
///////////////////////////////
// Verilog-HDL model for the
// 3-input logic circuit
///////////////////////////////

module gates (A, B, C, D, OUT);

input A;
input B;
input C;
input D;

output OUT;

or    I0 (X1, A, B);
and   I1 (X2, X1, C);
nand  I2 (OUT, X2, D);

endmodule

///////////////////////////////
// End of File
///////////////////////////////
```

**Fig. B.2.** Design module

```
/////////////////////////////////////////////////////////////////
// Testfixture for Verilog model for the 3-Input logic circuit
/////////////////////////////////////////////////////////////////

///////////////////////////////////////////////////////////
// Set timescales
///////////////////////////////////////////////////////////

`timescale 1ns / 1ps

///////////////////////////////////////////////////////////
// Define testfixture module
///////////////////////////////////////////////////////////

module testfixture();

/////////////////////////////////////
// Inputs and Output
/////////////////////////////////////

    reg A;
    reg B;
    reg C;
    reg D;

    wire OUT;

/////////////////////////////////////
// Instantiate the UUT
/////////////////////////////////////

gates UUT (.A(A), .B(B), .C(C), .C(C), .D(D), .OUT(OUT));

/////////////////////////////////////////////////////////////
// Initialize Inputs and set end of simulation time
/////////////////////////////////////////////////////////////

initial

begin

$write("\n\nStarting simulation run at time ...", $time, "\n\n");

$write("\t\t\tTime\tA\tB\tC\tD\tOUT\n\n");

A = 1'b0; B = 1'b0; C = 1'b0; D = 1'b0;

#160 $write("\n\nCompleting simulation run at time ...", $time, "\n\n");

$finish;

end

/////////////////////////////////////////////////
// Stimulus as clocks using 'always' statement
/////////////////////////////////////////////////

always # 10 D = ~D; always # 20 C = ~C;

always # 40 B = ~B; always # 80 A = ~A;

always #5 $write($time, "\t", A, "\t", B, "\t", C, "\t", D, "\t",OUT, "\n");

endmodule

/////////////////////////////////////////////////////////////
// End of File
/////////////////////////////////////////////////////////////
```

**Fig. B.3.** Testfixture file

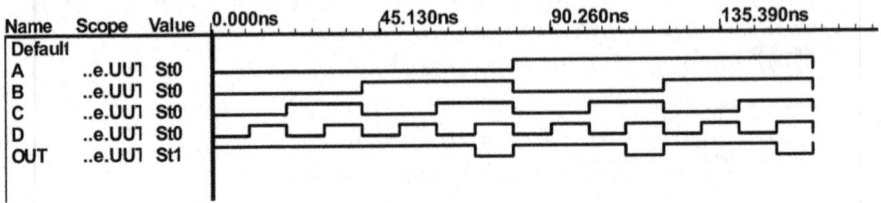

**Figure B.4.** Simulation results

# B.7 References

[1]   Smith M., "Application Specific Integrated Circuits", Addison-Wesley, 1999, ISBN 0-201-50022-1

[2]   Smith D., "HDL Chip Design", Doone Publications, USA, 1996, ISBN 0-9651934-3-8

[3]   IEEE 1364-1995, IEEE Standard Verilog Hardware Description Language, IEEE

# Appendix C
# Introduction to Spice

## C.1 Introduction

Within Integrated Circuit (IC) design and test activities, a great deal of the work is performed using Computer Aided Design (CAD) and simulation tools running on either a PC (with a Windows or LINUX operating system) or workstation (with a UNIX operating system) platform. For design activities, circuit simulation studies are part of the normal design process. Increasingly in test development, the use of simulation tools underlies the development and verification of the required test programs. Digital, analogue or mixed-simulation capability would be required dependent on the nature of the design. At one level, a simulation study would aim to be undertaken in order to perform the types of tests that would be undertaken with the physical hardware. However, an advantage of simulation is the capability of accessing nodes within a circuit that would not be accessible with the physical hardware. This is particularly important for Integrated Circuit (IC) based circuits where direct access to internal nodes of the physical circuit is not possible.

For analogue circuit simulation, the use of Spice [1-3] is dominant:

SPICE

<u>S</u>imulation <u>P</u>rogram with <u>I</u>ntegrated <u>C</u>ircuit <u>E</u>mphasis

Spice originated from the University of California, at Berkeley, in the early 1970s and has dominated the world of analogue simulation. Apart from a few company specific proprietary languages and commercially available simulators such as Spectre® from Cadence Design Systems Inc. (which itself supports Spice based circuit descriptions), simulators based on Spice are the most widely used, varying in their performance and supporting software such as schematic capture and results waveform viewers. However sophisticated the simulation tool, the underlying requirements for a textural description of the circuit, definition of the input stimuli and the type of analysis to perform, are common and adhere to the Spice syntax requirements.

The range of Spice simulators currently available include:

- Spice3 simulator, UC Berkeley.
- AIM-spice, AIMS-Software, Norway.
- Pspice® (part of the OrCAD product line), Cadence Design Systems, Inc.
- HSPICE® (part of the Discovery AMS simulation suite), Synopsys Inc.
- T-Spice, Tanner Research Inc.

## C.2 Scope of Discussions

The following introduction to Spice, whilst not exhaustive in its scope, is aimed to provide an introduction to relevant aspects of the language in order to start using the language for test development purposes. A comprehensive discussion to a particular simulator should be obtained from the relevant simulator documentation.

In the following description, the development of example *"Spice Input Files"* for the HSPICE® circuit simulator are provided. Where HSPICE® specific commands are included, they are noted. However, the syntax for the circuit and analysis type are non-simulator specific.

Whatever approach is taken to modelling and simulation, it must be considered with care. Whilst simulation studies do allow for the functionality of a design to be understood prior to a physical prototyping and final design production, simulation results are only ever as good as the underlying models and the accuracy of the simulator, along with the expertise of the operator. It is quite possible to realise a working design in simulation that will not work in reality!

## C.3 Input File Format

The simulator will analyse a circuit defined in an input file. This file is an ASCII text file that describes the circuit to simulate, the input stimuli (voltages and currents) and the type of simulation to perform, along with commands required to control the simulator. Whilst the format of the input file is flexible in the construction (for example, Spice is not case-sensitive), it would be good practice to adopt a particular format and always follow that format. The input file format adopted here will be as follows in Fig. C.1.

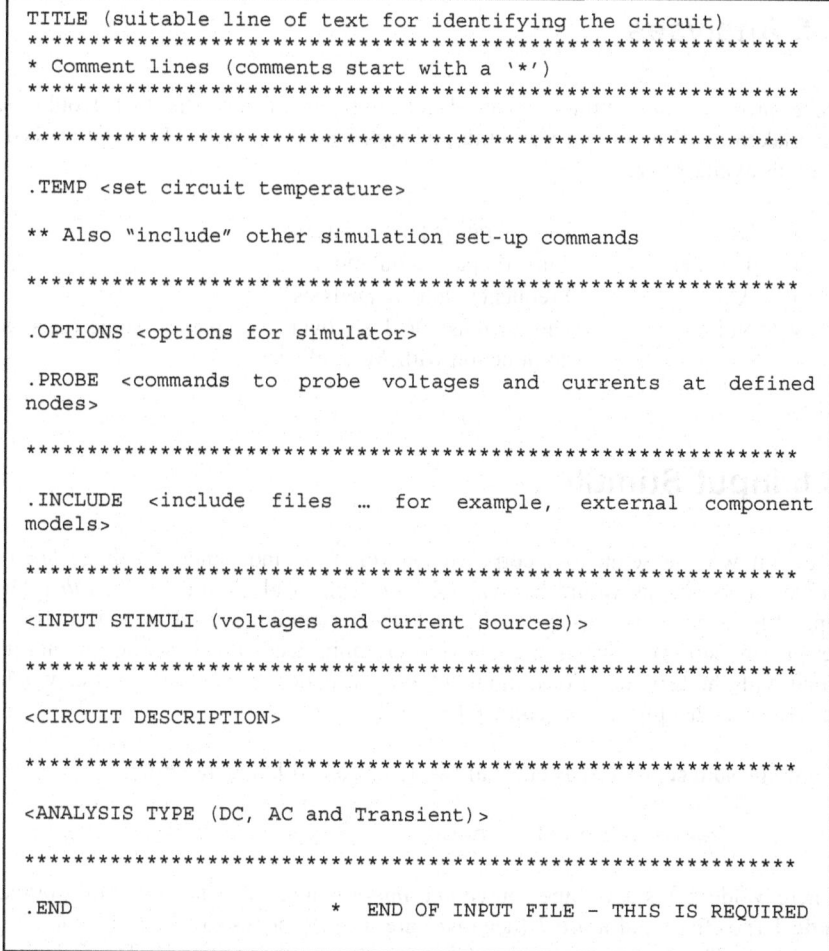

```
TITLE (suitable line of text for identifying the circuit)
******************************************************************
* Comment lines (comments start with a `*')
******************************************************************

******************************************************************

.TEMP <set circuit temperature>

** Also "include" other simulation set-up commands

******************************************************************

.OPTIONS <options for simulator>

.PROBE  <commands  to  probe  voltages  and  currents  at  defined
nodes>

******************************************************************

.INCLUDE  <include  files  …  for  example,  external  component
models>

******************************************************************

<INPUT STIMULI (voltages and current sources)>

******************************************************************

<CIRCUIT DESCRIPTION>

******************************************************************

<ANALYSIS TYPE (DC, AC and Transient)>

******************************************************************

.END                     *   END OF INPUT FILE - THIS IS REQUIRED
```

**Fig. C.1.** Spice input file format

# C.4 Commenting the Input File

Commenting any piece of software code is essential to providing a running commentary of the code author, date of creation, modification summary and code operation, etc. Comments in Spice begin with a single asterix (*) and can start at the beginning of a line, or in-line after the defined code. Additionally, in the Spice Input File, the first line of the file is for commenting (that is, it is used to enter a suitable title for the tool to use). The first line cannot be used for circuit code, as it will be treated only as a comment.

# C.5 Analyses

Spice provides the ability to set different types of analysis that would be undertaken as if the circuit was built and tested on the bench. The basic analysis methods available are:

- DC          DC operating point analysis.
- Transient    Time domain simulation.
- AC          Frequency domain analysis.
- Noise        Circuit noise analysis over a frequency range (used in conjunction with AC analysis).

# C.6 Input Stimuli

A circuit will be setup to incorporate one or more independent voltage and/or current sources – as would be available for a physical circuit "*in the lab*". The input file is be treated as a "*testbench*" (the terminology used in VHDL based design simulation) – that is it contains everything needed to describe the circuit, along with the test conditions, input stimuli and nodes to monitor – just as would be created with a physical circuit "*in the lab*".

A **voltage source** for a transient analysis definition will take the form

```
Vname    node1    node2    <type of source>
```

Where **V** identifies a voltage source and ***name*** is a suitable name for the source. **node 1** (positive) and **node 2** (negative) are the two nodes to which the source is connected. The type of source can take one of a number of forms. The following identifies the **DC, PWL, PULSE** and **SIN** sources. For example, a 5V DC voltage source between nodes "*IN*" and "*0*" (*0* defines the circuit common node) would be defined as:

```
Vin  IN  0  5
```

A **current source** can be defined in the same manner as a voltage source, except now the identifier is **I** instead of **V**. For example, a 10µA DC current source between nodes "*X*" and "*0*" (*0* defines the circuit common node) would be defined as:

```
Iin  X  0  10u
```

# C.7 Defining Components in Spice

The basic (primitive) passive and active circuit elements include:

- Resistor
- Capacitor
- Inductor
- Magnetic elements
- Bipolar Junction Transistor (BJT)
- Metal Oxide Semiconductor Field Effect Transistor (MOSFET)
- Junction Field Effect Transistor (JFET)

In addition, signal source (voltage and current) and behavioural models (analogue and to a certain extent, digital) circuit elements are utilised.

# C.8 Scale Factors in Spice

For component and input source value definition, writing out large or small numbers would be laborious. Scaling values by a power of ten (using the letter "E" – e.g. e+6 (times ten to the power of 6)) can be used. For convenience, the following suffix to express a value can be used, see Table C.1.

**Table C1.** Scale factors in Spice

| Suffix | Prefix | Power of 10 | Example use |
|--------|--------|-------------|-------------|
| F | femto- | $10^{-15}$ | 10 femto-farads |
| P | pico- | $10^{-12}$ | 10 pico-farads |
| N | nano- | $10^{-9}$ | 5 nano-henries |
| U | micro- | $10^{-6}$ | 5 micro-volts |
| M | milli- | $10^{-3}$ | 1 milli-ohm |
| K | kilo- | $10^{+3}$ | 10 kilo-ohm |
| MEG | mega- | $10^{+6}$ | 5 mega-volts |
| G | giga- | $10^{+9}$ | 1 giga-hertz |
| T | tera- | $10^{+12}$ | 1 tera-ohm |
| MIL | (0.001") | $2.54 \times 10^{-6}$ | --- |

## C.9 Dealing with Design Hierarchy

In all but the simplest of circuits, designs will incorporate several levels of design hierarchy. This would be required to define multiple instances of a cell design, along with the need to develop parts of the overall design and for these parts to be integrated. In Spice, the definition and use of subcircuits allows for this scenario. When subcircuits are utilised, care must be taken in order to identify and define "*global*" nodes (*i.e.* nodes that are required throughout the entire design such as power supply nodes).

## C.10 References

[1]      *SPICE*: Simulation Program with Integrated Circuit Emphasis, Version 3f5, University of California, Berkeley, USA

[2]      Tuinenga P., "SPICE, A Guide to Circuit Simulation and Analysis using Pspice", 3rd Edition, Prentice Hall, 1995, ISBN 0-13-158775-7

[3]      HSpice®, Synopsys Inc., USA

# Appendix D

# Introduction to MATLAB®

## D.1 Introduction

The MATLAB® [1, 2] software tool from *The Mathworks Inc.* provides for both a programming language and a powerful computational environment that works with complex arithmetic and matrix operations, and integrates mathematical computing and data visualisation tasks. These tasks are underpinned with the tool using its own modelling language. MATLAB® is accompanied with a range of Toolboxes, Blocksets and other tools that allow for a range of engineering and scientific applications. In such an approach, various ideas can be investigated as part of an overall design process in order to arrive at a final and optimal solution. The Toolboxes and Blocksets are utilised for:

- Data acquisition
- Data analysis and exploration
- Visualisation and image processing
- Algorithm prototyping and development
- Modelling and simulation
- Programming and application development

Examples of the currently available Toolboxes and Blocksets are shown in Table D.1. MATLAB® is widely used in the design of complex algorithms to be implemented in either electronic hardware or software. For example, in the design of data converters (A/D and D/A), the converter algorithm to be implemented can be designed and verified at the algorithm level prior to any detailed circuit design work. Both time domain and frequency domain operation can be quickly ascertained early in a design process. The same concepts can be extended to the test development area, with both DfT operation and BIST algorithms, along with

291

test program and tester hardware, modelled and simulated alongside the circuit under test model.

**Table D.1.** Toolboxes and blockets

| SIMULINK® | An interactive tool allowing for the modelling, simulation and analysis of dynamic, multidomain systems. A graphical, block diagram approach is adopted |
|---|---|
| Control system yoolbox | A collection of algorithms that implement common control system design, analysis, and modeling techniques |
| Signal processing toolbox | A collection of MATLAB® functions that provides a customisable framework for analogue and digital signal processing |
| Communications blockset | A blockset that builds on the SIMULINK® system level design environment for the modelling of the physical layer of a communication system |
| Communications toolbox | A library of MATLAB® functions that supports the design of communication system algorithms and components. It builds on the powerful capabilities of MATLAB® and the Signal Processing Toolbox by providing functions to model the physical layer of a communication system |
| Fuzzy logic toolbox | Provides a graphical user interface to support the steps involved in fuzzy logic design |

# D2. Interactive and Batch Modes

MATLAB® allows for two different methods in which to execute commands:

- **Interactive mode**:

    In the interactive mode, commands are typed (or cut-and-pasted) into the MATLAB® *Command Window.* In this mode, each step to be performed is explicitly typed into the command window by the user. Whilst this provides for a great deal of flexibility and user control, it is time consuming and can involve repetitive tasks.

- **Batch mode**:

    In the batch mode, a series of commands are saved in a text file which has a '.m' extension. The batch commands in the file (referred to as an 'm file') are then executed by typing the name of the file at the MATLAB® command prompt. The advantage to using an 'm file' is that small changes can be made to the code (and also used in different MATLAB® sessions) without having to recall and re-enter the entire set of commands.

# D3. Scalar Variables and Arithmetic Operators

The following provides an example use of MATLAB® by using the Command Prompt that appears in the *Command Window*. The >> is the command prompt and, by default, MATLAB® echoes back the value of the assignment that has just been entered. The "echo" can be turned off by placing a semicolon at the end of a statement.

Scalar variables can be assigned as follows:

```
>> x = (2 + 3) * 15

x =

    75

>>


>> x = (2 + 3) * 15;

>>
```

A range of functions are provided for scalar variable manipulation. Additionally, by creating an 'm file', user-defined functions can be created.

# D.4 Useful Commands

Variables in MATLAB® follow the usual variable naming conventions - any combination of letters, numbers and underscore symbols ('_') can be used, provided that the first character is a letter. Additionally, variable names are case sensitive ('X' is different from 'x').

Two useful MATLAB® commands are **who** and **whos**. These report the names and dimensions of all variables currently in the MATLAB® workspace. Help can be obtained at the command prompt by typing the keyword help followed by the command to get the help on. For example:

```
>> help who

>> help fft
```

## D.5 Matrix Variables and Arithmetic Operators

The following example is a 2 x 2 matrix (variable name a) which is entered into the MATLAB® workspace and then multiplied by itself to create a new variable (b).

```
>> a = [ 2 3; 3 2 ]

a =
      2      3
      3      2

>> b = a*a

ans =
     13     12
     12     13

>>
```

A range of functions are provided for matrix manipulation. Additionally, by creating an 'm file', user-defined functions can be created.

## D.6 Signals and Results Plotting

Results can be viewed in text form in the *Command Window*, either stored in a text file, or viewed graphically using the plot commands. The following example shows the creation of a sine wave (1V peak, 1kHz frequency, no DC offset) (variable input) and a mathematical manipulation of the waveform (variable output). Both waveforms are then plotted, see Fig. D.1.

```
>> time_step = [0:0.00001:0.005];
>> amplitude = 1.0;
>> frequency = 1000;
>> input = amplitude * sin(2 * pi * frequency * time_step);
>> plot(time_step, input);
>> hold on;
>> grid;
>> output = 1.0 + (0.5 * input);
>> plot(time_step, output, 'r');
```

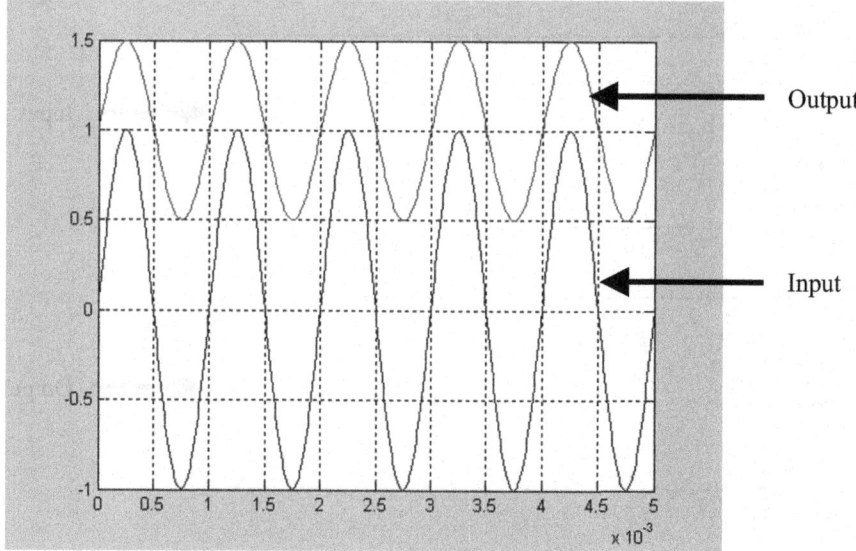

**Fig. D.1.** MATLAB® plot example

The two plots can be drawn separately using sub-plots:

```
>> time_step = [0:0.00001:0.005];
>> amplitude = 1.0;
>> frequency = 1000;
>> input = amplitude * sin(2 * pi * frequency * time_step);
>> subplot(2,1,1);
>> plot(time_step, input);
>> grid;
>> output = 1.0 + (0.5 * input);
>> subplot(2,1,1);
>> plot(time_step, output);
```

In this example, see Fig. D.2, the input is placed on the top sub-plot and the output on the bottom sub-plot. The number of sub-plots to draw is set by the user with the command:

$$subplot(a, b, c);$$

that subdivides the current plot into an a x b array of areas for plotting. The $c^{th}$ area in the array is selected to be active (*i.e.* to be drawn in).

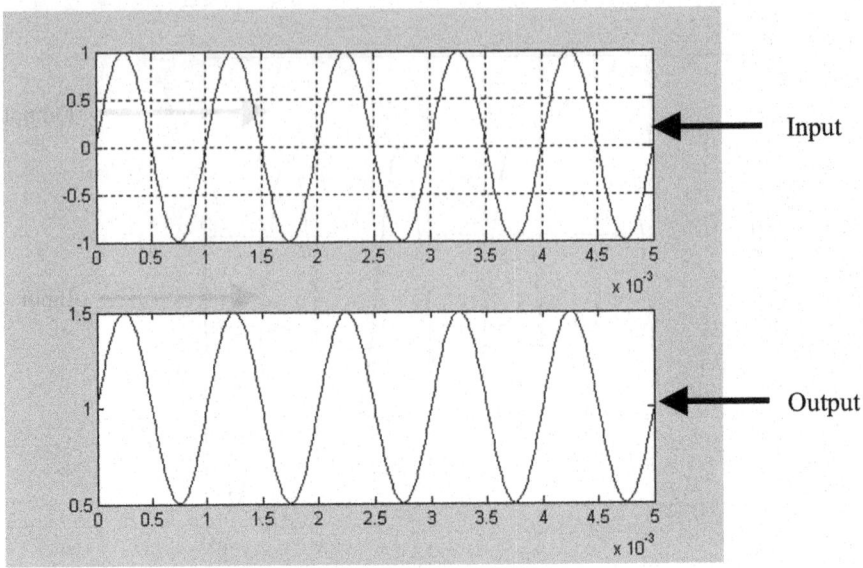

Input

Output

**Fig. D.2.** MATLAB® plot example

# D.7 References

[1]      The Mathworks Inc., http://www.themathworks.com

[2]      Hanselman D. and Littlefield B., "Mastering Matlab 6 – A Comprehensive Tutorial and Reference", Prentice Hall, USA, 2001, ISBN 0-13-019468-9

# Appendix E

# Hardware Experimentation

## E.1 Introduction

Test engineering concepts can be understood and developed through practical experience. Undertaking both simulation exercises and practical hardware experiments are essential to this achieving this understanding. Both simulation and hardware experiments have their uses:

- **Simulation exercises** provide for a means to investigate circuits and test techniques through the development of suitable circuit models and testfixtures. Typically, for digital circuits and systems, these would typically be undertaken using Verilog®-HDL and/or VHDL model simulation. The circuit operation can be determined through the use of a simulator and the insertion of specific logical faults, or by using a specific fault simulation tool. For analogue circuits, this would typically be undertaken using Spice simulation.

- **Hardware experiments** provide a means in which to investigate the test program development and application, along with understanding the test and measurement equipment issues. Both local and remote (Internet based) experiments can be undertaken.

Simulation can be used where access to hardware is not deemed useful in a particular learning scenario, or where access to test and measurement equipment is limited. A number of simulation tools are freely available for use, with specific licensing agreements that must be adhered to, and can be helpful in the understanding and interpretation of specific test concepts. Hardware experiments are useful where simulation has limitations, and for the user to gain invaluable experience of "real world" issues such as measurement limitations and the effect of the measurement equipment on the signal being measured. In fact, the basic principles of an ATE system can be demonstrated on a modest, PC based, tester arrangement that can be built in the laboratory.

# E.2 Generic Tester Arrangement

In the following discussions, the basic idea is to develop a PC based tester that may demonstrate IC test principles and test program development. The tester is based around a PC running a test program interfacing to a tester hardware arrangement and then to the device or circuit under test. In relation to the DUT/CUT, this would be an IC, or circuit comprising a number of ICs mounted on a PCB, in order to demonstrate test principles for the following classes of circuit:

- Digital logic
- Memory (RAM and ROM)
- Analogue circuit
- Mixed-signal circuit

The basic construction of the tester is shown in Fig. E.1. Here, a PC/Workstation is connected through one of the available ports (parallel or serial) to the tester hardware. The main controller circuitry physical interface to the PC/Workstation is via a dedicated communications hardware/software arrangement. Both analogue and digital I/O is provided for with suitably sized ADC, DAC and digital I/O buffer circuitry. The analogue output is passed through a low-pass filter before application to the DUT/CUT. The choices in tester design would be based on a number of issues, including:

- **PC or UNIX workstation**.

- **Operating System**:
  Would be dependent on the PC (Microsoft® Windows® or Linux) or workstation (UNIX®) to use.

- **Software**:
  A suitable programming language would be required to develop the test program and provide for interfacing between the PC/Workstation and tester hardware. For PC based applications, the C/C++, Java, Visual Basic™, *etc.* languages could be utilised.

- **Tester Hardware**:
  The tester hardware will need to provide for a great level of flexibility, suited for ease of maintenance and provide the ability for readily updating the system. A

processor would typically be used, although a suitably designed PLD (CPLD or FPGA) can also provide for a great level of flexibility.

- **I/O Capability**:

   Speed (rate of data transfer between the tester and DUT/CUT), and capacity (number of digital I/O, number of analogue I/O and data converter (A/D and D/A) resolution and conversion rates).

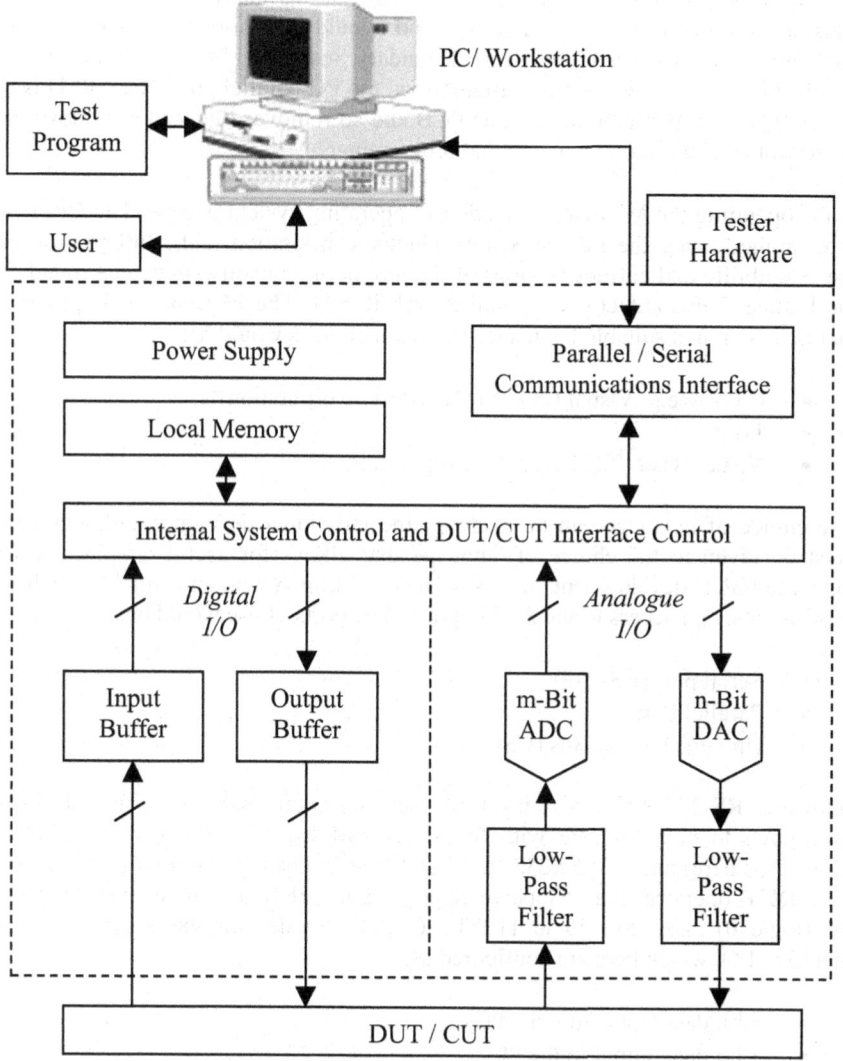

**Fig. E.1.** Generic tester arrangement

# E.3 CPLD Based Logic and Memory Tester

A specific implementation of the generic tester is shown in Fig. E.2. Here, a Lattice Semiconductor Corporation [1] CPLD is utilised – the Mach4A5–64/32 is a suitably sized device that would allow for a range of small to medium sized logic and memory (logic implementation) circuits to be implemented. The device operates on a +5V power supply voltage.

The CPLD is configured to act as the tester electronics (signal generator and signal capture, along with a serial (RS-232) communications interface to/from the PC). This is configured prior to first use and would not normally change in an application. However, a JTAG (IEEE Boundary Scan Std 1149.1) interface on the CPLD allows the device to be configured via the PC parallel port. The CPLD is to be configured whilst mounted on the PCB and so removes the need to remove the device for configuration in an external programmer.

A PC operating the Microsoft® Windows® operating system is used. The CPLD is programmed using the Lattice Semiconductor Corporation tools. Information on the availability and current versions of the programming software can be found on the Lattice Semiconductor Corporation website [1]. The PC runs a test program that is written in a suitable language. The main choices would be:

- C/C++ (e.g. Visual C++™ (Microsoft® Corporation))
- Java
- Visual Basic™ (Microsoft® Corporation)

The choice of language can be made by the individual developer, although care must be given to the choice of language that allows for useful Graphical User Interface (GUI) development, access to the operating system and suitable database applications, and access to the PC I/O ports. The ports of use would be:

- Serial port (RS-232)
- Parallel Port
- Universal Serial Bus (USB)

Here, the RS-232 serial port is to be used for communications. The CPLD is configured to communicate with the serial port via a UART IC (the 6402 is considered here) and a MAX232 (level shifter IC). The overall circuit, as does the CUT/DUT, operates on a +5V power supply. Hence all digital logic levels are to be 0V (logic 0) and +5V (logic 1). The CPLD provides for the ability for user specified I/O, which here are configured as:

- 8 bit data input from the PC
- 8 bit data output to the PC
- $m$ bit data output to the CUT/DUT
- $n$ bit data input from the CUT/DUT

- UART control signals
- JTAG connections for device configuration (fixed pins on the device)

The exact configuration would be determined by the individual user and should be based on the range of DUT/CUT circuits to be tested.

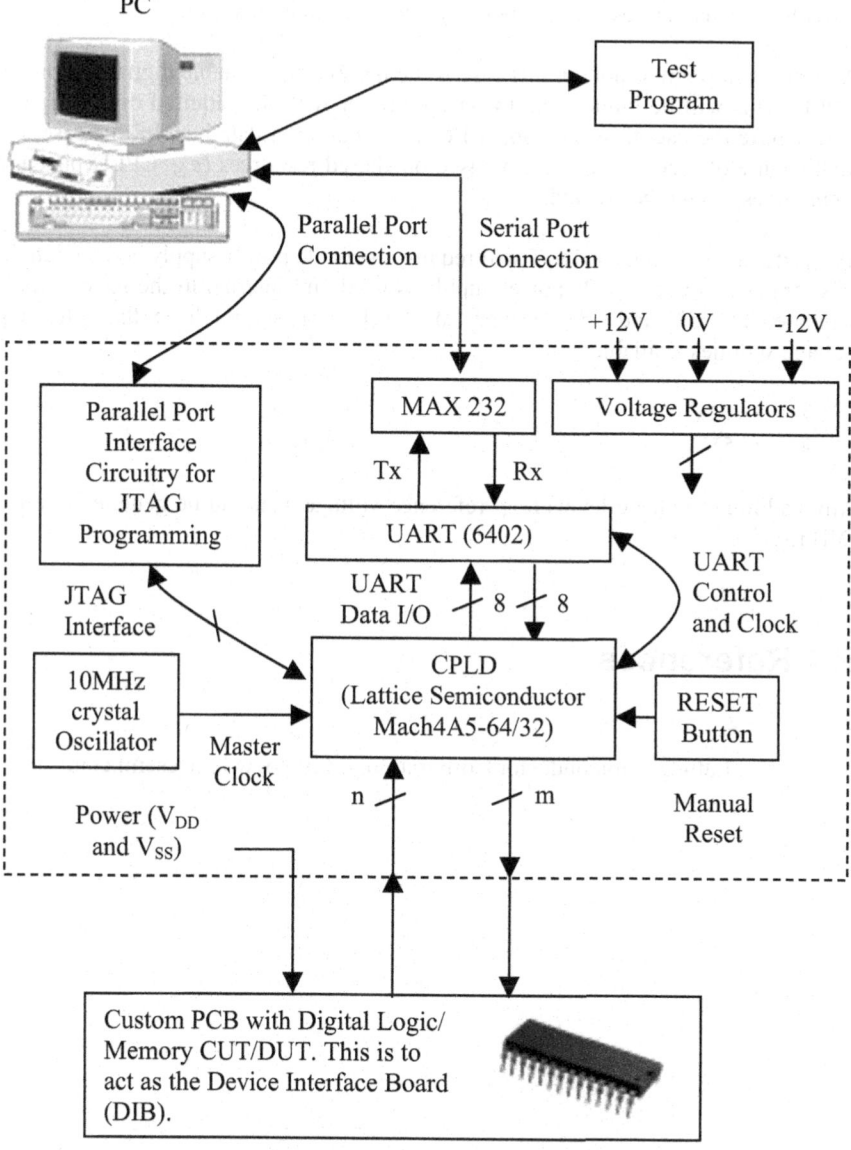

**Fig. E.2.** CPLD implementation of the generic tester

The circuit is run under the control of a 10MHz master clock signal, which can be created using a suitable crystal oscillator circuit. Internal to the CPLD, then clock division would be built into the design. For example, the serial communications link would require a clock that would be 16 times the baud rate of the link. A suitable baud rate would be set (to be chosen by the tester developer) and the UART control signals generated accordingly. A manual reset input (push-button) is provided for manual reset and temporary circuit disable operations.

An important point to note in this case is the number of available digital I/O on the CPLD. This will be limited and to set up a tester with the required digital signals will require the use of bi-directional I/O and a common data bus. In this case, the architecture of a computer system based on shared resources (*e.g.* CPLD pins and signal lines) would be created.

The tester board and CUT/DUT will require a suitable power supply connection. In this arrangement, a +/-12V power supply is used and internal to the tester, this is regulated to +5V and −5V. Hence, the CUT/DUT should have the following voltage supplies available:

- +-/12V
- +/-5V

Any additional voltage levels (*e.g.* reference voltages) should be generated on the DIB itself.

# E.4 References

[1]      Lattice Semiconductor Corporation, http://www.latticesemi.com

# Appendix F

# VHDL Simulation Examples

## F.1 Introduction

This appendix provides a set of digital logic and memory circuit models in VHDL, along with associated testbenches. These examples are identified in Table F.1.

**Table F.1.** VHDL Simulation Examples

| Identifier | Description |
|---|---|
| Combinational logic | |
| F.2 | 3-Input combinational logic circuit |
| F.3 | 8-Input priority encoder |
| Sequential logic | |
| F.4 | 3-bit binary counter |
| F.5 | 3-bit linear feedback shift register (LFSR) |
| Read Only memory | |
| F.6 | Simple 4 x 8-bit ROM |
| Random Access memory | |
| F.7 | Simple 4 x 8-bit RAM |

The first two examples are simple combinational logic circuits and are provided with their own testbench examples. The second two examples are small sequential logic circuits provided as the design descriptions only. These require suitable testbenches to be created. The third two examples are memory cells and are provided with their own testbench examples. The first memory is a ROM cell with an example memory content, whilst the second memory is a RAM cell.

These examples may be run as they are presented using a suitable VHDL simulator. However, they may also be modified and this allows for an investigation into "what if" scenarios. For example, the insertion of logical fault models (e.g. the stuck-at-fault and the bridging fault (wired-AND and wired-OR)), may be undertaken by either modification to the circuit description or to the testbench.

In the examples, where the design files contain references to logic gates, these will be used from a reference library called "COMPONENTS"

```
LIBRARY COMPONENTS;
USE COMPONENTS.LOGIC.ALL;
```

and the following logic gates will be used:

- INV     Inverter
- AND2    2-input AND gate
- AND3    3-input AND gate
- OR2     2-input OR gate
- OR3     3-input OR gate
- XOR2    Exclusive-OR gate
- DFF     D-Type bistable, active **high** asynchronous reset
- DFFs    D-Type bistable, active **high** asynchronous set

These logic gates would need to exist in the COMPONENTS library, or a suitable library with an alternative name if the reference to the appropriate library was used.

# F.2 3-Input Combinational Logic Circuit

The following circuit is a 3-input logic gate (A, B, C) with 1-primary output (Z), see Fig. F.1. The truth-table for this design is shown in Table F.2

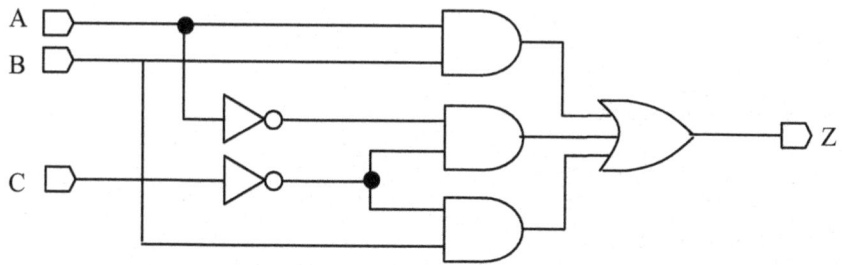

**Fig. F.1.** Circuit schematic

**Table F.2.** Truth Table

| A | B | C | Z |
|---|---|---|---|
| 0 | 0 | 0 | 1 |
| 0 | 0 | 1 | 0 |
| 0 | 1 | 0 | 1 |
| 0 | 1 | 1 | 0 |
| 1 | 0 | 0 | 0 |
| 1 | 0 | 1 | 0 |
| 1 | 1 | 0 | 1 |
| 1 | 1 | 1 | 1 |

The design is described in the following VHDL source files:

- **gates_vhdl.vhd**:
  The VHDL entity and architecture describing the design.

- **test_gates_vhdl.vhd**:
  The VHDL testbench which increments the logic circuit providing a binary count through all possible input codes.

In the VHDL code for the design, the functionality of the logic gates themselves are obtained from the Library "COMPONENTS".

## Design (gates_vhdl.vhd)

```
-------------------------------------------------------------------
-- VHDL description for combinational logic circuit.
-------------------------------------------------------------------


-------------------------------------------------------------------
-- Library Definitions
-------------------------------------------------------------------


LIBRARY ieee;
USE ieee.std_logic_1164.ALL;
USE ieee.numeric_std.ALL;
LIBRARY COMPONENTS;
USE COMPONENTS.LOGIC.ALL;


-------------------------------------------------------------------
-- Entity Definition
-------------------------------------------------------------------


ENTITY gates_vhdl IS
    PORT ( A        :       IN      STD_LOGIC;
           B        :       IN      STD_LOGIC;
           C        :       IN      STD_LOGIC;
           Z        :       OUT     STD_LOGIC);
end gates_vhdl;


-------------------------------------------------------------------
-- Architecture Definition
-------------------------------------------------------------------


ARCHITECTURE SCHEMATIC OF gates_vhdl IS
    SIGNAL node1  :         STD_LOGIC;
    SIGNAL node2  :         STD_LOGIC;
    SIGNAL node3  :         STD_LOGIC;
    SIGNAL node4  :         STD_LOGIC;
    SIGNAL node5  :         STD_LOGIC;

    COMPONENT AND2
        PORT ( I0 :         IN      STD_LOGIC;
               I1 :         IN      STD_LOGIC;
               O  :         OUT     STD_LOGIC);
    END COMPONENT;

    COMPONENT INV
        PORT ( I  :         IN      STD_LOGIC;
               O  :         OUT     STD_LOGIC);
    END COMPONENT;

    COMPONENT OR3
        PORT ( I0 :         IN      STD_LOGIC;
               I1 :         IN      STD_LOGIC;
               I2 :         IN      STD_LOGIC;
               O  :         OUT     STD_LOGIC);
    END COMPONENT;

BEGIN
    I1 : AND2 PORT MAP (I0=>A, I1=>B, O=>node3);
    I2 : AND2 PORT MAP (I0=>node1, I1=>node2, O=>node4);
    I3 : AND2 PORT MAP (I0=>node2, I1=>B, O=>node5);
    I4 : INV  PORT MAP (I=>A, O=>node1);
    I5 : INV  PORT MAP (I=>C, O=>node2);
    I6 : OR3  PORT MAP (I0=>node3, I1=>node4, I2=>node5, O=>Z);
END SCHEMATIC;

-------------------------------------------------------------------
-- End of File
-------------------------------------------------------------------
```

## Design Test Bench (test_gates_vhdl.vhd)

```vhdl
--------------------------------------------------------------------
-- VHDL Test Bench Created to simulate source file gates.vhd
--------------------------------------------------------------------


--------------------------------------------------------------------
-- Library Definitions
--------------------------------------------------------------------

LIBRARY ieee;
USE ieee.std_logic_1164.ALL;
USE ieee.numeric_std.ALL;

LIBRARY COMPONENTS;
USE COMPONENTS.LOGIC.ALL;


--------------------------------------------------------------------
-- Entity Definition
--------------------------------------------------------------------

ENTITY testbench IS
END testbench;


--------------------------------------------------------------------
-- Architecture Definition
--------------------------------------------------------------------

ARCHITECTURE behavioral OF testbench IS

    COMPONENT gates_vhdl
    PORT( A        :        IN        STD_LOGIC;
          B        :        IN        STD_LOGIC;
          C        :        IN        STD_LOGIC;
          Z        :        OUT       STD_LOGIC);
    END COMPONENT;

    SIGNAL A        :        STD_LOGIC;
    SIGNAL B        :        STD_LOGIC;
    SIGNAL C        :        STD_LOGIC;
    SIGNAL Z        :        STD_LOGIC;

BEGIN

    UUT: gates_vhdl PORT MAP(
                A => A,
                B => B,
                C => C,
                Z => Z);

    test_stimulus : PROCESS
    BEGIN
       WAIT for 0 ns;    A <= '0'; B <='0'; C <= '0';
       WAIT for 100 ns; A <= '0'; B <='0'; C <= '1';
       WAIT for 100 ns; A <= '0'; B <='1'; C <= '0';
       WAIT for 100 ns; A <= '0'; B <='1'; C <= '1';
       WAIT for 100 ns; A <= '1'; B <='0'; C <= '0';
       WAIT for 100 ns; A <= '1'; B <='0'; C <= '1';
       WAIT for 100 ns; A <= '1'; B <='1'; C <= '0';
       WAIT for 100 ns; A <= '1'; B <='1'; C <= '1';
       WAIT for 100 ns;
    END PROCESS;
END;


--------------------------------------------------------------------
-- End of File
--------------------------------------------------------------------
```

# F.3 8-Input Priority Encoder

The priority encoder is a combinational logic circuit that gives an order of importance to the inputs; see Fig. F.2. The truth-table for an 8-input priority encoder is shown in Table F.2. In this the 8 inputs (A7 – A0) are encoded to produce 3 outputs (B2 – B0) which is a binary up-count defined by the shifting of the logic '1' in the input signal.

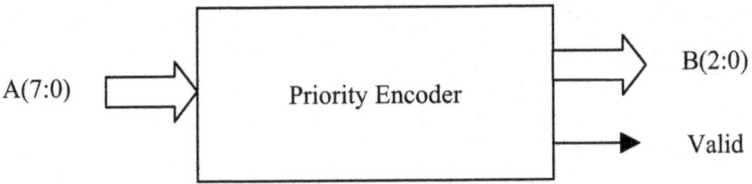

**Fig. F.2.** 8-input priority encoder

**Table F.3.** 8-input priority encoder

| Inputs | | | | | | | | Outputs | | |
|---|---|---|---|---|---|---|---|---|---|---|
| A7 | A6 | A5 | A4 | A3 | A2 | A1 | A0 | B2 | B1 | B0 |
| 0 | 0 | 0 | 0 | 0 | 0 | 0 | 1 | 0 | 0 | 0 |
| 0 | 0 | 0 | 0 | 0 | 0 | 1 | 0 | 0 | 0 | 1 |
| 0 | 0 | 0 | 0 | 0 | 1 | 0 | 0 | 0 | 1 | 0 |
| 0 | 0 | 0 | 0 | 1 | 0 | 0 | 0 | 0 | 1 | 1 |
| 0 | 0 | 0 | 1 | 0 | 0 | 0 | 0 | 1 | 0 | 0 |
| 0 | 0 | 1 | 0 | 0 | 0 | 0 | 0 | 1 | 0 | 1 |
| 0 | 1 | 0 | 0 | 0 | 0 | 0 | 0 | 1 | 1 | 0 |
| 1 | 0 | 0 | 0 | 0 | 0 | 0 | 0 | 1 | 1 | 1 |

The design is described in the following VHDL source files:

- **encoder.vhd**:

    The VHDL entity and architecture describing the design.

- **test_encoder.vhd**:

    The VHDL testbench which increments the encoder providing both expected (as defined in the truth-table) and a selected number of unexpected inputs (the remainder of the 256 possible inputs).

In the VHDL code for the design, an additional output (**Valid**) is provided and is a logic '1' when the expected input codes are applied, and a logic '0' whenever any unexpected input is encountered.

# Design (encoder.vhd)

```
------------------------------------------------------------------
-- VHDL model for a Priority Encoder
------------------------------------------------------------------


------------------------------------------------------------------
-- Library Definitions
------------------------------------------------------------------

library IEEE;
use IEEE.STD_LOGIC_1164.ALL;
use IEEE.STD_LOGIC_ARITH.ALL;
use IEEE.STD_LOGIC_UNSIGNED.ALL;

------------------------------------------------------------------
-- Entity Definition
------------------------------------------------------------------

entity encoder is
    Port (A :     in    std_logic_vector(7 downto 0);
          B :     out   std_logic_vector(2 downto 0);
          Valid : out   std_logic);
end entity encoder;

------------------------------------------------------------------
-- Architecture Definition
------------------------------------------------------------------

architecture simple of encoder is

begin

    B <=    "000"    when    A = "00000001"  else
            "001"    when    A = "00000010"  else
            "010"    when    A = "00000100"  else
            "011"    when    A = "00001000"  else
            "100"    when    A = "00010000"  else
            "101"    when    A = "00100000"  else
            "110"    when    A = "01000000"  else
            "111"    when    A = "10000000"  else
            "000";

Valid <= '1'    when    (A = "00000001" or  A = "00000010"
                or       A = "00000100" or  A = "00001000"
                or       A = "00010000" or  A = "00100000"
                or       A = "01000000" or  A = "10000000")
                else     '0';

end architecture simple;

------------------------------------------------------------------
-- End of File
------------------------------------------------------------------
```

## Design Test Bench (test_encoder.vhd)

```
------------------------------------------------------------------
-- VHDL Test Bench Created to simulate source file encoder.vhd
------------------------------------------------------------------


------------------------------------------------------------------
-- Library Definitions
------------------------------------------------------------------

LIBRARY ieee;
USE ieee.std_logic_1164.ALL;
USE ieee.numeric_std.ALL;

------------------------------------------------------------------
-- Entity Definition
------------------------------------------------------------------

ENTITY testbench IS
END testbench;

------------------------------------------------------------------
-- Architecture Definition
------------------------------------------------------------------

ARCHITECTURE behavior OF testbench IS

        COMPONENT encoder
        PORT(   a : IN std_logic_vector(7 downto 0);
                b : OUT std_logic_vector(2 downto 0);
                valid : OUT std_logic);
        END COMPONENT;

        SIGNAL a :  std_logic_vector(7 downto 0);
        SIGNAL b :  std_logic_vector(2 downto 0);
        SIGNAL valid :  std_logic;

BEGIN

uut: encoder PORT MAP(a => a, b => b, valid => valid);

    test_stimulus : PROCESS
    BEGIN
  -- Normal (expected) inputs
        wait for 0 ns;    a <= "00000001";
        wait for 100 ns;  a <= "00000010";
        wait for 100 ns;  a <= "00000100";
        wait for 100 ns;  a <= "00001000";
        wait for 100 ns;  a <= "00010000";
        wait for 100 ns;  a <= "00100000";
        wait for 100 ns;  a <= "01000000";
        wait for 100 ns;  a <= "10000000";
  -- Selected unexpected inputs
        wait for 100 ns;  a <= "00000000";
        wait for 100 ns;  a <= "00000011";
        wait for 100 ns;  a <= "00001100";
        wait for 100 ns;  a <= "00110000";
        wait for 100 ns;  a <= "11000000";
        wait for 100 ns;  a <= "11111111";
    END PROCESS;

END;


------------------------------------------------------------------
-- End of File
------------------------------------------------------------------
```

# F.4 8-3-Bit Binary Counter

The binary counter is a sequential logic circuit that has a state change on a clock edge. The circuit, see Fig. F.3, changes state in a binary increment step of 1.

The circuit is active low reset (asynchronous) and the D-Type bistables used are positive edge triggered.

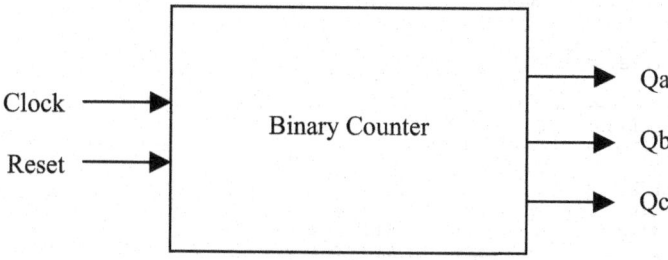

**Fig. F.3.** Binary counter

The design is described in the following VHDL source file:

- **counter.vhd**:
  The VHDL entity and architecture describing the design.

The counter design requires a suitable testbench to be created.

## Design (counter.vhd)

```
------------------------------------------------------------------
-- VHDL description for 3-bit counter circuit.
------------------------------------------------------------------


------------------------------------------------------------------
-- Library Definitions
------------------------------------------------------------------


LIBRARY ieee;
USE ieee.std_logic_1164.ALL;
USE ieee.numeric_std.ALL;

LIBRARY COMPONENTS;
USE COMPONENTS.LOGIC.ALL;


------------------------------------------------------------------
-- Entity Definition
------------------------------------------------------------------


ENTITY counter IS
    PORT ( Clk    :       IN      STD_LOGIC;
           Reset  :       IN      STD_LOGIC;
           A      :       OUT     STD_LOGIC;
           B      :       OUT     STD_LOGIC;
           C      :       OUT     STD_LOGIC);

end counter;


------------------------------------------------------------------
-- Architecture Definition
------------------------------------------------------------------


ARCHITECTURE structural OF counter IS
    SIGNAL An   :         STD_LOGIC;
    SIGNAL Bn   :         STD_LOGIC;
    SIGNAL Cn   :         STD_LOGIC;
    SIGNAL Db   :         STD_LOGIC;
    SIGNAL Dc   :         STD_LOGIC;
    SIGNAL X1   :         STD_LOGIC;
    SIGNAL X2   :         STD_LOGIC;
    SIGNAL X3   :         STD_LOGIC;

    COMPONENT AND2
        PORT ( I0  :      IN      STD_LOGIC;
               I1  :      IN      STD_LOGIC;
               O   :      OUT     STD_LOGIC);
    END COMPONENT;

    COMPONENT AND3
        PORT ( I0  :      IN      STD_LOGIC;
               I1  :      IN      STD_LOGIC;
               I2  :      IN      STD_LOGIC;
               O   :      OUT     STD_LOGIC);
    END COMPONENT;

    COMPONENT DFF
        PORT ( C   :      IN      STD_LOGIC;
               CLR :      IN      STD_LOGIC;
               D   :      IN      STD_LOGIC;
               Q   :      OUT     STD_LOGIC);
    END COMPONENT;

    COMPONENT INV
        PORT ( I   :      IN      STD_LOGIC;
               O   :      OUT     STD_LOGIC);
    END COMPONENT;
```

```
COMPONENT OR3
    PORT ( I0  :        IN      STD_LOGIC;
           I1  :        IN      STD_LOGIC;
           I2  :        IN      STD_LOGIC;
           O   :        OUT     STD_LOGIC);
END COMPONENT;

COMPONENT XOR2
    PORT ( I0  :        IN      STD_LOGIC;
           I1  :        IN      STD_LOGIC;
           O   :        OUT     STD_LOGIC);
END COMPONENT;


BEGIN


    I0 : AND2       PORT MAP (I0=>Cn, I1=>A, O=>X1);
    I1 : AND2       PORT MAP (I0=>Bn, I1=>A, O=>X2);
    I2 : AND3       PORT MAP (I0=>C, I1=>B, I2=>An, O=>X3);


    I3 : DFF        PORT MAP (C=>Clk, CLR=>Reset, D=>Db, Q=>B);
    I4 : DFF        PORT MAP (C=>Clk, CLR=>Reset, D=>An, Q=>A);
    I5 : DFF        PORT MAP (C=>Clk, CLR=>Reset, D=>Dc, Q=>C);


    I6 : INV        PORT MAP (I=>A, O=>An);
    I7 : INV        PORT MAP (I=>B, O=>Bn);
    I8 : INV        PORT MAP (I=>C, O=>Cn);


    I9 : OR3        PORT MAP (I0=>X1, I1=>X2, I2=>X3, O=>Dc);


    I10 : XOR2      PORT MAP (I0=>C, I1=>B, O=>Db);


END structural;

------------------------------------------------------------------
-- End of File
------------------------------------------------------------------
```

# F.5 3-Bit Linear Feedback Shift Register (LFSR)

The LFSR is a counter circuit that changes state on the edge of a clock. The output pattern is considered pseudorandom in that the output pattern appears random in nature, but repeats itself after a set number of clock pulses. The 3-bit LFSR in Fig. F.4 is a maximal count circuit in that it repeats the sequence after 7 clock pulses ($2^n - 1$, where n = 3).

The forbidden state (all bistable outputs are '0' is avoided by setting one bistable rather than resetting it).

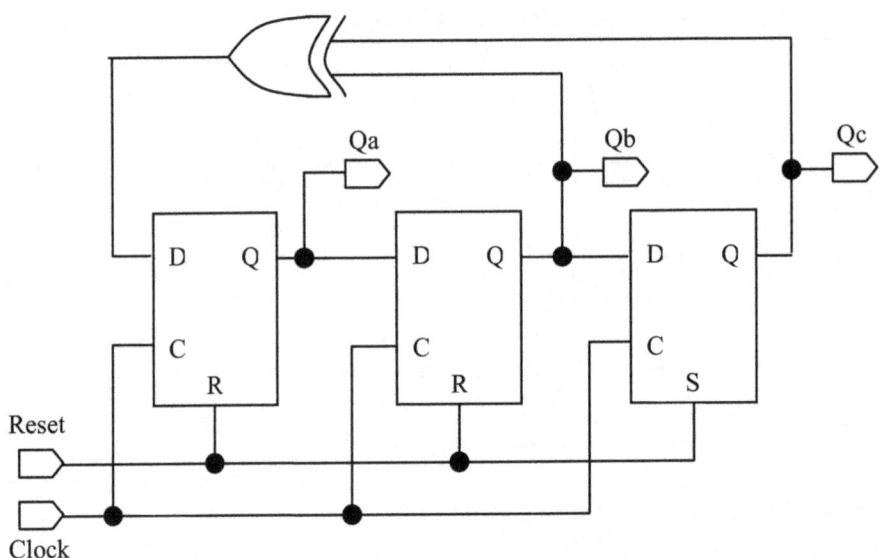

**Fig. F.4.** Linear feedback shift register

The design is described in the following VHDL source file:

- **lfsr.vhd**:

  The VHDL entity and architecture describing the design.

The LFSR design requires a suitable testbench to be created.

# Design (lfsr.vhd)

```
------------------------------------------------------------------
-- VHDL description for 3-bit counter circuit.
------------------------------------------------------------------


------------------------------------------------------------------
-- Library Definitions
------------------------------------------------------------------

LIBRARY ieee;
USE ieee.std_logic_1164.ALL;
USE ieee.numeric_std.ALL;

LIBRARY COMPONENTS;
USE COMPONENTS.LOGIC.ALL;

------------------------------------------------------------------
-- Entity Definition
------------------------------------------------------------------

ENTITY lfsr IS
    PORT ( Clk    :      IN      STD_LOGIC;
           Reset  :      IN      STD_LOGIC;
           Qa     :      OUT     STD_LOGIC;
           Qb     :      OUT     STD_LOGIC;
           Qc     :      OUT     STD_LOGIC);

end lfsr;

------------------------------------------------------------------
-- Architecture Definition
------------------------------------------------------------------

ARCHITECTURE structural OF lfsr IS

    SIGNAL X1    :        STD_LOGIC;

    COMPONENT DFF
        PORT ( C  :      IN      STD_LOGIC;
               D  :      IN      STD_LOGIC;
               R  :      IN      STD_LOGIC;
               Q  :      OUT     STD_LOGIC);
    END COMPONENT;

    COMPONENT DFFs
        PORT ( C  :      IN      STD_LOGIC;
               D  :      IN      STD_LOGIC;
               S  :      IN      STD_LOGIC;
               Q  :      OUT     STD_LOGIC);
    END COMPONENT;

    COMPONENT XOR2
        PORT ( I0 :      IN      STD_LOGIC;
               I1 :      IN      STD_LOGIC;
               O  :      OUT     STD_LOGIC);
    END COMPONENT;

BEGIN
    I0 : DFF      PORT MAP (C=>Clk, D=>X1, R=>Reset, Q=>Qa);
    I1 : DFF      PORT MAP (C=>Clk, D=>Qa, R=>Reset, Q=>Qb);
    I2 : DFFs     PORT MAP (C=>Clk, D=>Qb, S=>Reset, Q=>Qc);
    I3 : XOR2     PORT MAP (I0=>Qb, I1=>Qc, O=>X1);
END structural;

------------------------------------------------------------------
-- End of File
------------------------------------------------------------------
```

# F.6 Simple 4 x 8-Bit ROM

This design model is a simple input-output value mapping, see Fig. F.5. There are no control signals and the output is defined to create a logic '0' or '1' at all times. The output creates a RAMP signal that if fed to an 8-bit DAC will create a coarse ramp output (voltage or current).

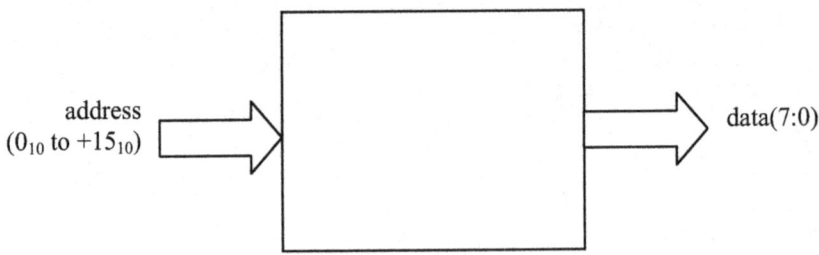

address
$(0_{10}$ to $+15_{10})$

data(7:0)

**Fig. F.5.** ROM cell

The design is described as follows:

- **rom.vhd**:

  The VHDL entity and architecture describing the design.

- **test_rom.vhd**:

  The VHDL testbench which increments the ROM address by 1 starting at 0.

The input *address* is identified using decimal number notation for a 4-bit address, whilst the output *data* is identified using binary number notation. In the final logic circuit implementation, then all signals will be in binary.

## Design (rom.vhd)

```
------------------------------------------------------------------
-- VHDL model for a 4 x 8-bit ROM
------------------------------------------------------------------
-- The model is a simple input-output value mapping. There are no
-- control signals and the output is defined to create a logic '0'
-- or '1' at all times. The output creates a RAMP signal which if
-- fed to an 8-bit DAC will create a coarse ramp output(voltage
-- or current).
------------------------------------------------------------------


------------------------------------------------------------------
-- Library Definitions
------------------------------------------------------------------

library IEEE;
use IEEE.STD_LOGIC_1164.ALL;
use IEEE.STD_LOGIC_ARITH.ALL;
use IEEE.STD_LOGIC_UNSIGNED.ALL;


------------------------------------------------------------------
-- Entity Definition
------------------------------------------------------------------

entity ROM is
    Port ( address :      in      Integer range 0 to 15;
           data :         out     std_logic_vector(7 downto 0));
end entity ROM;


------------------------------------------------------------------
-- Architecture Definition
------------------------------------------------------------------

architecture Simple of ROM is

type rom_array is array (0 to 15) of std_logic_vector(7 downto 0);

constant ROM: rom_array := (
                                "00000000",
                                "00010000",
                                "00100000",
                                "00110000",
                                "01000000",
                                "01010000",
                                "01100000",
                                "01110000",
                                "10000000",
                                "10010000",
                                "10100000",
                                "10110000",
                                "11000000",
                                "11010000",
                                "11100000",
                                "11110000");

begin
        data <= rom(address);

end architecture Simple;

------------------------------------------------------------------
-- End of File
------------------------------------------------------------------
```

## Design Test Bench (test_rom.vhd)

```
-----------------------------------------------------------------
-- VHDL Test Bench Created to simulate source file ramp.vhd
-----------------------------------------------------------------


-----------------------------------------------------------------
-- Library Definitions
-----------------------------------------------------------------

LIBRARY ieee;
USE ieee.std_logic_1164.ALL;
USE ieee.numeric_std.ALL;

-----------------------------------------------------------------
-- Entity Definition
-----------------------------------------------------------------

ENTITY testbench IS
END testbench;

-----------------------------------------------------------------
-- Architecture Definition
-----------------------------------------------------------------

ARCHITECTURE behavior OF testbench IS

        COMPONENT ROM
        PORT(    address : IN Integer range 0 to 15;
                 data :    OUT std_logic_vector(7 downto 0));
        END COMPONENT;

        SIGNAL address :  Integer range 0 to 15;
        SIGNAL data :     std_logic_vector(7 downto 0);

BEGIN
        uut: ROM PORT MAP(address => address, data => data);


test_stimulus : PROCESS

        BEGIN

                wait for 0 ns;      address <= 0;
                wait for 100 ns;    address <= 1;
                wait for 100 ns;    address <= 2;
                wait for 100 ns;    address <= 3;
                wait for 100 ns;    address <= 4;
                wait for 100 ns;    address <= 5;
                wait for 100 ns;    address <= 6;
                wait for 100 ns;    address <= 7;
                wait for 100 ns;    address <= 8;
                wait for 100 ns;    address <= 9;
                wait for 100 ns;    address <= 10;
                wait for 100 ns;    address <= 11;
                wait for 100 ns;    address <= 12;
                wait for 100 ns;    address <= 13;
                wait for 100 ns;    address <= 14;
                wait for 100 ns;    address <= 15;
                wait for 100 ns;

        END PROCESS;

END;

-----------------------------------------------------------------
-- End of File
-----------------------------------------------------------------
```

# F.7 Simple 4 x 8-Bit RAM

This design model is a simple random access memory cell, see Fig. F.6.

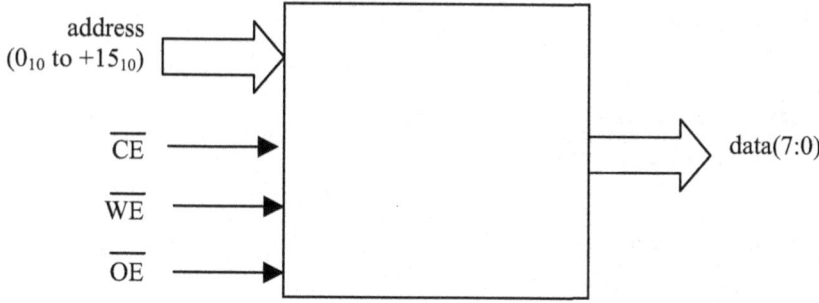

**Fig. F.6**. RAM cell

The design is described as follows:

- **ram.vhd**:

    The VHDL entity and architecture describing the design.

- **test_ram.vhd**:

    The VHDL testbench that writes values to addresses 1 and 1 and then reads back the contents.

The input *address* is identified using decimal number notation for a 4-bit address, whilst the output *data* is identified using binary number notation. In the final logic circuit implementation, then all signals will be in binary.

The RAM is written to when the $\overline{\text{CE}}$ and $\overline{\text{WE}}$ signals are low (logic 0). The RAM is read from when the $\overline{\text{CE}}$ and $\overline{\text{OE}}$ signals are low (logic 0).

In this design, no timing values for the signals are identified. However, in the circuit implementation, there would be timing constraints linked to the circuit.

## Design (ram.vhd)

```
--------------------------------------------------------------------
-- VHDL description for 4x4 RAM circuit.
--------------------------------------------------------------------


--------------------------------------------------------------------
-- Library Definitions
--------------------------------------------------------------------


library IEEE;
use IEEE.STD_LOGIC_1164.ALL;
use IEEE.STD_LOGIC_ARITH.ALL;
use IEEE.STD_LOGIC_UNSIGNED.ALL;


--------------------------------------------------------------------
-- Entity Definition
--------------------------------------------------------------------


entity ram is
    Port ( address        : in    Integer range 0 to 15;
           ce             : in    std_logic;
           we             : in    std_logic;
           oe             : in    std_logic;
           data           : inout std_logic_vector(7 downto 0));
end entity ram;


--------------------------------------------------------------------
-- Architecture Definition
--------------------------------------------------------------------


architecture simple of ram is

begin

memory:process(address, ce, we, oe) is

type ram_array is array (0 to 15) of std_logic_vector(7 downto 0);

variable mem: ram_array;

begin
        data <= (others => 'Z');

        if (ce = '0') then

                if (we = '0') then
                        mem(address) := data;

                elsif (oe = '0') then
                        data <= mem(address);
                end if;

        end if;

end process memory;

end architecture simple;


--------------------------------------------------------------------
-- End of File
--------------------------------------------------------------------
```

# Design Test Bench (test_ram.vhd)

```
-------------------------------------------------------------------
-- VHDL Test Bench Created to simulate source file ram.vhd
-------------------------------------------------------------------


-------------------------------------------------------------------
-- Library Definitions
-------------------------------------------------------------------


LIBRARY ieee;
USE ieee.std_logic_1164.ALL;
USE ieee.numeric_std.ALL;


-------------------------------------------------------------------
-- Entity Definition
-------------------------------------------------------------------

ENTITY testbench IS
END testbench;


-------------------------------------------------------------------
-- Architecture Definition
-------------------------------------------------------------------

ARCHITECTURE behavior OF testbench IS

        COMPONENT ram
          Port ( address : in Integer range 0 to 15;
                 ce : in std_logic;
                 we : in std_logic;
                 oe : in std_logic;
                 data : inout std_logic_vector(7 downto 0));
          END COMPONENT;
signal address : Integer range 0 to 15;
signal ce : std_logic;
signal we : std_logic;
signal oe : std_logic;
signal data : std_logic_vector(7 downto 0);


BEGIN

        uut: ram PORT MAP(address => address,
                        ce => ce, oe => oe, we => we, data => data);

    testbench : PROCESS
    BEGIN


-- Set initial conditions
        wait for 0 ns;
        address <= 0; data <= "ZZZZZZZZ";
        ce <= '1'; we <= '1'; oe <= '1';


-- Write to memory address 0
        wait for 10 ns; data <= "01010101";
        wait for 10 ns; ce <= '0';
        wait for 10 ns; we <= '0';
        wait for 10 ns; ce <= '1'; we <= '1'; data <= "ZZZZZZZZ";
```

```
-- Write to memory address 1
        wait for 0 ns; address <= 1;
        wait for 10 ns; data <= "10101010";
        wait for 10 ns; ce <= '0';
        wait for 10 ns; we <= '0';
        wait for 10 ns; ce <= '1'; we <= '1'; data <= "ZZZZZZZZ";


-- Read from memory address 0
        wait for 0 ns; address <= 0;
        wait for 10 ns; ce <= '0';
        wait for 10 ns; oe <= '0';
        wait for 10 ns; ce <= '1'; oe <= '1';


-- Read from memory address 1
        wait for 0 ns; address <= 1;
        wait for 10 ns; ce <= '0';
        wait for 10 ns; oe <= '0';
        wait for 10 ns; ce <= '1'; oe <= '1';


        wait for 1 ms;


    END PROCESS;

END;

--------------------------------------------------------------------
-- End of File
--------------------------------------------------------------------
```

# Appendix G

# HSPICE® Simulation Examples

## G.1 Introduction

The following Spice examples are provided in netlist format for analogue circuit simulation, see Table G.1. These simulation models contain both the circuit description and simulation commands. These files are written for the HSPICE® simulator, but can be used with any Spice simulator with suitable modifications. Essentially, the only HSPICE® specific command is the .OPTIONS POST line.

**Table G.1.** HSPICE® simulation examples

| Identifier | Description |
|---|---|
| Analogue circuit | |
| G.2 | Series resonant LCR circuit |
| G.3 | MOS transistor amplifier |
| | |
| Digital circuit | |
| G.4 | CMOS inverter |
| | |
| Mixed-signal circuit | |
| G.5 | 3-Bit current weighted digital to analogue converter |

The first two examples are analogue circuits and the simulation is set-up to perform a transient analysis. The third example is the CMOS inverter logic gate that was discussed in the main text of the book. The fourth circuit is a mixed-signal design. This is a digital to analogue converter (DAC) design and the simulation is set-up to perform a transient analysis.

# G.2 Series Resonant LCR Circuit

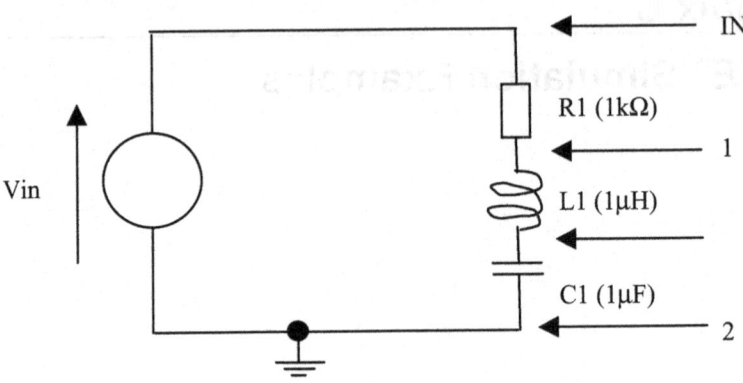

**Fig. G.1.** Series resonant LCR circuit

The HSPICE® input file is as follows:

```
** Example Input File

********************************************************************
** Set temperature

.TEMP 25                    * set operating temperature to 25C

********************************************************************
***** set options (for HSpice simulator, set results post-processing)

.OPTIONS POST               * options post

********************************************************************
** Set input stimuli - Input voltage PWL (between nodes IN and 0)

Vin  IN 0  PWL(0mS,0V 1mS,0V 1.01mS,1V)

********************************************************************
** Define netlist

R1  IN  1  1k    * Resistor R1, value 1kohm (between nodes IN and 1)
L1  1   2  1uH   * Inductor L1, value 1uH (between nodes 1 and 2)
C1  2   0  1uF   * Capacitor C1, value 1uF (between nodes 2 and 0)

********************************************************************
** Set Analysis - Run a transient analysis for 10mS,

.tran 1mS 10mS

********************************************************************
* End of Spice File

.end
```

# G.3 MOS Transistor Amplifier

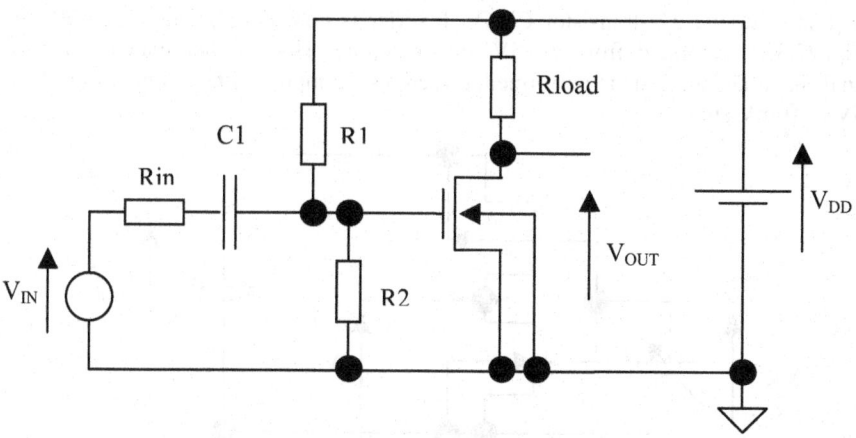

**Fig. G.2.** MOS transistor amplifier

The HSPICE® input file is as follows:

```
** Common source amplifier: transient analysis for 5mS

.OPTIONS POST
.TEMP 25

*********************************************************************

Vdd   Vdd  0   +5V
Vin   IN   0   SIN(0 1mV 1kHz 0 0 0)

Rin      IN    1    1k
R1       Vdd   2    61.8k
R2       2     0    100k
Rload    Vdd   OUT  5k

C1       1     2    10uF

Mn1      OUT   2    0  0    nmod    w=70um    l=10um

*********************************************************************

.model nmod nmos (Level=1 Vto=0.7V Kp=25e-6 Lambda=0.0  Gamma=1)

*********************************************************************

.tran 1mS  5mS

*********************************************************************

.END                    *  END OF INPUT FILE
```

# G.4 CMOS Inverter

In this example, a 2-transistor CMOS Inverter is considered. The logic cell, see Fig. G.3, is operated from a +5V power supply. Here, rather than a transient analysis, a DC analysis is performed, sweeping the input voltage (Vin) from 0V to 5V in 10mV steps.

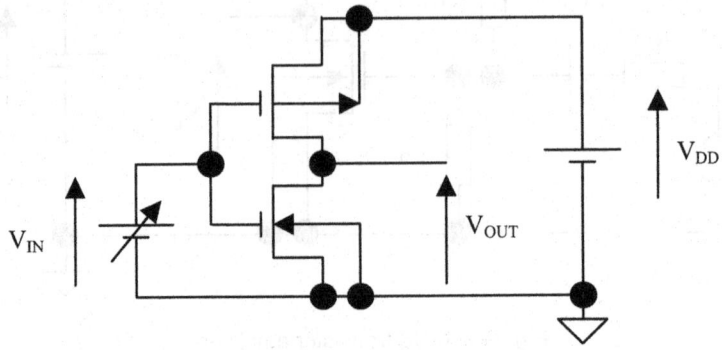

**Fig. G.3.** CMOS inverter

The transistor lengths are both set to 1μm and the pMOS width 2.5 times that of the nMOS transistor. The transistor models (level 1) are defined in the input file.

The HSPICE® input file is as follows:

```
*Inverter DC analysis

.OPTIONS POST

.TEMP 25


Vdd   vdd   0   5V

Vin   in    0


Mn1   out   in   0    0    MODN   W=4um    L=1um

Mp1   out   in   vdd  vdd  MODP   W=10um   L=1um


.model MODN   nMOS (LEVEL=1  VTO=0.7V   KP=25e-6  LAMBDA=0.00)

.model MODP   pMOS (LEVEL=1  VTO=-0.7V  KP=10e-6  LAMBDA=0.00)


.DC   Vin   0   5   0.01

.end
```

# G.5 3-Bit Current Weighted DAC

The weighted current Digital to Analogue Converter (DAC) creates binary weighted currents using current mirrors. This type of DAC can be created using bipolar or MOS transistors. In the circuit, an array of current mirrors is used and these are switched by each of the digital logic inputs. The switches are implemented using suitably sized transistors. By directing current to flow through one of two resistors connected around an op-amp, an output voltage dependent on a digital input can be created. The current is directed through the transistor switches where the switch position is dependent on the logic value of the digital input.

Considering a 3-bit DAC, see Fig. G.4, which uses nMOS transistor current mirrors, where:

- The transistors have their source and substrate connections joined
- All transistors have the same dimensions
- When the transistors conduct, they are to operate in saturation. (i.e. ($V_{GS}$ >=$V_T$) and ($V_{DS}$ >= ($V_{GS}$ -$V_T$)) and the current flow is given by:

$$I_D = \frac{\mu_n.C_{ox}}{2}.\frac{W}{L}.\left[(V_{GS}-V_T)^2.(1+\lambda.V_{DS})\right]$$

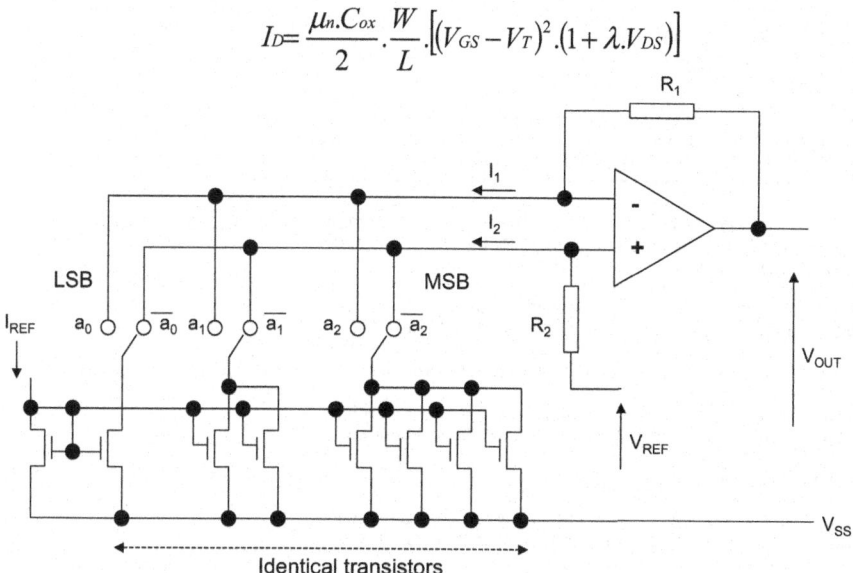

**Fig. G.4.** 3-bit current weighted DAC

The current flowing each of the resistors will vary from 0 to 7x $I_{REF}$. Considering $V_{SS}$ to be 0V, then the output voltage is given by:

$$V_{OUT} = (V_{REF} - I_2.R_2) + (I_1.R_1)$$

## The HSPICE® input file is as follows:

```
* Example HSPICE Input file to simulate a CMOS Weighted current DAC
*************************************************************************
* Set Temperature and HSPICE Options
*************************************************************************

.TEMP 25
.OPTIONS POST

*************************************************************************
* Probe transistor drain currents
*************************************************************************

.PROBE I(M1)
.PROBE I(M2)
.PROBE I(M3)
.PROBE I(M4)
.PROBE I(M5)
.PROBE I(M6)
.PROBE I(M7)
.PROBE I(M8)

*************************************************************************
* Set current and voltage reference sources
*************************************************************************

Iref  Vdd  Gate  10uA
Vref  REF  0     2.5V
Vdd   Vdd  0     5V

*************************************************************************
* Set digital input voltages (0V/+5V) in binary count
* from 000 to 111 (+ inverters for complement signal)
*************************************************************************

Vd0   in0   0  PWL(0us,0V 1us,0V 1.01us,5V 2us,5V 2.01us,0V 3us,0V 3.01us,5V
+ 4us,5V 4.01us,0V 5us,0V 5.01us,5V 6us,5V 6.01us,0V
+ 7us,0V 7.01us,5V)

Minv0n  in0n  in0  0    0    MODN  W=1.0um  L=0.6um
Minv0p  in0n  in0  Vdd  Vdd  MODP  W=2.5um  L=0.6um

Vd1   in1   0  PWL(0us,0V 2us,0V 2.01us,5V 4us,5V 4.01us,0V 6us,0V 6.01us,5V)

Minv1n  in1n  in1  0    0    MODN  W=1.0um  L=0.6um
Minv1p  in1n  in1  Vdd  Vdd  MODP  W=2.5um  L=0.6um

Vd2   in2   0  PWL(0us,0V  4us,0V 4.01us,5V)

Minv2n  in2n  in2  0    0    MODN  W=1.0um  L=0.6um
Minv2p  in2n  in2  Vdd  Vdd  MODP  W=2.5um  L=0.6um

*************************************************************************
* Unit sized transistors (each transistor to carry 10uA)
*************************************************************************

M1  Gate  Gate  0  0  MODN  W=12.5um  L=10um
M2  D0    Gate  0  0  MODN  W=12.5um  L=10um
M3  D1    Gate  0  0  MODN  W=12.5um  L=10um
M4  D1    Gate  0  0  MODN  W=12.5um  L=10um
M5  D2    Gate  0  0  MODN  W=12.5um  L=10um
M6  D2    Gate  0  0  MODN  W=12.5um  L=10um
M7  D2    Gate  0  0  MODN  W=12.5um  L=10um
M8  D2    Gate  0  0  MODN  W=12.5um  L=10um
```

```
****************************************************************************
* Use CMOS transmission gates for switches
****************************************************************************

Msw0n1  NEG  in0   D0  0    MODN  W=2.5um  L=0.6um
Msw0p1  NEG  in0n  D0  Vdd  MODP  W=7.5um  L=0.6um
Msw0n2  POS  in0n  D0  0    MODN  W=2.5um  L=0.6um
Msw0p2  POS  in0   D0  Vdd  MODP  W=7.5um  L=0.6um
Msw1n1  NEG  in1   D1  0    MODN  W=2.5um  L=0.6um
Msw1p1  NEG  in1n  D1  Vdd  MODP  W=7.5um  L=0.6um
Msw1n2  POS  in1n  D1  0    MODN  W=2.5um  L=0.6um
Msw1p2  POS  in1   D1  Vdd  MODP  W=7.5um  L=0.6um
Msw2n1  NEG  in2   D2  0    MODN  W=2.5um  L=0.6um
Msw2p1  NEG  in2n  D2  Vdd  MODP  W=7.5um  L=0.6um
Msw2n2  POS  in2n  D2  0    MODN  W=2.5um  L=0.6um
Msw2p2  POS  in2   D2  Vdd  MODP  W=7.5um  L=0.6um


****************************************************************************
* OP-AMP Model and Resistors
****************************************************************************

E1  OUT  0    POS  NEG  MAX=+5V  MIN=0V  1e+5
R1  OUT  NEG  20k
R2  REF  POS  20k

****************************************************************************
* Set transistor models
****************************************************************************

.model MODN nMOS (LEVEL=1 VTO = 0.7  KP=25e-6 LAMBDA = 0)
.model MODP pMOS (LEVEL=1 VTO = -0.7 KP=10e-6 LAMBDA = 0)

****************************************************************************
* Run transient analysis
****************************************************************************

.Tran  0.1uS  8uS

****************************************************************************
* End Simulation file
****************************************************************************

.END

****************************************************************************
```

# Appendix H
# MATLAB® Simulation Examples

## H.1 Introduction

The following system models are provided as MATLAB® examples, see Table H.1. Two examples of mixed-signal designs are provided:

- **ADC example**:
  A MATLAB® model of an 8-Bit ADC is provided and analysed within the MATLAB® environment.

- **DAC example**:
  Results from the testing of an 8-Bit DAC are provided and analysed within the MATLAB® environment.

**Table H.1.** MATLAB® simulation examples

| Identifier | Description |
|---|---|
| Mixed-signal circuit | |
| H.2 | 8-Bit analogue to digital converter |
| H.3 | 8-Bit digital to analogue converter |

# H.2 8-Bit Analogue to Digital Converter

The following example is a MATLAB® simulation model and analysis commands for the model of an ideal 8-bit ADC, see Fig. H.1.

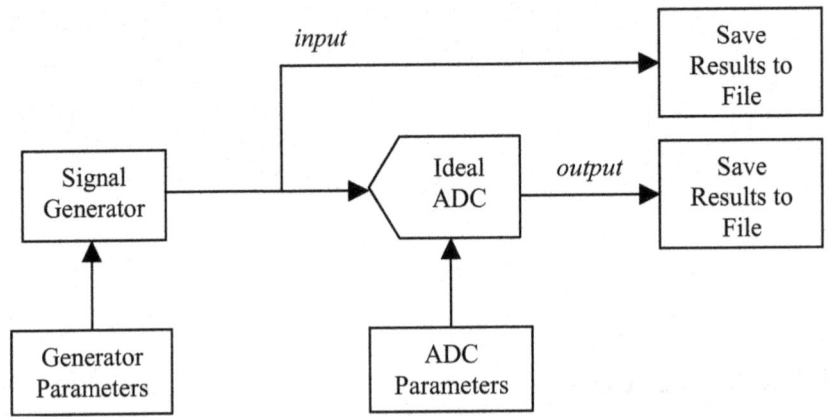

**Fig. H.1.** ADC model and simulation set-up in MATLAB®

In this model, a sine wave is generated and applied directly to the ADC model. The time steps, input and output signals are saved to a text file for later analysis. In this arrangement, the following parameters are set up:

- Sine wave:
  - i. Amplitude       2.5V
  - ii. Frequency      1kHz
  - iii. Offset        +2.5V

- The simulation is run for 32 cycles of the sine wave input.

- The ADC coherently samples the input signal.

- The output code increments every 1LSB change in the input voltage. The reference voltage ($V_{REF}$) is the same value as the full-scale voltage ($V_{FS}$) and is set to +5V. The unipolar input signal is to vary from 0V to +5V.

- 1024 samples of the input signal are taken for every cycle of the input signal. For a 1kHz sine wave, there will be (ideally) 1024 samples taken every 1ms, if the sampling period is correctly and accurately set-up.

The code provided does not undertake any analysis of the signals. However, analysis routines such as FFT and Histogram tests can be readily undertaken with the in-built, or user defined, functions. The input sine wave is defined as follows:

*The time step at which the input signal value is calculated has been chosen in order to calculate the value of the sine wave at the sample point of the ADC model. The sample time is rounded to every 20ns to model a finite time step of 20ns for the digital timing of the ADC. This will however create an error in the resulting calculations that would need to be taken into account.*

The MATLAB® code for the input signal definition is:

```
%%%%%%%%%%%%%%%%%%%%%%%%%%%%%%%%%%%%%%%%%%%%%%%%%%%%%%%%%%%%%%%
% DEFINE INPUT SIGNAL
%%%%%%%%%%%%%%%%%%%%%%%%%%%%%%%%%%%%%%%%%%%%%%%%%%%%%%%%%%%%%%%
time_step = [0:0.00000098:0.032];
amplitude = 2.5;
offset = 2.5;
frequency = 1000;
input = amplitude * sin(2 * pi * frequency * time_step) + offset;
plot(time_step, input);
grid;
xlabel('time (secs)');
ylabel('input signal (volts)');
%%%%%%%%%%%%%%%%%%%%%%%%%%%%%%%%%%%%%%%%%%%%%%%%%%%%%%%%%%%%%%%
```

The input signal plot is shown in Fig. H.2.

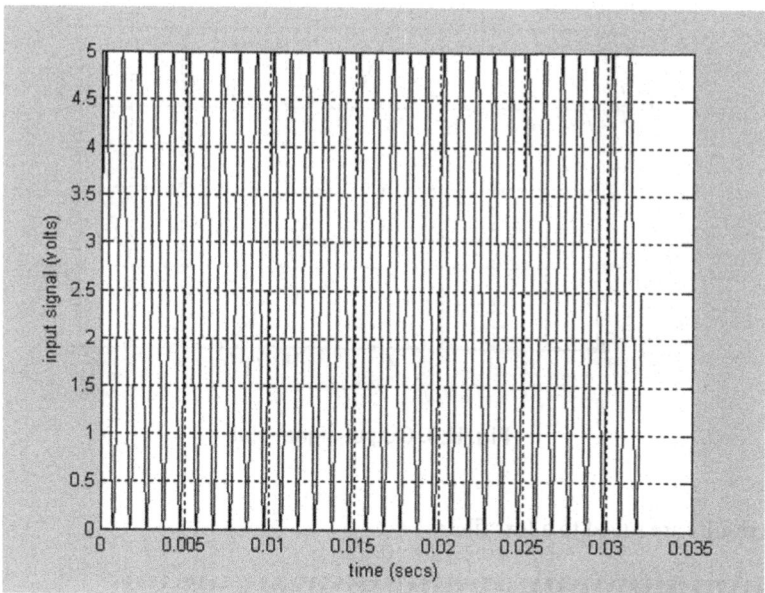

**Fig. H.2.** Input signal plot

The ADC output is a sample of the analogue input signal at the sample time (every 980ns) and the value rounded to every Least Significant Bit (LSB) of the converter.

The MATLAB® code for the output signal definition is:

```
%%%%%%%%%%%%%%%%%%%%%%%%%%%%%%%%%%%%%%%%%%%%%%%%%%%%%%%%%%%
% CALCULATE IDEAL OUTPUT (time domain)
%%%%%%%%%%%%%%%%%%%%%%%%%%%%%%%%%%%%%%%%%%%%%%%%%%%%%%%%%%%
no_of_bits = 8;
Vref = 5.0;
Vfs = Vref;
LSB = Vfs / (2^(no_of_bits));
output = round((input / LSB));
plot(time_step, output);
grid;
xlabel('time (secs)');
ylabel('output signal (bits)');
%%%%%%%%%%%%%%%%%%%%%%%%%%%%%%%%%%%%%%%%%%%%%%%%%%%%%%%%%%%
```

The output signal plot for 1 cycle of the input is shown in Fig. H.3. This varies from a code value of $0_{10}$ ($00000000_2$) to $+255_{10}$ ($11111111_2$). The y-axis numbers represent the decimal value of the binary output signal.

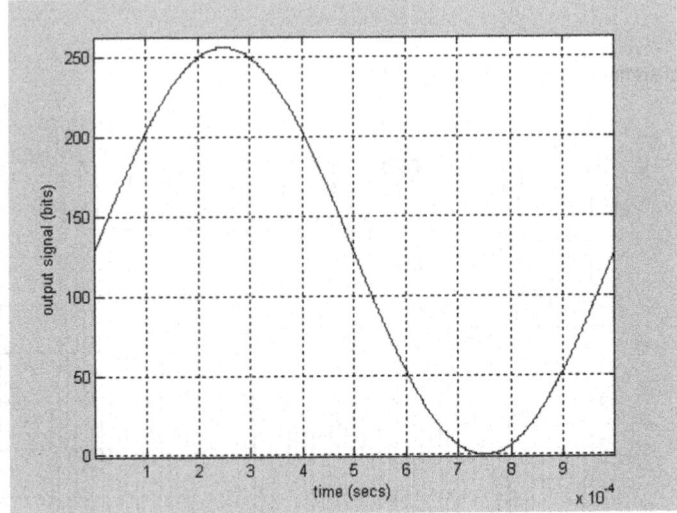

**Fig. H.3.** Output signal plot

The results are saved to a text file:

```
%%%%%%%%%%%%%%%%%%%%%%%%%%%%%%%%%%%%%%%%%%%%%%%%%%%%%%%%%%%
% Save signals to text file for later analysis
%%%%%%%%%%%%%%%%%%%%%%%%%%%%%%%%%%%%%%%%%%%%%%%%%%%%%%%%%%%
dlmwrite('c:\time_step.mat', time_step, '\n');
dlmwrite('c:\input.mat', input, '\n');
dlmwrite('c:\output.mat', output, '\n');
%%%%%%%%%%%%%%%%%%%%%%%%%%%%%%%%%%%%%%%%%%%%%%%%%%%%%%%%%%%
```

# H.3 8-Bit Digital to Analogue Converter

An 8-Bit DAC is manually tested using the following arrangement, see Fig. H.4. Here, the output of the DAC is a current that is converted to a voltage. The DAC is set-up within a simple test circuit in order to produce a unipolar output.

The digital inputs are manually selected (logic 0 or 1) by a switch array and an output voltage (the output of an external op-amp) is monitored. The digital inputs are generated by switching to $V_{DD}$ (logic 1 via a resistor) and $V_{SS}$ (logic 0). A binary count from $00_{10}$ to $255_{10}$ can be manually created. The DAC used in the AD7524 8-bit Buffered Multiplying DAC The expected output for a particular digital input may be obtained from the device datasheet.

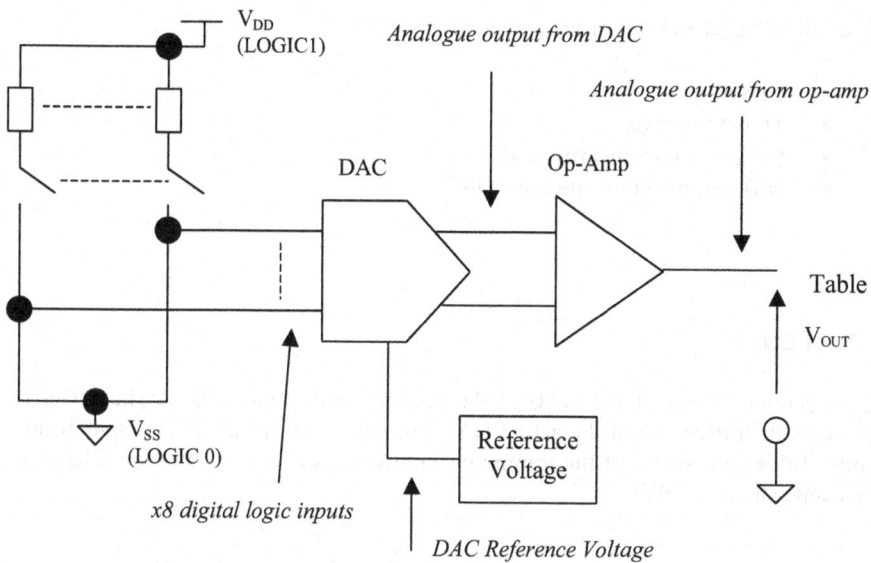

**Fig. H.4.** DAC test circuit

The reference voltage is set to $-5.0V$. Table H.2 identifies the measured voltage for each input code. The results are (voltage values) are stored in a text file:

$$c:\backslash results.mat$$

assuming that a PC is used in the analysis. For example, the following results example, see Fig. H.5, contains 8 numbers representing the output of an example 3-bit DAC. In the 8-bit DAC circuit, the output file would contain 256 values.

```
0.05
0.3125
1.25
1.875
2.7
3
4
4.375
```

**Fig. H.5.** Example output file format

The test results for the 8-bit DAC are tabulated in Table H.2.

The following MATLAB® routines are provided:

- Offset Error
- Transfer Curve
- Integral Non-Linearity (INL)
- Differential Non-Linearity (DNL)

## Offset Error

Consider the output of the DAC at the minimal code input ($00_{10}$). The offset is given as the difference of the actual DAC output and the ideal DAC output at this code. Taking the offset at the minimum input code and that the output voltage at this code should be 0V:

```
Offset = DACout(1)/ (Vfs/no_of_codes)
```

(Offset in LSBs). For the converter under test, the offset error is 0.0512 LSBs.

```
>> DACout = dlmread('c:\results.mat')
>> no_of_bits = 8
>> no_of_codes = 2^no_of_bits
>> Vfs = 5.0
>> Offset = DACout(1)/ (Vfs/no_of_codes)

Offset =

0.0512

>>
```

## Table H.2. 8-Bit DAC test results

| Input Code | o/p voltage | Input Code | o/p voltage | Input Code | o/p voltage | Input Code | o/p voltage |
|---|---|---|---|---|---|---|---|
| 0 | 0.0010 | 66 | 1.283 | 132 | 2.6 | 198 | 3.91 |
| 1 | 0.0034 | 67 | 1.303 | 133 | 2.62 | 199 | 3.93 |
| 2 | 0.0285 | 68 | 1.323 | 134 | 2.64 | 200 | 3.95 |
| 3 | 0.0433 | 69 | 1.342 | 135 | 2.66 | 201 | 3.97 |
| 4 | 0.0630 | 70 | 1.362 | 136 | 2.68 | 202 | 3.99 |
| 5 | 0.0828 | 71 | 1.382 | 137 | 2.70 | 203 | 4.01 |
| 6 | 0.1027 | 72 | 1.401 | 138 | 2.72 | 204 | 4.03 |
| 7 | 0.1222 | 73 | 1.421 | 139 | 2.74 | 205 | 4.05 |
| 8 | 0.1417 | 74 | 1.441 | 140 | 2.76 | 206 | 4.07 |
| 9 | 0.1617 | 75 | 1.46 | 141 | 2.78 | 207 | 4.09 |
| 10 | 0.1817 | 76 | 1.48 | 142 | 2.8 | 208 | 4.11 |
| 11 | 0.2010 | 77 | 1.499 | 143 | 2.82 | 209 | 4.12 |
| 12 | 0.2200 | 78 | 1.519 | 144 | 2.84 | 210 | 4.14 |
| 13 | 0.2400 | 79 | 1.539 | 145 | 2.86 | 211 | 4.16 |
| 14 | 0.2590 | 80 | 1.558 | 146 | 2.88 | 212 | 4.18 |
| 15 | 0.279 | 81 | 1.578 | 147 | 2.90 | 213 | 4.20 |
| 16 | .298 | 82 | 1.598 | 148 | 2.92 | 214 | 4.22 |
| 17 | .318 | 83 | 1.618 | 149 | 2.94 | 215 | 4.24 |
| 18 | .338 | 84 | 1.637 | 150 | 2.96 | 216 | 4.26 |
| 19 | .358 | 85 | 1.657 | 151 | 2.98 | 217 | 4.28 |
| 20 | .378 | 86 | 1.677 | 152 | 3.02 | 218 | 4.30 |
| 21 | .397 | 87 | 1.696 | 153 | 3.04 | 219 | 4.32 |
| 22 | .417 | 88 | 1.715 | 154 | 3.06 | 220 | 4.34 |
| 23 | .437 | 89 | 1.735 | 155 | 3.08 | 221 | 4.36 |
| 24 | .456 | 90 | 1.755 | 156 | 3.1 | 222 | 4.38 |
| 25 | .476 | 91 | 1.774 | 157 | 3.12 | 223 | 4.40 |
| 26 | .496 | 92 | 1.794 | 158 | 3.14 | 224 | 4.42 |
| 27 | .515 | 93 | 1.814 | 159 | 3.16 | 225 | 4.44 |
| 28 | .535 | 94 | 1.833 | 160 | 3.18 | 226 | 4.46 |
| 29 | .554 | 95 | 1.852 | 161 | 3.20 | 227 | 4.48 |
| 30 | .574 | 96 | 1.872 | 162 | 3.21 | 228 | 4.5 |
| 31 | .593 | 97 | 1.892 | 163 | 3.23 | 229 | 4.52 |
| 32 | .612 | 98 | 1.912 | 164 | 3.25 | 230 | 4.54 |
| 33 | .632 | 99 | 1.932 | 165 | 3.27 | 231 | 4.56 |
| 34 | .652 | 100 | 1.951 | 166 | 3.29 | 232 | 4.58 |
| 35 | .672 | 101 | 1.971 | 167 | 3.31 | 233 | 4.6 |
| 36 | .692 | 102 | 1.991 | 168 | 3.33 | 234 | 4.62 |
| 37 | .711 | 103 | 2.01 | 169 | 3.35 | 235 | 4.64 |
| 38 | .731 | 104 | 2.05 | 170 | 3.37 | 236 | 4.66 |
| 39 | .751 | 105 | 2.07 | 171 | 3.39 | 237 | 4.68 |
| 40 | .770 | 106 | 2.09 | 172 | 3.41 | 238 | 4.7 |
| 41 | .79 | 107 | 2.11 | 173 | 3.43 | 239 | 4.71 |
| 42 | .81 | 108 | 2.13 | 174 | 3.45 | 240 | 4.73 |
| 43 | .83 | 109 | 2.15 | 175 | 3.47 | 241 | 4.75 |
| 44 | .849 | 110 | 2.17 | 176 | 3.49 | 242 | 4.77 |
| 45 | .869 | 111 | 2.19 | 177 | 3.51 | 243 | 4.79 |
| 46 | .888 | 112 | 2.21 | 178 | 3.53 | 244 | 4.81 |
| 47 | .908 | 113 | 2.23 | 179 | 2.64 | 245 | 4.83 |
| 48 | .927 | 114 | 2.25 | 180 | 3.55 | 246 | 4.85 |
| 49 | .947 | 115 | 2.27 | 181 | 3.57 | 247 | 4.87 |
| 50 | .967 | 116 | 2.29 | 182 | 3.59 | 248 | 4.89 |
| 51 | .987 | 117 | 2.31 | 183 | 3.61 | 249 | 4.91 |
| 52 | 1.006 | 118 | 2.32 | 184 | 3.63 | 250 | 4.93 |
| 53 | 1.026 | 119 | 2.34 | 185 | 3.65 | 251 | 4.95 |
| 54 | 1.046 | 120 | 2.36 | 186 | 3.67 | 252 | 4.97 |
| 55 | 1.065 | 121 | 2.38 | 187 | 3.69 | 253 | 4.99 |
| 56 | 1.084 | 122 | 2.40 | 188 | 3.71 | 254 | 5.01 |
| 57 | 1.104 | 123 | 2.42 | 189 | 3.73 | 255 | 5.03 |
| 58 | 1.124 | 124 | 2.44 | 190 | 3.75 | | |
| 59 | 1.144 | 125 | 2.46 | 191 | 3.77 | | |
| 60 | 1.163 | 126 | 2.48 | 192 | 3.79 | | |
| 61 | 1.183 | 127 | 2.50 | 193 | 3.81 | | |
| 62 | 1.203 | 128 | 2.53 | 194 | 3.83 | | |
| 63 | 1.222 | 129 | 2.54 | 195 | 3.85 | | |
| 64 | 1.241 | 130 | 2.56 | 196 | 3.87 | | |
| 65 | 1.263 | 131 | 2.58 | 197 | 3.89 | | |

**Note:-** there are significant rounding errors in the measurements taken.

## Transfer Curve (Input-Output relationship) plot

The aim of this test is to create the converter transfer curve plot.

**Table H.3.** Transfer curve plot routine

| Procedure | Details | MATLAB® code |
|---|---|---|
| 1 | Read in data from file (assuming ASCII text file called 'results.mat' on C: drive. The data is read into a variable called 'DACout' using the read ASCII delimited file command. The data is stored in a column vector | `DACout = dlmread('c:\results.mat')` |
| 2 | Create a row vector (called 'X') to contain the input code (in decimal number format) | `X = 0:(length(DACout) - 1)` |
| 3 | Create a column vector (called 'Code') that is the transposed 'X' vector | `Code = X'` |
| 4 | Plot the input-output relationship points | `plot(Code, DACout, 'o')` |
| 5 | Hold the current plot | `hold on` |
| 6 | Draw a grid | `grid` |
| 7 | Draw a straight line between the points on the plot | `plot(Code, DACout)` |
| 8 | Create a title for the plot | `title('DAC Test Transfer Curve Test')`<br>`xlabel('Input Code')`<br>`ylabel('Output Voltage')` |
| 9 | Create the ideal straight line plot for the 8-bit ADC | `Ideal = Code * (5/256)` |
| 10 | Plot the ideal straight line plot | `plot(Code, Ideal, 'r')` |

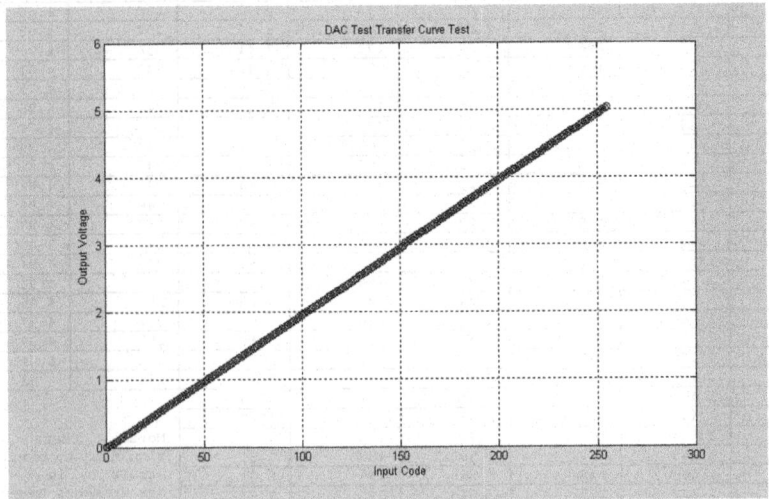

**Fig. H.6.** Transfer curve plot

## INL Test

For each input code, the INL is a measure of the deviation of the actual converter output from the straight-line approximation. In this case, the Ideal Converter characteristic is used as the straight-line approximation.

**Table H.4.** INL test

| Procedure | Details | MATLAB® code |
|---|---|---|
| 1 | Read in data from file (assuming ASCII text file called 'results.mat' on C: drive. The data is read into a variable called 'DACout' using the read ASCII delimited file command. The data is stored in a column vector | `DACout = dlmread('c:\results.mat')` |
| 2 | Create a row vector (called 'X') to contain the input code (in decimal number format) | `X = 0:(length(DACout) - 1)` |
| 3 | Create a column vector (called 'Code') that is the transposed 'X' vector | `Code = X'` |
| 4 | Set the DAC full scale output voltage (at 5V) | `Vfs = 5` |
| 5 | Set the DAC resolution (8-bit DAC = 256 codes) | `no_of_codes = 2.^8` |
| 6 | Create the Ideal Converter Output | `Ideal = Code * (Vfs/ no_of_codes)` |
| 7 | Create the INL for each code (in volts) | `INLvolts = DACout - Ideal` |
| 8 | Create the INL in terms of LSBs | `INL = INLvolts/(Vfs/ no_of_codes)` |
| 9 | Plot the INL points | `plot(Code, INL, 'o')` |
| 10 | Hold the current plot | `hold on` |
| 11 | Draw a grid | `grid` |
| 12 | Draw a straight line between the points on the plot | `plot(Code, INL)` |
| 13 | Create a title for the plot | `title('DAC Test INL')`<br>`xlabel('Input Code')`<br>`ylabel('INL (LSBs)')` |

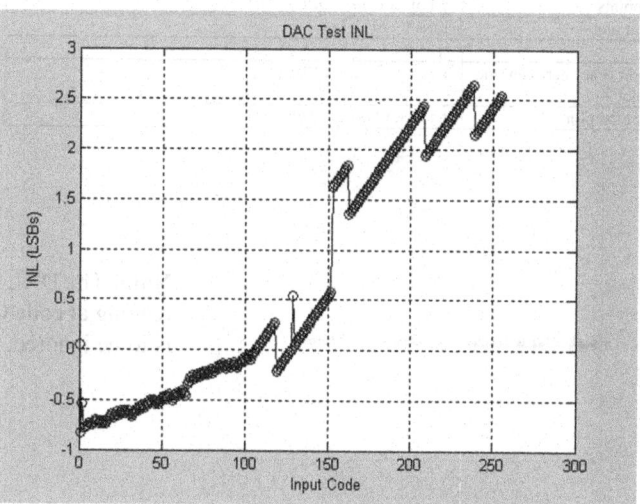

**Note:** In this set of results, there is a trend for the INL to increase as the input code increases. The INL should be expected to vary around 0 LSBs, although here, the trend in INL increase would need to be investigated.

**Fig. H.7.** INL plot

## DNL Test

Where a binary input code change of 1 bit occurs, the output should change by 1 LSB. The DNL is the maximum deviation between each output step size of the converter and an ideal step size of 1 LSB. For a given input code the output step size is taken between the current input code and the previous code.

**Table H.5.** DNL test

| Procedure | Details | MATLAB® code |
|---|---|---|
| 1 | Read in data from file (assuming ASCII text file called 'results.mat' on C: drive. The data is read into a variable called 'DACout' using the read ASCII delimited file command. The data is stored in a column vector | `DACout = dlmread('c:\results.mat')` |
| 2 | Create a row vector (called 'X') to contain the input code (in decimal number format) | `X = 0:(length(DACout) - 1)` |
| 3 | Create a column vector (called 'Code') that is the transposed 'X' vector | `Code = X'` |
| 4 | Set the DAC full scale output voltage (at 5V) | `Vfs = 5` |
| 5 | Set the DAC resolution (8-bit DAC = 256 codes) | `no_of_codes = 2.^8` |
| 6 | Create the Ideal Converter Output | `Ideal = Code * (Vfs/ no_of_codes)` |
| 7 | Create a vector to store DNL values of 0 (in volts) | `l = length(DACout)`<br>`for n = 1:l,`<br>`d(n) = 0;`<br>`end`<br>`DNLvolts = d'` |
| 8 | Determine output change steps between previous and current code. Note the first value remains at zero | `for n = 2:l,`<br>`DNLvolts(n) = DACout(n) - DACout(n-1);`<br>`end` |
| 9 | Calculate DNL for each code (in LSBs) | `DNL = (DNLvolts / (Vfs/no_of_codes)) - 1` |
| 9 | Plot the DNL points | `plot(Code, DNL, 'o')` |
| 10 | Hold the current plot | `hold on` |
| 11 | Draw a grid | `grid` |
| 12 | Draw a straight line between the points on the plot | `plot(Code, DNL)` |
| 13 | Create a title for the plot | `title('DAC Test DNL')` |

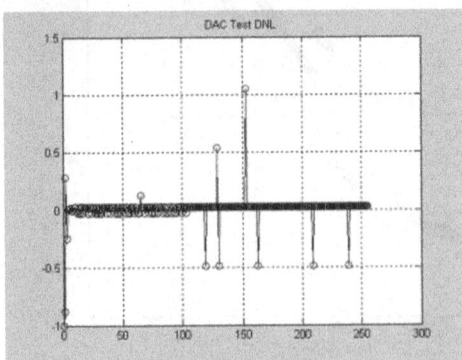

**Note:** The DNL reading at code 0 is to be ignored.

**Fig. H.8.** DNL plot

# Appendix I

# Journals, Conferences and Organisations

## Journals and Publications

IEEE Transactions on Computer-Aided Design of Integrated Circuits and Systems, IEEE Circuits and Systems Society
http://www.ieee-cas.org

IEEE Design and Test of Computers, IEEE Computer Society
http://computer.org/dt

IEEE Circuits and Systems Magazine, IEEE

IEEE Spectrum, IEEE

IEE Proceedings, Circuits, Devices & Systems, IEE

IEE Proceedings, Computer and Digital Techniques, IEE

Journal of Electronic Testing Theory and Applications (JETTA), Kluwer Academic Publishers.

## IEEE Standards

| | |
|---|---|
| IEEE Std 746-1984 | IEEE standard for performance measurements of A/D and D/A converters for PCM television video circuits |
| IEEE Std 829-1998 | IEEE standard for software test methods |
| IEEE Std 1076-1987 | IEEE standard VHDL language reference manual |
| IEEE 1076-CONC-1990 | The Sense of VASG |
| IEEE Std 1076/INT-1991 | IEEE standards interpretations: IEEE Std 1076-1987, IEEE standard VHDL language reference manual |
| ANSI/IEEE Std 1076-1993 | IEEE standard VHDL language reference manual |
| IEEE Std 1076-2000 | IEEE standard VHDL language reference manual |
| IEEE Std 1076-2002 | (Revision of IEEE Std 1076, 2002 Edition) IEEE standard VHDL language reference manual |
| IEEE Std 1076.1-1999 | IEEE standard VHDL analog and mixed-signal extensions |
| IEEE Std 1076.2-1996 | IEEE standard VHDL mathematical packages |
| IEEE Std 1076.3-1997 | IEEE standard VHDL synthesis packages |
| IEEE Std 1076.4-1995 | IEEE standard for VITAL Application-Specific Integrated Circuit (ASIC) modeling specification |
| IEEE Std 1076.4-2000 | IEEE standard for VITAL Application-Specific Integrated Circuit (ASIC) modeling specification |
| IEEE Std 1076.6-1999 | IEEE standard for VHDL Register Transfer Level (RTL) synthesis |
| IEEE Std 1076.6-2004 | Revision of IEEE Std 1076.6-1999 |

| | |
|---|---|
| IEEE Std 1149.1-1990 | IEEE standard test access port and boundary - scan architecture |
| IEEE Std 1149.1a-1993 | IEEE standard test access port and boundary-scan architecture |
| IEEE Std 1149.1b-1994 | Supplement to IEEE Std 1149.1-1990, IEEE standard test access port and boundary-scan architecture |
| IEEE Std 1149.1-2001 | IEEE standard test access port and boundary-scan architecture |
| IEEE Std 1149.4-1999 | IEEE standard for a mixed-signal test bus |
| IEEE Std 1149.5-1995 | IEEE standard for module test and maintenance bus (MTM-Bus) protocol |
| IEEE Std 1149.6-2003 | IEEE standard for boundary-scan testing of advanced digital networks |
| IEEE Std 1241-2000 | IEEE standard for terminology and test methods for analog-to-digital converters |
| IEEE 1364-1995 | IEEE Standard Verilog® Hardware Description Language |
| IEEE 1364-2001 | IEEE Standard Verilog® Hardware Description Language |
| IEEE 1364.1-2002 | IEEE Standard for Verilog® Register Transfer Level Synthesis |
| IEEE Std 1450-1999 | IEEE standard for standard test interface language (STIL) |
| IEEE Std 1532-2001 | IEEE standard for in-system configuration of programmable devices |
| IEEE Std 1532-2002 | Revision of IEEE standard 1532-2001 |
| P1500 | Standard testability method for embedded core-based integrated circuits |

## Conferences, Workshops and Symposiums

| | |
|---|---|
| ITC | International Test Conference |
| DATE | Design and Test in Europe Conference |
| IMSTW | International Mixed-Signal Test Workshop |
| ETS | European Test Symposium |
| VTS | IEEE VLSI Test Symposium |
| LATW | IEEE Latin American Test Workshop |
| IOLTS | IEEE International On-Line Test Symposium |
| ISSCC | IEEE International Solid-State Circuits Conference |
| ISCAS | IEEE International Symposium on Circuits and Systems |
| DAC | IEEE Design Automation Conference |
| ICCAD | International Conference on Computer Aided Design |
| MTDT | IEEE International Workshop on Memory Technology, Design, and Testing |

## Organisations

| | |
|---|---|
| IEE | Institution of Electrical Engineers |
| IEEE | Institute of Electrical and Electronics Engineers |
| IMAPS | International Microelectronics and Packaging Society |
| OVI | Open Verilog International |
| TTTC | Test Technology Technical Council |
| STC | Semiconductor Test Consortium |
| JEDEC | Joint Electronic Device Engineering Council |

# Bibliography

## Books

Bellaouar A. and Elmasry M., "Low-Power Digital VLSI Design Circuits and Systems", Kluwer Academic Publishers, The Netherlands, 1995, ISBN 0-7923-9587-5

Burns M. and Roberts G.W., "An Introduction to Mixed-Signal IC Test and Measurement", Oxford University Press, New York, 2001, ISBN 0-19-514016-8

Bushnell M. and Agrawal V., "Essentials of Electronic Testing for Digital, Memory & Mixed-Signal VLSI Circuits", Kluwer Academic Publishers, 2000, ISBN 0-7923-7991-8

Chang C.Y. and Sze S.M., "ULSI Technology", McGraw-Hill International Editions, Singapore, 1996, ISBN 0-07-114105-7

Cheung V. and Luong H., "Design of Low-voltage CMOS Switched-opamp Switched-capacitor Systems", Kluwer Academic Publishers, 2003, ISBN 1-4020-7466-2

Diaz C.H., Kang S. and Duvvury C., "Modeling of Electrical Overstress in Integrated Circuits", Kluwer Academic Publishers, 1994, ISBN 792395050

Doane D. A. and Franzon P.D., "Multichip Module Technologies and Alternatives, The Basics", Van Nostrand Reinhold, New York, 1993, ISBN 0-442-01236-5

Franco S., "Design with Operational Amplifiers and Analog Integrated Circuits", McGraw-Hill International Editions, Singapore, 1988, ISBN 0-07-100435-1

Hanselman D. and Littlefield B., "Mastering Matlab 6 – A Comprehensive Tutorial and Reference", Prentice Hall, USA, 2001, ISBN 0-13-019468-9

Haskard M.R. and May I.C., "Analog VLSI Design nMOS and CMOS", Prentice Hall Silicon Systems Engineering Series, Australia, 1988, ISBN 0-7248-0027-1

345

Hurst S., "VLSI Testing digital and mixed analogue/digital techniques", IEE, 1998, ISBN 0-85296-901-5

Ifeachour E. and Jervis B., "Digital Signal Processing, A Practical Approach", Prentice Hall, UK, 2002, ISBN 0-201-59619-9

Jespers P., "Integrated Converters, D to A and A to D Architectures, Analysis and Simulation", Oxford University Press, USA, 2001, ISBN 0-19-856446-5

Kang S. and Leblebici Y., "CMOS Digital Integrated Circuits Analysis and Design", McGraw-Hill International Editions, Singapore, 1996, ISBN 0-07-114423-4

Laker K.R. and Sansen W.M.C. "Design of Analog Integrated Circuits and Systems", McGraw-Hill International Editions, Singapore, 1994, ISBN 0-07-113458-1

Navabi Z., "VHDL Analysis and Modeling of Digital Systems", McGraw-Hill International Editions, 1993, ISBN 0-07-112732-1

Needham W., "Designer's Guide to Testable ASIC Devices", Van Nostrand Reinhold, 1991, ISBN 0-442-00221-1

O'Connor P., "Test Engineering, A Concise Guide to Cost-effective Design, Development and Manufacture", John Wiley & Sons Ltd., England, 2001, ISBN 0-471-49882-3

Parker K., "The Boundary-Scan Handbook, Analog and Digital", 2nd Edition, Kluwer Academic Publishers, USA, 2000, ISBN 0-7923-8277-3

Prince B., "High Performance Memories, New architecture DRAMs and SRAMs evolution and function", Wiley, England, 1996, ISBN 0-471-95646-5

Rajsuman, R., "System-on-a-Chip Design and Test", Artech House Publishers, USA, 2000, ISBN 1-58053-107-5

Roberts G. and Lu A., "Analog Signal Generation for Built-In Self-Test of Mixed-Signal Integrated Circuits", Kluwer Academic Publishers, USA, 1995, ISBN 0-7923-9564-6

Sanjay Dabral S. and Maloney T., "Basic ESD and I/O Design", Wiley, 1999, ISBN 0-471-25359-6

Schaumann R. and Valkenburg M., "Design of Analog Filters", Oxford University Press, USA, 2001, ISBN 0-19-511877-4

Sharma A. K., "Semiconductor Memories: Technology, Testing, and Reliability", Wiley-IEEE Press, 2002, ISBN 0-7803-1000-4

Singh R., Modest M.M. and Oprysko D.H., "Silicon Germanium: Technology, Modeling, and Design", Wiley-IEEE Press, 2003, ISBN: 0-471-44653-X

Smith D., "HDL Chip Design", Doone Publications, USA, 1996, ISBN 0-9651934-3-8

Smith M., "Application Specific Integrated Circuits", Addison-Wesley, 1999, ISBN 0-201-50022-1

Sze S.M., "Semiconductor devices Physics and Technology", Wiley, New York, 1985, ISBN 0-471-83704-0

Tocci R.J., Widmer N.S. and Moss G.LK., "Digital Systems 9th Edition", Pearson Education International, USA, 2004, ISBN 0-13-121931-6

Toumazou C., Lidgey F. and Haigh D., "Analogue IC design: the current-mode approach", IEE Circuits and Systems Series 2, IEE, UK, 1993, ISBN 0-86341-297-1

Toumazou C., Hughes J. and Battersby N., "Switched-Currents an analogue technique for digital technology", IEE Circuits and Systems Series 5, IEE, UK, 1993, ISBN 0-86341-294-7

Tuinenga P., "SPICE, A Guide to Circuit Simulation and Analysis using Pspice®", 3rd Edition, Prentice Hall, 1995, ISBN 0-13-158775-7

van de Goor, A. J., "Testing Semiconductor Memories", Wiley, New York, 1991, ISBN 0-4719-2587-x

Wilkins B.R. "Testing Digital Circuits An Introduction", Van Nostrand Reinhold (UK), UK, 1986, ISBN 0-442-31748-4

Zwolinski M., "Digital System Design with VHDL", Pearson Education Limited, 2000, England, ISBN 0-201-36063-2

# Journals

Al-Ars Z. and van de Goor A., "Test Generation and Optimization for DRAM Cell Defects Using Electrical Simulation", IEEE Transactions on Computer-Aided Design of Integrated Circuits and Systems, Vol. 22, No. 10, October 2003, pp1371-1384

Barth J. et al., "Embedded DRAM design and architecture for the IBM 0.11um ASIC offering", IBM Journal of Research and Development, Vol. 46, No. 6, November 2002, pp675-689

Bertozzi D. and Benini L., "Xpipes: A Network-on-Chip Architecture for Gigascale Systems-on-Chip", IEEE Circuits and Systems Magazine, Vo.4, No. 2, 2004, pp18-31

Burbidge M. et al., "Motivations towards BIST and DfT for embedded charge-pump phase-locked loop frequency synthesisers", IEE Proceedings on Circuits, Devices and Systems, Vol. 151, Issue 4, August 2004, pp337-348

Cassol L. et al., "The $\Sigma\Delta$-BIST Method Applied to Analog Filters", Journal of Electronic Testing: Theory and Applications, No. 19, 2003, pp13-20

Chakrabarty K., Iyengar V. and Krasniewski M., "Test Planning for Modular Testing of Hierarchical SOCs", IEEE Transactions on Computer-Aided Design of Integrated Circuits and Systems, Vol. 24, No. 3, March 2005, pp435-448

Champac V. Rubio A. and Figueras J., "Electrical Model of the Floating Gate Defect in CMOS ICs: Implications on IDDQ Testing", IEEE Transactions on Computer-Aided Design of Integrated Circuits and Systems, Vol. 13, No. 3, March 1994

Cheng K., Tsai M. and Wu C., "Neighbourhood Pattern Sensitive Fault Testing and Diagnostics for Random-Access Memories", IEEE Transactions on Computer-Aided Design of Integrated Circuits and Systems, Vol. 21, No. 11, November 2002, pp1328-1336

Chess B. et al., "Logic Testing of Bridging Faults in CMOS Integrated Circuits", IEEE Transactions on Computers, Vol. 47, No. 3, March 1998, pp338-345

Christie P. and Stroobandt D., "The Interpretation and Application of Rent's Rule", IEEE Transactions on Very Large Scale Integration (VLSI) Systems, Vol. 8, No. 6, December 2000, pp639-648

D. C. Huang and W. B. Jone, "A Parallel Transparent BIST Method for Embedded Memory Arrays by Tolerating Redundant Operations", IEEE Transactions on Computer-Aided Design of Integrated Circuits and Systems, Vol. 21, No. 5, May 2002, pp617-628

Der-Cheng Huang and Wen-Ben Jone, "A Parallel Built-In Self-Diagnostic Method for Embedded Memory Arrays", IEEE Transactions on Computer-Aided Design of Integrated Circuits and Systems, Vol. 21, No. 4, April 2002, pp449-465

Diaz C., Kang S.M. and Duvvury C., "Electrical overstress and electrostatic discharge", IEEE Transactions on Reliability, Vol. 44, Issue 1, March 1995, pp2-5

Dufort B. and Roberts G., "On-Chip Analog Signal Generation for Mixed-Signal Built-In Self-Test", IEEE Journal of Solid-State Circuits, Vol. 34, No. 3, March 1999, pp318-330

Edenfeld D. et al., "2003 Technology Roadmap for Semiconductors", Computer, IEEE Computer Society, January 2004, pp47-56

Edwards C., "Questions hover over the package path to Integration", IEE Electronics Systems and Software, August-September 2004, pp30-31

Edwards C., "Speeding up is hard to do", IEE Review, September 2004, pp44-46

Eklow B., Barnhart C. and Parker K., "IEEE 1149.6: A Boundary-Scan Standard for Advanced Digital Networks", IEEE Design and Test of Computers, September-October 2003, pp76-80

Evans-Pughe C., "Got to get a packet or two", IEE Review, December 2004, pp40-43

Ferguson F.J. and Shen J.P., "A CMOS Fault Extractor for Inductive Fault Analysis", IEEE Transactions on Computer Aided Design, Vol, 7, No. 11, November 1988, pp1181-1194

Flaherty N., "In the chip or on the fly", IEE Review, September 2004, pp48-51

Gaitonde D. and Walker D., "Hierarchical Mapping of Spot Defects to Catastrophic Faults – Design and Applications", IEEE Transactions on Semiconductor Manufacturing, Vol. 8, No. 2, May 1995, pp167-177

Geppert L., "Sun's Big Splash", IEEE Spectrum magazine, January 2005, pp50-54

Harlow J., "Overview of Popular Benchmark Sets", IEEE Design and Test of Computers, Vol. 17, No. 3, July-September 2000, pp15-17

Hoffmann C. and Ohletz M., "Feasibility Study for the Hybrid-Built-In Self-Test (HBIST) for Mixed-Signal Integrated Circuits", IEEE Design & Test of Computers, July-September 2000, pp106-115

Hogan T. and Heffernan D., "Virtual Test reduces semiconductor product development time", IEE Electronics and Communication Engineering Journal, April 2001, pp77-83

Huertas G. et al., "Practical Oscillation-Based Test of Integrated Filters", IEEE Design & Test of Computers, November-December 2002, pp64-72

Huertas G. et al., "Testing Mixed-Signal Cores: A Practical Oscillation-Based Test in an Analog Macro", IEEE Design & Test of Computers, November-December 2002, pp73-82

Koranne S., "Design of reconfigurable access wrappers for embedded core based SoC test", IEEE Transactions on Very Large Scale Integration (VLSI) Systems, Vol. 11, Issue 5, October 2003, pp955-960

Landman B. and Russo R., "On a pin versus block relationship for partitions of logic graphs", IEEE Transactions on Computers, C-20, 1971, pp1469-1479

Li J. et al., "A hierarchical test methodology for systems on chip", IEEE Micro, Vol. 22, Issue 5, September-October 2002, pp69-81

Liu J. and Lin X., "Equalization in High-Speed Communication Systems, IEEE Circuits and Systems Magazine, Vo.4, No. 2, 2004, pp4-17

Lu S. "Built-in self-repair techniques for embedded RAMs", IEE Proceedings on Computers and Digital Techniques, Vol. 150, Issue 4, July 2003, pp201-208

MacMillen D. et al., "An Industrial View of Electronic Design Automation", IEEE Transactions on Computer Aided Design of Integrated Circuits and Systems, Vol. 19, No. 12, December 2000, pp1428-1448

Mallarapu S. and Hoffman A., "IDDQ Testing on a Custom Automotive IC", IEEE Journal of Solid-State Circuits, Vol. 30, No. 3, March 1995, pp295-299

Meyerson B.S., "High speed silicon-germanium electronics," Scientific American, vol. 270, no. 3, pp. 42-47, 1994

Pennino T.P. and Potechin J., "Design for Manufacture", IEEE Spectrum, September 1993, pp51-53

Perkins A. et al., "Fault Modeling And Simulation Using VHDL-AMS", Analog Integrated Circuits and Signal Processing, Vol. 16, No. 2, 1998, pp141-155.

Piotr R. Sidorowicz and Janusz A. Brzozowski, "A Framework for Testing Special-Purpose Memories", IEEE Transactions on Computer-Aided Design of Integrated Circuits and Systems, Vol. 21, No. 12, December 2003, pp1459-1468

Provost B. and Sanchez-Sinencio E., "On-Chip Ramp Generators for Mixed-Signal BIST and ADC Self-Test", IEEE Journal of Solid-State Circuits," Vol. 38, No. 2, February 2003, pp263-273

Psarakis M. Gizopoulos D. and Paschalis A., "Built-In Sequential Fault Self-Testing of Array Multipliers", IEEE Transactions on Computer-Aided Design of Integrated Circuits and Systems, Vol. 24, No. 3, March 2005, pp449-460

Raab W. et al., "A 100GOPS Programmable Processor for Vehicle Vision Systems", IEEE Design & Test of Computers, January-February 2003, pp8-16

Rosinger P. et al., "Analysing trade-offs in scan power and test data compression for systems-on-a-chip", IEE Proceedings on Computers and Digital Techniques, Vol. 149, No. 4, July 2002, pp188-196

Russell G. and Learmouth D., "Systematic approaches to testing embedded analogue circuit functions", Microelectronics Journal, No. 25, 1994, pp133-138

Sabade S. and Walker D., "IDDQ Test: Will It Survive the DSM Challenge?", IEEE Design and Test of Computers, September-October 2002, pp8-16

Sadiku M.N.O. and Akujuobi C.M., "Electrostatic discharge (ESD)", IEEE Potentials, Vol. 22, Issue 5., December 2003 – January 2004, pp39-41

Shen J., Maly W. and Ferguson F., "Inductive Fault Analysis of MOS Integrated Circuits", IEEE Design and Test of Computers, Vol. 2, No. 12, December 1985, pp13-26

Singh R., Modest M.M. and Oprysko D.H., "Silicon Germanium: Technology, Modeling, and Design", Wiley-IEEE Press, 2003, ISBN: 0-471-44653-X

Soden J.M., Hawkins C.F., Gulati R.K. and Weiwei M., "$I_{DDQ}$ Testing: A Review", Journal of Electronic Testing, Theory and Applications, No. 3, 1992, pp291-303

Soma M. et al., "Hierarchical ATPG for Analog Circuits and Systems", IEEE Design and Test of Computers, January-February 2001, pp72-81

Sunter S., "Cost/benefit analysis of the P1149.4 mixed-signal test bus" IEE Proceedings on Circuits, Devices and Systems, Vol. 143, No. 6, December 1996, pp393-398

Takach A., "Turning C into hardware", IEE Electronics Systems and Software, December/January 2004/05, pp20-23

Uros K. et al., "Extending IEEE Std. 1149.4 Analog Boundary Modules to Enhance Mixed-Signal Test", IEEE Design & Test of Computers, March-April 2003, pp32-39

van de Goor, "Using March Tests to Test SRAMs", IEEE Design & Test of Computers, Vol. 10, Issue 1, March 1993, pp8-14 van de Goor A., "An Industrial Evaluation of DRAM Tests", IEEE Design and Test of Computers, September-October 2004, pp430-440

Van Spaandonk J. and Kevenaar T., "Selecting Measurements to Test the Functional Behavior of Analog Circuits", Journal of Electronic Testing: theory and Applications, No. 9, 1996, pp9-18

Van treuren B. and Miranda J., "Embedded Boundary Scan", IEEE Design and Test of Computers, March-April 2003, pp20-25

Vermeulen B. and Goel S. K., "Design for Debug: Catching Design Errors in Digital Chips", IEEE Design and Test of Computers, Vol. 19, No. 3, May-June 2002, pp37-45

Vollrath J., "Testing and Characterization of SDRAMs", IEEE Design and Test of Computers, January-February 2003, pp42-50

Walker H. and Stephen W., "VLASIC: A Catastrophic Fault Yield Simulator for Integrated Circuits", IEEE Transactions on Computer-Aided Design, Vol.CAD-5, No. 4, October 1986, pp541-556

Wang S. and Gupta S., "ATPG for heat dissipation minimization during test application", IEEE Transactions on Computers, 1998, Vol. 47, No.2, pp256-262

Xijiang L. et al., "High-Frequency, At-Speed Scan Testing", IEEE Design and Test of Computers, September-October 2003, pp17-25

Yazdani M., Ferry D.K. and Akers L.A, "Microprocessor Pin Predicting", IEEE Circuits and Devices Magazine, Vol. 13, No. 2, March 1997, pp28-31

Zaid Al-Ars and Ad J. van de Goor, "Test Generation and Optimization for DRAM Cell Defects Using Electrical Simulation", IEEE Transactions on Computer-Aided Design of Integrated Circuits and Systems, Vol. 22, No. 10, October 2003, pp1371-1384

Zorian Y. and Shoukourian S., "Embedded-Memory Test and Repair: Infrastructure IP for SoC Yield", IEEE Design and Test of Computers, May-June 2003, pp58-67

Zorian Y. and Gizopoulos D., "Guest Editors' Introduction: Design for Yield and Reliability", IEEE Design and Test of Computers, Vol. 21, No. 3, May-June 2004, pp177-182

# Conferences, Workshops and Symposiums

Acevedo G. and Ramirez-Angulo J., "Built-in self-test scheme for on-chip diagnosis, compliant with the IEEE 1149.4 mixed-signal test bus standard", Proceedings of the IEEE International Symposium on Circuits and Systems, Vol. 1, 2002, pp I-149 - I-152

Ahmed N., Tehranipour, M and Nourani, M, "Extending JTAG for testing signal integrity in SoCs", Proceeding of the Design, Automation and Test in Europe Conference and Exhibition, 2003, pp218-223

Aitken R., "Finding defects with fault models", Proceedings of the International Test Conference, 1995, pp498-505

Baker K. et al., "Development of a Class 1 QTAG Monitor", Proceedings of the International Test Conference, 1994, pp213-222

Baosheng W. et al., "Yield, overall test environment timing accuracy, and defect level trade-offs for high-speed interconnect device testing", Proceedings of the 12th Asian Test Symposium, 2003, pp348-353

Barth R., "Selective Optimization of Test for Embedded Flash Memory", Proceedings of the International Test Conference, 2002, pp1222

Bennetts B., "Status of IEEE testability standards 1149.4, 1532 and 1149.6", Proceedings of the Design, Automation and Test in Europe Conference and Exhibition, 2004, Vol. 2 , 2004, pp1184 – 1189

Bradford J. et al., "Simulating Realistic Bridging and Crosstalk Faults in an Industrial Setting", Proceedings of the 7th IEEE European Test Symposium, 2002, pp75-80

Chess B., Roth C. and Larrabee T., "On Evaluating Competing Bridge Fault Models for CMOS ICs", Proceedings of the 12th IEEE VLSI Test Symposium, 1994, pp446-451

Dance D., "Modeling Test Cost Ownership", Proceedings of the 3rd International Conference on the Economics of Design, Test and Manufacturing, 1994, pp13-17

Datta R., Sebastine A. and Abraham J., "Delay Fault Testing and Silicon Debug Using Scan Chains", Proceedings of the 9th IEEE European Test Symposium, May 2004, pp111,116

Dervisoglu B., "A unified DFT architecture for use with IEEE 1149.1 and VSIA/IEEE P1500 compliant test access controllers", Proceedings of the Design Automation Conference, 2001, pp53-58

de Vries R. et al., "Built-in self-test methodology for A/D converters", Proceedings of the European Design and Test Conference, 1997, pp353-358

Fang L., Kerkhoff H. and Gronthoud G., "Reducing Analogue Fault-Simulation Time by Using High-Level Modelling in Dotss for an Industrial Design", Proceedings of the IEEE European Test Workshop (ETW'01), 2001, pp61-67

Godambe N. and Shi C., "Behavioral Level Noise Modeling and Jitter Simulation of Phase-Locked Loops with Faults using VHDL-AMS, Proceedings of the 15th IEEE VLSI Test Symposium, 1997, pp177-182

Goel S. and Marinissen E., "Effective and efficient test architecture design for SOCs", Proceedings of the International Test Conference, 2002, pp529-538

Grout I., "An Analogue and Mixed-Signal Fault Simulation Tool based on Tcl/Tk and HSpice", Proceedings of the Iberchip 2002 Workshop, 2002

Grout I. and Santana J., "Mechatonic System Fault Simulation study with Spectre and Verilog-A", Proceedings of the Mechatronics Forum Conference, The Netherlands, 2002

Guanglin W. et al., "Implementation of a BIST scheme for ADC test", Proceedings of the 5th International Conference on ASIC, 2003, Vol. 2, 2003, pp1128-1131

Hawkins C. Soden J., Righter A. and Ferguson F., "Defect Classes – An Overdue Paradigm for CMOS IC Testing", Proceedings of International Test Conference, USA, 1994, pp 413-425

Hawkins C. and Segura J., "Failure Modes in Nanometer Technologies", Tutorial D2, Design and Automation in Europe Conference (DATE), 2003

Hunter C. Vida-Torku E. and LeBlanc J., "Balancing Structured and Ad-hoc Design for Test: Testing of the PowerPC 603[TM] Microprocessor", Proceedings of the International Test Conference, 1994, pp76-83

Iyengar V., Goel S., Marinissen E. and Chakrabarty K., "Test Resource Optimization for Multi-Site Testing of SOCs Under ATE Memory Depth Constraints", Proceedings of the International Test Conference, 2002, pp115-1168

Jannesari S., "HP94000 Multi_site Mixed Signal Test Development Considerations", Proceedings of the 16th IEEE Instrumentation and Measurement Technology Conference, Vol. 3, 1999, pp1563 – 1568

Jee A. and Ferguson F.J., "Carafe: An Inductive Fault Analysis Tool for CMOS VLSI Circuits", IEEE VLSI Test Symposium, 1993, pp92-98

Kilic Y. and Zwolinski M., "Speed-up Techniques for Fault-based Analogue Fault Simulation, Proceedings of the European Test Workshop, 2001

Kuijstermans, F., Sachdev, M. and Thijssen, A., "Defect-oriented test methodology for complex mixed-signal circuits", Proceedings of the European Design and Test Conference, 1995, pp18 – 23

Lubaszewski, M., et al., "A built-in multi-mode stimuli generator for analogue and mixed-signal testing", Proceedings of the XI Brazilian Symposium on Integrated Circuit Design, 1998, pp175 – 178

Majernik, D. et al., "Using simulation to improve fault coverage of analog and mixed-signal test program sets", Proceedings of the IEEE Autotestcon, 1997, pp371-375

Marinissen E. et al., "Towards a standard for embedded core test: an example", Proceedings of the International Test Conference, 1999, pp616-627

Maroufi W. et al., "Solving the I/O bandwidth problem in system on a chip testing", Proceedings of the 13th Symposium on Integrated Circuits and Systems Design, 2000, pp 9-14

Miettinen, J., Mantysalo, M., Kaija, K. and Ristolainen, E.O. "System design issues for 3D system-in-package (SiP)", Proceedings of the Electronic Components and Technology Conference (ECTC), 2004, Vol. 1, pp610-614

Miara A. and Giambiasi N., "Dynamic and deductive fault simulation", Proceedings of the 15th conference on Design Automation Conference, USA, 1978, pp439-443

Mir S. et al., "SWITTEST: Automatic Switch-level Fault Simulation and Test Evaluation of Switched-Capacitor Systems2, Proceedings of the 34th Design Automation Conference, 1997, pp281-286

Montanes R.R., Bruls E.M. and Figueras J., "Bridging Defects Resistance Measurements in a CMOS Process", Proceedings of the International Test Conference, 1992, pp892-899

Nicolaidis, M, Achouri, N. and Anghel, L., "Memory built-in self-repair for nanotechnologies", Proceedings of the 9th IEEE On-Line Testing Symposium, 2003, pp94-98

Oakland S., "Considerations for implementing IEEE 1149.1 on system-on-a-chip integrated circuits", Proceedings of the International Test Conference, 2000, pp628-637

Olbrich T. et al, "A New Quality Estimation Methodology for Mixed-Signal and Analogue ICs", Proceedings of the European Design & Test Conference, France, 1997, pp573-580

Olbrich T. et al., "Defect-Oriented Vs Schematic-Level Based Fault Simulation for Mixed-Signal ICs", Proceedings of International Test Conference USA, 1996, pp 511-520

Ohletz M., "Hybrid Built-In Self-Test (HBIST) for Mixed Analogue/Digital Integrated Circuits", Proceedings of the European Test Conference, 1991, pp307-316

Park S., et al., "Designing Built-In Self-Test Circuits for Embedded Memories Test", Proceedings of the Second IEEE Asia Pacific Conference on ASICs, 2000, pp315-318

Parnas B., Pramanick A., Elston M. and Adachi T., "Software Development for an Open Architecture Test System", Proceedings of the 9th IEEE European Test Symposium, 2004, pp81-86

Powell T. et al., "BIST for Deep Submicron ASIC Memories with High Performance Application", Proceedings of the International Test Conference, 2003, pp386-392

Rajsuman R., "Rambist builder: a methodology for automatic built-in self-test design of embedded RAMs", Records of the 1996 IEEE International Workshop on Memory Technology, Design and Testing, 1996, pp50-56

Riedel M. and Rajski J., "Fault coverage analysis of RAM test algorithms", Proceedings of the 13th IEEE VLSI Test Symposium, 1995, pp227-234

Rivoir J., "Lowering cost of test: parallel test or low-cost ATE?" Proceedings of the 12th Asian Test Symposium, 2003, pp360-363

Roberts G., "Metrics, Techniques and Recent Developments in Mixed-Signal Testing", Proceedings of the IEEE/ACM International Conference on Computer Aided Design, 1996, pp1-8

Sachdev M. and Atzema B., "Industrial Relevance of Analog IFA: A Fact or a Fiction", Proceedings of the International Test Conference, 1995, pp61-70

Sauer C. et al. "Developing a Flexible Interface for RapidIO, Hypertransport, and PCI-Express", Proceedings of the International Conference on Parallel Computing in Electrical Engineering, 2004, pp129-134

Schrader M. and McConnell R., "SoC Design and Test Considerations", Proceedings of the Design, Automation and Test in Europe Conference and Exhibition, 2003, pp202-207

Song Y. et al., "The reliability issues on ASIC/memory integration by SiP (system-in-package) technology", Proceedings of the IEEE International SOC Conference, 2003, pp7-10

Spinks S. et al., "Generation and Verification of Tests for Analogue Circuits Subject to Process Parameter Deviations", Proceedings of the IEEE International Symposium on Defect and Fault Tolerance in VLSI Systems, 1997, pp100-108

Stoffels R., "Cost effective frequency measurement for production testing: new approaches on PLL testing", Proceedings of the International Test Conference, 1996, pp708-716

Straube B., Vermeiren W., Müller B., "Using an Analogue Fault Simulator for Microsystem Fault Simulation", Proceedings of the 4th IEEE International Mixed Signal Testing Workshop, The Netherlands, 1998

Sunter S. and Nadeau-Dostie B., "Complete, Contactless I/O Testing – Reaching the Boundary in Minimizing Digital IC Testing Cost", Proceedings of the International Test Conference, 2002, pp446-455

Tai K.L., "System-In-Package (SIP): challenges and opportunities", Proceedings of the Asia and South Pacific Design Automation Conference, 2000, pp191-196

Tehranipour M. and Navabi Z., "Zero-Overhead BIST for Internal SRAM Testing", Proceedings of the 12th International Conference on Microelectronics, 2000, pp109-112

van de Goor A. and Offernan A., "Towards a Uniform Notation for Memory Tests", Proceedings of the European Design and Test Conference, 1996, pp420-427

Volkerink E., Khoche A., Kamas L., Rivoir J. and Kerkhoff H., "Tackling Test Trade-offs from Design, Manufacturing to Market using Economic Modeling", Proceedings of the International Test Conference, 2001, pp1098-1107

Volkerink E., Khoche A., Rivoir J. and Hilliges K., "Test Economics for Multi-site Test with Modern Cost Reduction Techniques", Proceedings of the 20th IEEE VLSI Test Symposium, 2002

Vranken H.P.E. and Segers M.T.M., "Design-for-Debug in Hardware/Software Co-Design", Proceedings of the 5th International Workshop on Hardware/Software Co-Design, 1997, pp35-39

Wallquist K., Richter A. and Hawkins C., "A General Purpose IDDQ Measurement Circuit", Proceedings of the International Test Conference, 1993, pp642-651

West B., "Highly Reconfigurable ATE for Rapidly Evolving Test Requirements", Proceedings of the 9th IEEE European Test Symposium, 2004, pp87-93

Yong-Ha Song, Soon-Gon Kim, Kwang-Joon Rhee, Dong-Soo Cho and Taek-Soo Kim, "The reliability issues on ASIC/memory integration by SiP (system-in-package) technology", Proceedings of the IEEE International SOC Conference, 2003, pp7-10

Zorian Y., "Testing semiconductor chips: trends and solutions", Proceedings of the XII Symposium on Integrated Circuits and Systems Design, 1999, pp226-233

Zorian Y., "Test requirements for embedded core-based systems and IEEE P1500", Proceedings of the International Test Conference, 1997, pp191 – 199

Zorian Y. et al, "Testing embedded-core based system chips", Proceedings of the International Test Conference, 1998, pp130-143

Zorian Y., "Test requirements for embedded core-based systems and IEEE P1500", Proceedings of the International Test Conference, 1997, pp191-199

# Articles, Reports, Datasheets and Standards

ADA4851 Low Cost, High Speed Rail-to-Rail Output Op Amp, dataheet, Analog Devices Inc., USA

Advanced Micro Devices, Inc., USA, "AMD Athlon™ 64 FX Product Data Sheet"

Advanced Micro Devices, Inc., USA, "AMD Duron™ Processor Model 8 Data Sheet"

"Dual Positive Edge Triggered D-Type Flip-Flops with Clear and Preset", datasheet, Texas Instruments Inc., USA

EIA/JEDEC Test Method A114-A, "Electrostatic Discharge (ESD) Sensitivity Testing Human Body Model (HBM)", Electronic Industries Association, 1997

ESD-STM5.1-1998, "ESD Association Standard Test Method for Electrostatic Discharge Sensitivity Testing: Human Body Model (HBM) – Component Level", ESD Association, 1998

Garcia R., "Rethink fault models for submicron-IC test", Test & Measurement World, October 2001

IEEE Std 1076-2002, IEEE Standard VHDL Language Reference Manual, IEEE, USA

IEEE standard VHDL analog and mixed-signal extensions, Std 1076.1-1999, IEEE

IEEE standard test access port and boundary - scan architecture, IEEE Std 1149.1-2001, IEEE, USA

IEEE Standard for a Mixed-Signal Test Bus, IEEE Std 1149.4-2000, IEEE, USA

IEEE 1364-1995, IEEE Standard Verilog® Hardware Description Language, IEEE, USA

IEEE Std 1532-2002, IEEE Standard for In-System Configuration of Programmable Devices, IEEE, USA

Intel Corporation, USA, "Intel® Pentium® 4 Processor 660, 650, 640, and 630Δ and Intel® Pentium® 4 Processor Extreme Edition Datasheet"

Intel Corporation, USA, "Intel® Celeron® M Processor Datasheet"

International Technology Roadmap for Semiconductors, 2003 Edition, "Executive Summary"

International Technology Roadmap for Semiconductors, 2003 Edition, "Test and Test Equipment"

International Technology Roadmap for Semiconductors, 2003 Edition, "Design"

International Technology Roadmap for Semiconductors, 2003 Edition, "Assembly and Packaging"

International Technology Roadmap for Semiconductors, 2003 Edition, "Lithography"

"Logic Selection Guide", Texas Instruments Inc., USA

MIL-STD-833 Method 3015.7, "Electrostatic Discharge Sensitivity Classification"

Moore G., "Cramming more components onto integrated circuits", Electronics, Vol. 38, No. 8, April 1965

Pennino T.P. and Potechin J., "Design for Manufacture", IEEE Spectrum, September 1993, pp51-53

SPICE: Simulation Program with Integrated Circuit Emphasis, Version 3f5, University of California, Berkeley, USA

Star-Hspice®, Synopsys Inc., USA

United States Department of Defense, MIL-STD-883F, "Test Method Standard Microchips", 18th June 2004

# Internet Resources

Antics Analogue Fault Simulation Software, University of Hull, UK, http://www.eng.hull.ac.uk/research/ee_vlsi/antics.htm

Cadence Design Systems Inc., USA, http:://www.cadence.com

DFT Compiler®, Synopsys Inc., USA

IBM Research, http://www.research.ibm.com/

ITC 1999 benchmark circuits, http://www.cerc.utexas.edu/itc99-benchmarks/bench.html

ITC 2002 benchmark circuits, http://www.extra.research.philips.com/itc02socbenchm/

Moore's Law, Intel, http://www.intel.com

Open Verilog International (OVI), Verilog Analog Mixed-Signal Group, http://www.eda.org/verilog-ams/

P1500 Working Group, http://www.grouper.ieee.group/1500

Synopsys Inc., USA, http://www.synopsys.com

TetraMAX® ATPG, Synopsys Inc., USA

Texas Instruments Inc., USA. http://www.ti.com

The Mathworks Inc., http://www.themathworks.com

The Semiconductor Test Consortium (STC), http://www.semitest.org

Verifault-XL® Fault Simulator User Guide, Cadence Design Systems Inc., USA

Verilog-XL® User Guide, Cadence Design Systems Inc., USA

Xilinx Inc. http://www.xilinx.com

# Index